Corrosion and Protection of Metals

Corrosion and Protection of Metals

Editor

David M. Bastidas

MDPI • Basel • Beijing • Wuhan • Barcelona • Belgrade • Manchester • Tokyo • Cluj • Tianjin

Editor
David M. Bastidas
The University of Akron
USA

Editorial Office
MDPI
St. Alban-Anlage 66
4052 Basel, Switzerland

This is a reprint of articles from the Special Issue published online in the open access journal *Metals* (ISSN 2075-4701) (available at: https://www.mdpi.com/journal/metals/special_issues/corrosion_protection_metals).

For citation purposes, cite each article independently as indicated on the article page online and as indicated below:

LastName, A.A.; LastName, B.B.; LastName, C.C. Article Title. *Journal Name* **Year**, *Article Number*, Page Range.

ISBN 978-3-03943-152-6 (Hbk)
ISBN 978-3-03943-153-3 (PDF)

© 2020 by the authors. Articles in this book are Open Access and distributed under the Creative Commons Attribution (CC BY) license, which allows users to download, copy and build upon published articles, as long as the author and publisher are properly credited, which ensures maximum dissemination and a wider impact of our publications.

The book as a whole is distributed by MDPI under the terms and conditions of the Creative Commons license CC BY-NC-ND.

Contents

About the Editor vii

David M. Bastidas
Corrosion and Protection of Metals
Reprinted from: *Metals* 2020, *10*, 458, doi:10.3390/met10040458 1

Ulises Martin, Jacob Ress, Juan Bosch and David M. Bastidas
Evaluation of the DOS by DL−EPR of UNSM Processed Inconel 718
Reprinted from: *Metals* 2020, *10*, 204, doi:10.3390/met10020204 7

Juan J. Santana, Víctor Cano, Helena C. Vasconcelos and Ricardo M. Souto
The Influence of Test-Panel Orientation and Exposure Angle on the Corrosion Rate of Carbon Steel. Mathematical Modelling
Reprinted from: *Metals* 2020, *10*, 196, doi:10.3390/met10020196 23

Ricardo Galván-Martínez, Ricardo Orozco-Cruz, Andrés Carmona-Hernández, Edgar Mejía-Sánchez, Miguel A. Morales-Cabrera and Antonio Contreras
Corrosion Study of Pipeline Steel under Stress at Different Cathodic Potentials by EIS
Reprinted from: *Metals* 2019, *9*, 1353, doi:10.3390/met9121353 35

Juan J. Santana, Alejandro Ramos, Alejandro Rodriguez-Gonzalez, Helena C. Vasconcelos, Vicente Mena, Bibiana M. Fernández-Pérez and Ricardo M. Souto
Shortcomings of International Standard ISO 9223 for the Classification, Determination, and Estimation of Atmosphere Corrosivities in Subtropical Archipelagic Conditions—The Case of the Canary Islands (Spain)
Reprinted from: *Metals* 2019, *9*, 1105, doi:10.3390/met9101105 51

Asunción Bautista, Francisco Velasco and Manuel Torres-Carrasco
Influence of the Alkaline Reserve of Chloride-Contaminated Mortars on the 6-Year Corrosion Behavior of Corrugated UNS S32304 and S32001 Stainless Steels
Reprinted from: *Metals* 2019, *9*, 686, doi:10.3390/met9060686 65

Jamal Choucri, Federica Zanotto, Vincenzo Grassi, Andrea Balbo, Mohamed Ebn Touhami, Ilyass Mansouri and Cecilia Monticelli
Corrosion Behavior of Different Brass Alloys for Drinking Water Distribution Systems
Reprinted from: *Metals* 2019, *9*, 649, doi:10.3390/met9060649 85

Nining Purwasih, Naoya Kasai, Shinji Okazaki, Hiroshi Kihira and Yukihisa Kuriyama
Atmospheric Corrosion Sensor Based on Strain Measurement with an Active Dummy Circuit Method in Experiment with Corrosion Products
Reprinted from: *Metals* 2019, *9*, 579, doi:10.3390/met9050579 105

Najmeh Ahledel, Robert Schulz, Mario Gariepy, Hendra Hermawan and Houshang Alamdari
Electrochemical Corrosion Behavior of Fe_3Al/TiC and Fe_3Al-Cr/TiC Coatings Prepared by HVOF in NaCl Solution
Reprinted from: *Metals* 2019, *9*, 437, doi:10.3390/met9040437 117

Yong-Sang Kim, In-Jun Park and Jung-Gu Kim
Simulation Approach for Cathodic Protection Prediction of Aluminum Fin-Tube Heat Exchanger Using Boundary Element Method
Reprinted from: *Metals* 2019, *9*, 376, doi:10.3390/met9030376 127

Federica Zanotto, Vincenzo Grassi, Andrea Balbo, Cecilia Monticelli and Fabrizio Zucchi
Resistance of Thermally Aged DSS 2304 against Localized Corrosion Attack
Reprinted from: *Metals* **2018**, *8*, 1022, doi:10.3390/met8121022 143

Lucien Veleva, Mareny Guadalupe Fernández-Olaya and Sebastián Feliu Jr.
Initial Stages of AZ31B Magnesium Alloy Degradation in Ringer's Solution: Interpretation of EIS, Mass Loss, Hydrogen Evolution Data and Scanning Electron Microscopy Observations
Reprinted from: *Metals* **2018**, *8*, 933, doi:10.3390/met8110933 161

Kyu-Hyuk Lee, Seung-Ho Ahn, Ji-Won Seo and HeeJin Jang
Passivity of Spring Steels with Compressive Residual Stress
Reprinted from: *Metals* **2018**, *8*, 788, doi:10.3390/met8100788 177

Dan Song, Jinghua Jiang, Xiaonan Guan, Yanxin Qiao, Xuebin Li, Jianqing Chen, Jiapeng Sun and Aibin Ma
Effect of Surface Nanocrystallization on Corrosion Resistance of the Conformed Cu-0.4%Mg Alloy in NaCl Solution
Reprinted from: *Metals* **2018**, *8*, 765, doi:10.3390/met8100765 189

Heon-Young Ha, Tae-Ho Lee, Jee-Hwan Bae and Dong Won Chun
Molybdenum Effects on Pitting Corrosion Resistance of FeCrMnMoNC Austenitic Stainless Steels
Reprinted from: *Metals* **2018**, *8*, 653, doi:10.3390/met8080653 201

M.A. Mohtadi-Bonab
Effects of Different Parameters on Initiation and Propagation of Stress Corrosion Cracks in Pipeline Steels: A Review
Reprinted from: *Metals* **2019**, *9*, 590, doi:10.3390/met9050590 215

About the Editor

David M. Bastidas is a professor at The University of Akron, Dept. Chemical, Biomolecular, and Corrosion Engineering. He is a faculty member and corrosion expert at the National Center for Education and Research on Corrosion & Materials Performance, NCERCAMP-UA (Akron, OH, US). Currently, his scientific research activity is focused on metallic materials corrosion and its inhibition. He is head of the research group in Corrosion and Electrochemistry of Materials. Prof. D.M. Bastidas is Chair of the NACE RIP symposium on Corrosion of Steel in Concrete, Research Committee Member of the National Association of Corrosion Engineers (NACE), Vice-Chairman of NACE RIP Symposium and Vice-Chairman of Corrosion of Steel in Concrete Research Committee of the European Federation of Corrosion (EFC). Prof D.M. Bastidas is the Trustee of The University of Akron NACE Section, and the Faculty Advisor of Corrosion Squad Student Association. Prof. David M Bastidas is a Guest-Editor for the journal Metals and an Associate Editor of Journal Revista Metalurgia. He also serves as an Editorial Board Member for Corrosion and Materials Degradation, Corrosion Engineering Science and Technology, and npj-Materials Degradation (Nature research journal). Prof. D.M. Bastidas was awarded the National Concrete and Corrosion Award in 2020.

Editorial

Corrosion and Protection of Metals

David M. Bastidas

National Center for Education and Research on Corrosion and Materials Performance, NCERCAMP-UA, Department of Chemical, Biomolecular, and Corrosion Engineering, The University of Akron, 302 E Buchtel Ave, Akron, OH 44325-3906, USA; dbastidas@uakron.edu; Tel.: +1-330-972-2968

Published: 1 April 2020

1. Introduction and Scope

During the last few decades, an enormous effort has been made to understand corrosion phenomena and their mechanisms, and to elucidate the causes that dramatically influence the service lifetime of metal materials. The performance of metal materials in aggressive environments is critical for a sustainable society. The failure of the material in service impacts the economy, the environment, health, and society. In this regard, corrosion-based economic losses due to maintenance, repair, and the replacement of existing structures and infrastructure account for up to 4% of gross domestic product (GDP) in well developed countries.

One of the biggest issues in corrosion engineering is estimating service lifetime. Corrosion prediction has become very difficult, as there is no direct correlation with service lifetime and experimental lab results, usually as a result of discrepancies between accelerated testing and real corrosion processes. It is of major interest to forecast the impact of corrosion-based losses on society and the global economy, since existing structures and infrastructure are becoming old, and crucial decisions now need to be taken to replace them.

On the other hand, environmental protocols seek to reduce greenhouse effects. Therefore, low emission policies, in force, establish regulations for the next generation of materials and technologies. Advanced technologies and emergent materials will enable us to get through the next century. Great advances are currently in progress for the development of corrosion-resistant metal materials for different sectors, such as energy, transport, construction, and health.

This Special Issue on the corrosion and protection of metals is focused on current trends in corrosion science, engineering, and technology, ranging from fundamental to applied research, thus covering subjects related to corrosion mechanisms and modelling, protection and inhibition processes, and mitigation strategies.

2. Contributions

This Special Issue comprises a large variety of interesting corrosion and protection of metal-based studies, including research on stainless steel, carbon steel, Inconel, copper, and magnesium alloys. Sensitization and surface modification, coatings, and processing influence on corrosion. The different corrosion mechanisms are also included in this collection, including intergranular corrosion, localized pitting corrosion, stress corrosion cracking, atmospheric corrosion, galvanic corrosion, numerical simulation, and modeling. In addition, a wide spread of electrochemical, microstructural and surface characterization techniques (CPP, EIS, DL-EPR, SEM, XPS, DRX, OM) are described in the aforementioned studies.

The work by Ha et al. found Mo to impart a positive effect on pitting corrosion resistance of high interstitial alloyed (HIA) FeCrMnMoNC austenitic stainless steel [1]. The alloyed Mo suppressed metastable pitting corrosion and raised both pitting and repassivation potential E_{pit} and E_{rp}, respectively. In addition, Mo reduced the critical dissolution rate of the HIA in acidified chloride solutions, and the HIA with higher Mo content was able to resist active dissolution in stronger acid.

In a study by Song and coauthors [2], the impact of high-speed rotating wire-brushing nanocrystal surface modification (SNC) on the corrosion of extruded Cu-0.4%Mg alloys was reported. Strain-induced grain refinement weakens the corrosion resistance of the SNC alloy during the initial corrosion period in 0.1 M NaCl solution, resulting in the lower E_{corr} value and higher I_{corr} values in polarization tests, a smaller capacitive loop and Rp value in EIS tests, higher mass-loss rate, and a partially corroded surface. The SNC sample with a smaller grain size has lower corrosion resistance, indicating that the increased crystal defects and higher surface roughness results in increased corrosion activity.

Lee et al. investigated the effect of shot peening on the corrosive behavior of spring steel electrochemical polarization tests and the Mott–Schottky analysis [3]. The passive current density of the specimens with stress was higher and showed fluctuation. It was found that compressive stress produced passive films with lower point defect density than non-stressed specimens, thus revealing that the growth mechanism of passive film and the transport of vacancies in the film on metals and alloys depend on the residual stress on the metallic surface.

Veleva et al. discussed the initial stages of corrosion of AZ31B magnesium alloy in Ringer's solution at 37 °C [4]. Among the main findings, the corrosion current densities estimated by hydrogen evolution are in good agreement with the time-integrated reciprocal charge transfer resistance values estimated by electrochemical impedance spectroscopy (EIS). Moreover, the formation of corrosion products with poorer protection properties and the increase in the tendency for pitting corrosion are promoted by the significant content of Cl^- in the form of aluminum oxychlorides salts. A marked decrease in the EIS inductive loop was found to reflect the dissolution of aluminum oxychloride salt, which is probably formed across the uniform corrosion layer during the initial stages, as suggested by the EDS analysis.

The study by Zanotto et al. found that microstructural modifications imparted by heat treatment produced sensitization and influenced the localized corrosion and stress corrosion cracking of 2304 duplex stainless steel [5]. Pitting and intergranular corrosion mainly initiated in Cr- and Mo-depleted regions (ferrite/austenite interphases), near to the $Cr_{23}C_6$ precipitates within the γ_2 and γ phases, then propagated in the ferrite matrix. Moreover, SCC failure initiated at the bottom of pits and was likely stimulated by hydrogen penetration.

In a study by Kim et al., the multi-galvanic effect of an Al fin-tube heat exchanger with cathodic or anodic joints was evaluated using polarization tests, numerical simulation, and the seawater acetic acid test (SWAAT) [6]. Determination of the polarization state using polarization curves was well correlated with numerical simulations using a high-conductivity electrolyte, thus envisaging a novel approach to improve the design of products subject to multi-galvanic corrosion. Results were verified by SWATT, and the leakage time of the Al fin-tube heat exchanger assembled with the anodic joint was 42% longer than that of the exchanger assembled with the cathodic joint.

The work by Ahledel et al concluded that TiC additions into a Fe_3Al matrix, prepared by high-velocity oxy-fuel (HVOF) spraying, increased the corrosion resistance of Fe_3Al/TiC composite coating in 3.5 wt.% NaCl solution [7]. The addition of Cr contributed to the decrease in the corrosion rate of Fe_3Al-Cr/TiC HVOF coating, three times lower than that of Fe_3Al/TiC. Furthermore, the addition of TiC particles into Fe_3Al matrix benefit the wear resistance while keeping corrosion-resistant properties.

A new atmospheric corrosion sensor utilizing strain measurements (ACSSM) was developed by Purwasih et al. [8]. The sensor fundamentals are based on the influence of the strain variations ($\Delta\varepsilon$) on the compressive surface of a low-carbon steel under a bending moment, considering the corrosion product layers formed, and assuming three different stages: stage I, free corrosion Fe surface ($\Delta\varepsilon = 0$); stage II, tight corrosion products ($\Delta\varepsilon < 0$); and stage III, porous corrosion products layer ($\Delta\varepsilon > 0$).

In the review paper by Mohtadi-Bonab [9], the important damage modes in pipeline steels including stress corrosion cracking (SCC) and hydrogen induced cracking (HIC) are reported. Based on a literature survey, it was concluded that many factors influence SCC, such as the microstructure of steel, residual stresses, chemical composition of steel, applied load, alternating current, surface texture, and grain boundaries, influencing the crack initiation and propagation in pipeline steels. Crystallographic texture plays a key role in crack propagation. Grain boundaries associated with {111} and {110} parallel to the rolling plane, coincident site lattice boundaries and low angle grain boundaries are recognized as crack resistant paths, while grains with high angle boundaries favor the SCC intergranular crack propagation.

Choucri et al. studied a corrosion failure in service of copper pipes in drinking water distribution systems, mostly related to their high β' phase content, which undergoes dezincification and selective dissolution attacks [10]. The corrosive behaviors of two representative α + β' brass components were compared to that of brass alloys with nominal compositions $CuZn_{36}Pb_2As$ and $CuZn_{21}Si_3P$, marketed as dezincification resistant. Analyses evidenced that the highest dezincification resistance was afforded by $CuZn_{36}Pb_2As$ (longitudinal section of extruded bar), exhibiting dealloying and subsequent oxidation of β', only at a small depth. Limited surface dealloying was also found in $CuZn_{21}Si_3P$, which underwent selective silicon and zinc dissolution and negligible inner oxidation of both α and κ constituent phases, likely due to galvanic effects.

A study by Bautista et al. observed the performance of lean duplex stainless steel (LDSS) reinforcements (UNS S32304 and S32001) after long term exposure to chloride contained corrosion environment [11]. The authors concluded that a decrease in the alkaline reserve of the mortars can affect the corrosive behavior of the LDSS exposed to environments with high chloride concentrations. In the pits formed in regions of the corrugated surface which were only moderately strained, the austenite phase was dissolved selectively, while ferrite tended to remain uncorroded. The higher tendency of ferrite to dissolve Cr can explain this observation. Ferrite should be more Cr-rich that austenite, and therefore more corrosion-resistant. The duplex structure of the stainless steel influences the selective dissolution of the phases, and austenite corrodes preferentially except in the most strained areas of the corrugated surface, where ferrite dissolves selectively.

Both works by Santana et al. reported on atmospheric corrosion of zinc, copper and carbon steel [12,13]. It was found that the most influential environmental parameter affecting the corrosion rates was the chloride deposition rate (S_d), and on the contrary, the environmental temperature (T) showed the smallest influence. The influence of test-coupon orientation and exposure angle on the time of wetness (*TOW*) was of major interest. The authors summarized that corrosivity mathematical models would need to be redefined, introducing the time of wetness and a new set of operation constants. Therefore, they concluded that atmospheric corrosion classification standards need to be revisited.

The work by Galván-Martínez et al. on X70 pipeline steel immersed in acidified and aerated synthetic soil solution [14] found a higher susceptibility to stress corrosion cracking (SCC) as the cathodic polarization increased (E_{cp}). Nevertheless, when the E_{cp} was subjected to the maximum cathodic potential (−970 mV), the susceptibility decreased; this behavior is attributed to the fact that the anodic dissolution was suppressed and the process of the SCC was dominated only by hydrogen embrittlement (HE). The EIS results showed that the cathodic process was influenced by the mass transport (hydrogen diffusion) due to the steel undergoing so many changes in the metallic surface as a result of the applied strain that it generated active sites at the surface.

Research studies by Martin and coauthors revealed the influence of ultrasonic nanocrystal surface modification (UNSM) on the degree of sensitization (DOS) in Inconel 718 [15]. The double-loop electrochemical potentiodynamic reactivation method (DL−EPR) showed that for UNSM processed samples with no thermal treatment, the DOS increased, while for UNSM treated samples that were post-annealed at 1000 °C and water quenched, the DOS notably decreased. It was found that the annealing at 1000 °C and the water quenching of the UNSM treated specimens promoted the transformation of γ" to form the δ phase on the grain boundaries, which reduces the intergranular corrosion susceptibility.

3. Conclusions and Caveats

This Special Issue on the corrosion and protection of metals presents a collection of research articles covering the relevant topics and the current state of the art in the field. As the guest editor, I hope that this collection of original research papers and reviews may be useful to researchers working in the field, promoting more research studies, debates, and discussions that will continue to shed light and bridge the gap in the understanding of corrosion and protection fundamentals and mechanisms.

Acknowledgments: As guest editor, I would like to especially thank Kinsee Guo, assistant editor, for his support and active role in the publication. I am also grateful to the entire staff of the Metal Editorial Office for the precious collaboration. Last but not least, I wish to express my gratitude to all the contributing authors and reviewers: without your excellent work, it would not have been possible to accomplish this Special Issue on Corrosion and Protection of Metals, that I hope will be a piece of interesting reading and reference literature.

Conflicts of Interest: The author declares no conflict of interest.

References

1. Ha, H.Y.; Lee, T.H.; Bae, J.H.; Chun, D.W. Molybdenum Effects on Pitting Corrosion Resistance of FeCrMnMoNC Austenitic Stainless Steels. *Metals* **2018**, *8*, 653. [CrossRef]
2. Song, D.; Jiang, J.; Guan, X.; Qiao, Y.; Li, X.; Chen, J.; Sun, J.; Ma, A. Effect of Surface Nanocrystallization on Corrosion Resistance of the Conformed Cu-0.4%Mg Alloy in NaCl Solution. *Metals* **2018**, *8*, 765. [CrossRef]
3. Lee, K.H.; Ahn, S.H.; Seo, J.W.; Jang, H.J. Passivity of Spring Steels with Compressive Residual Stress. *Metals* **2018**, *8*, 788. [CrossRef]
4. Veleva, L.; Fernández-Olaya, M.G.; Feliu, S., Jr. Initial Stages of AZ31B Magnesium Alloy Degradation in Ringer's Solution: Interpretation of EIS, Mass Loss, Hydrogen Evolution Data and Scanning Electron Microscopy Observations. *Metals* **2018**, *8*, 933. [CrossRef]
5. Zanotto, F.; Grassi, V.; Balbo, A.; Monticelli, C.; Zucchi, F. Resistance of Thermally Aged DSS 2304 against Localized Corrosion Attack. *Metals* **2018**, *8*, 1022. [CrossRef]
6. Kim, Y.S.; Park, I.J.; Kim, J.G. Simulation Approach for Cathodic Protection Prediction of Aluminum Fin-Tube Heat Exchanger Using Boundary Element Method. *Metals* **2019**, *9*, 376. [CrossRef]
7. Ahledel, N.; Schulz, R.; Gariepy, M.; Hermawan, H.; Alamdari, H. Electrochemical Corrosion Behavior of Fe_3Al/TiC and Fe_3Al-Cr/TiC Coatings Prepared by HVOF in NaCl Solution. *Metals* **2019**, *9*, 437. [CrossRef]
8. Purwasih, N.; Kasai, N.; Okazaki, S.; Kihira, H.; Kuriyama, Y. Atmospheric Corrosion Sensor Based on Strain Measurement with an Active Dummy Circuit Method in Experiment with Corrosion Products. *Metals* **2019**, *9*, 579. [CrossRef]
9. Mohtadi-Bonab, M.A. Effects of Different Parameters on Initiation and Propagation of Stress Corrosion Cracks in Pipeline Steels: A Review. *Metals* **2019**, *9*, 590. [CrossRef]
10. Choucri, J.; Zanotto, F.; Grassi, V.; Balbo, A.; Ebn Touhami, M.; Mansouri, I.; Monticelli, C. Corrosion Behavior of Different Brass Alloys for Drinking Water Distribution Systems. *Metals* **2019**, *9*, 649. [CrossRef]
11. Bautista, A.; Velasco, F.; Torres-Carrasco, M. Influence of the Alkaline Reserve of Chloride-Contaminated Mortars on the 6-Year Corrosion Behavior of Corrugated UNS S32304 and S32001 Stainless Steels. *Metals* **2019**, *9*, 686. [CrossRef]
12. Santana, J.J.; Ramos, A.; Rodríguez-González, A.; Vasconcelos, H.C.; Mena, V.; Fernández-Pérez, B.M.; Souto, R.M. Shortcomings of International Standard ISO 9223 for the Classification, Determination, and Estimation of Atmosphere Corrosivities in Subtropical Archipelagic Conditions—The Case of the Canary Islands (Spain). *Metals* **2019**, *9*, 1105. [CrossRef]
13. Santana, J.J.; Cano, V.; Vasconcelos, H.C.; Souto, R.M. The Influence of Test-Panel Orientation and Exposure Angle on the Corrosion Rate of Carbon Steel. Mathematical Modelling. *Metals* **2020**, *10*, 196. [CrossRef]
14. Galván-Martínez, R.; Carmona-Hernández, A.; Mejía-Sánchez, E.; Morales-Cabrera, M.A.; Contreras, A. Corrosion Study of Pipeline Steel under Stress at Different Cathodic Potentials by EIS. *Metals* **2019**, *9*, 1353. [CrossRef]

15. Martin, U.; Ress, J.; Bosch, J.; Bastidas, D.M. Evaluation of the DOS by DL–EPR of UNSM Processed Inconel 718. *Metals* **2020**, *10*, 204. [CrossRef]

 © 2020 by the author. Licensee MDPI, Basel, Switzerland. This article is an open access article distributed under the terms and conditions of the Creative Commons Attribution (CC BY) license (http://creativecommons.org/licenses/by/4.0/).

Article

Evaluation of the DOS by DL−EPR of UNSM Processed Inconel 718

Ulises Martin, Jacob Ress, Juan Bosch and David M. Bastidas *

National Center for Education and Research on Corrosion and Materials Performance, NCERCAMP-UA, Dept. Chemical, Biomolecular, and Corrosion Engineering, The University of Akron, 302 E Buchtel Ave, Akron, OH 44325-3906, USA; um11@zips.uakron.edu (U.M.); jtr45@zips.uakron.edu (J.R.); jb394@zips.uakron.edu (J.B.)
* Correspondence: dbastidas@uakron.edu; Tel.: +1-330-972-2968

Received: 31 December 2019; Accepted: 29 January 2020; Published: 1 February 2020

Abstract: In this work, influence of ultrasonic nanocrystal surface modification (UNSM) on the degree of sensitization (DOS) in Inconel 718 has been studied and correlated with the resulting microstructure. The UNSM processed samples decreased their grain size from 11.9 µm to 7.75 µm, increasing the surface of grain boundaries, and thus enhancing the area where δ phase and niobium carbides precipitate. The effect of the UNSM process on the DOS of Inconel 718 was studied by the double loop electrochemical potentiokinetic reactivation (DL−EPR) test. The DL−EPR showed that for UNSM processed samples with no thermal treatment, the DOS increased up to 59.6%, while for UNSM treated samples that were post-annealed at 1000 °C for 10 min and water quenched the DOS decreased down to 40.9%. The increase of grain boundaries surface area and triple junctions after the UNSM process enables the formation of twice the amount of δ phase compared to the as-received Inconel 718 bulk sample. The area fraction of the grain boundary covered by δ phase was of 9.87% in the UNSM region while in the bulk the area fraction was 4.09%. In summary, it was found that after UNSM process, the annealing at 1000 °C for 10 min and water quenching promoted the transformation of γ″ to form δ phase on the grain boundaries, which reduces the intergranular corrosion susceptibility.

Keywords: double loop electrochemical potentiokinetic reactivation (DL−EPR); sensitization; ultrasonic nanocrystal surface modification (UNSM); intergranular corrosion; Inconel 718

1. Introduction

Inconel 718 is a Ni-Fe-Cr superalloy widely used in the aerospace and nuclear industries due to its enhanced corrosion properties within extreme environment conditions [1–3]. Its mechanical and corrosion properties are maintained at temperatures as high as 700 °C, making the Inconel 718 alloy suitable for demanding working conditions [1,4–6]. Among the high strength, fatigue, creep and wear resistance properties, Inconel 718 also has favorable weldability [7–10]. The outstanding mechanical and corrosion properties of Inconel 718 are attributed to its microstructure, which is mainly constituted of austenite, γ phase. The composition of Inconel 718 presents a Ni equivalent value (Ni_{eq}) high enough to promote a single-phase microstructure (γ phase) as seen in the Schaeffler diagram [11]. Besides the γ phase, Inconel 718 precipitates other phases, the most common are: γ′ being a face centered cubic (FCC), with a Ni_3(Al, Ti) composition; γ″ being a body centered tetragonal (BCT) with a Ni_3Nb composition, and δ phase with an orthorhombic crystal structure having a Ni_3Nb composition [4,12–14]. The γ″ phase confers most of the hardening to the Inconel 718 γ-phase matrix; however, it is a metastable phase the more stable form of which is the δ phase. The increased amount of δ phase is at an expense of depleting the γ″ phase concentration in the γ-phase matrix; leading to the worsening of the mechanical properties, mainly the hardness. Nevertheless, δ phase can control the limiting grain growth during solution treatment when present in small amounts [4,15]. Moreover, the δ phase enhances the corrosion resistance as it is a more stable phase [16].

Although Inconel 718 excels with its mechanical properties at elevated temperatures, the depletion of the γ″ phase in exchange of the δ phase, promotes early failure by fatigue and creep [1]. Thus, further improvement of the mechanical properties of Inconel 718 can be achieved through the combination of mechanical and thermal treatments. Mechanical treatments like laser shock peening, shot peening or ultrasonic nanocrystal surface modification (UNSM) are among the most common surface modification treatments used to improve the tribological performance, wear and friction resistance, of Inconel 718 [17–19]. However, the improvement of the mechanical properties may produce some decrease in the corrosion performance. The UNSM process causes the grains on the surface and at the nearest surroundings layers to be crushed and hence the grain boundaries increase, enlarging the available surface coverage for the chromium and niobium carbides to precipitate which will raise the degree of sensitization (DOS) and therefore the intergranular corrosion susceptibility. This corrosion issue has not been considered in previous studies regarding the UNSM surface processing of Inconel and only few works mention it [17,20]. Previous works assert that the corrosion properties are improved after UNSM processing, showing micrographs of the top surface and its deformation by using TEM (transmission electron microscopy). However, there is a lack of electrochemical tests to prove the hypothesis based on the microstructure studies previously mentioned. A combined study of the mechanical performance, the microstructure and the electrochemistry should be carried out to better understand the tradeoff between the mechanical and the corrosion resistance properties.

In order to characterize the effect of thermal treatments on the DOS, previous researchers have used the double loop electrochemical potentiokinetic reactivation (DL–EPR) test. The DL–EPR enables a straightforward comparison between current peaks in the forward scan, activation scan; and backward scan, reactivation scan [21–23]. This method has been shown to produce reliable data from different thermal treatments to sensitized steels [21,24–29]. The DL–EPR has a higher reproducibility among results than the single loop EPR [30]. Previous studies on the Inconel family—mainly the 600 series—have provided promising results on the characterization of the DOS by the DL–EPR method. The Cr depleted areas due to thermal treatments or working conditions can be detected by the DL–EPR, due to the ability of the method to selectively attack the grain boundaries. Studies by DL–EPR on Inconel 600 have shown the enhancement of the susceptibility to intergranular corrosion due to the Cr depleted grain boundaries [31,32]. More recently, studies on grain boundary engineering on Inconel 600 have also been tested with the DL–EPR. They have shown the improvement on Inconel 600 by thermo-mechanical treatments on the protection against intergranular corrosion [33–35]. In addition to the thermal treatments, mechanical processing such as cold work also use DL–EPR to assess the degree of sensitization of steels [36–38].

This work studies the effect of thermo-mechanical processing, UNSM plus annealing, has on the intergranular corrosion susceptibility of Inconel 718. The grain size reduction induces larger grain boundary areas, which then are thermally activated promoting the growth of precipitates, mainly δ phase and Nb carbides. The effect of the UNSM on the DOS in Inconel 718 is studied by means of the DL–EPR. In addition to the electrochemical tests, a microstructural characterization was performed by optical, scanning electron microscopy (SEM) and X-ray diffraction (XRD) to support the DOS results.

2. Materials and Methods

2.1. Materials and Thermo-Mechanical Processing

The material used for this study was Inconel 718, the chemical composition of which is shown in Table 1. Samples were cut into squared sheets of 15 mm length size with a thickness of 3 mm. Before any thermo-mechanical treatment, the samples were polished up to grade 1200 with SiC sandpapers. The different samples studied can be identified in Table 2, where each sample abbreviation corresponds with its thermal and/or mechanical treatment. Four different thermo-mechanically treated samples were studied: Sample I1 was thermally treated in the furnace for 2 h at 675 °C and then water quenched; sample I2 was thermally treated in the furnace at 1000 °C for 10 min and then water quenched, this process was repeated three times; sample I3 was mechanically processed with the UNSM treatment three times; and sample I4 was mechanically processed with the UNSM, annealed at 1000 °C for 10 min and then water quenched. This thermo-mechanical process was repeated three times.

Table 1. Chemical composition of Inconel 718 (wt.%).

Element	Al	C	Cr	Fe	Mo	Nb	Ni	Ti
Content (wt.%).	0.2–1	0.1	17–21	Bal.	2.8–3.3	4.6–5.75	50–55	0.3–1.3

Table 2. UNSM (ultrasonic nanocrystal surface modification) and thermal processing details of Inconel 718.

Sample	Treatment
I1	Annealed at 675 °C for 2 h
I2	Annealed at 1000 °C for 10 min, water quenched, repeated 3 times
I3	UNSM treated, repeated 3 times
I4	UNSM treated and annealed at 1000 °C for 10 min, water quenched, 3 times

The processing parameters used for the UNSM treatment were a tungsten carbide ball with 2.4 mm tip diameter, a static load of 20 N, a scanning speed of 1000 mm/min, an amplitude of 16 μm and a spacing of 10 μm [39].

2.2. Electrochemical Characterization

Cyclic potentiodynamic polarization (CPP) tests were done for each sample in 3.5 wt.% NaCl solution (VWR Chemicals, LLC, Solon, OH, USA) at room temperature (25 °C). All electrochemical tests were conducted using a potentiostat/galvanostat Gamry Reference 600 (Gamry Instruments Inc., Warminster, PA, USA). A three-electrodes configuration cell setup was used, with a saturated calomel electrode (SCE) as the reference electrode (RE), a graphite rod as the counter electrode (CE) and the Inconel 718 samples as the working electrode (WE). The area exposed for the WE was 1 cm^2. The polarization scan was ±1.0 V$_{OCP}$ at a scan rate of 1.667 mV/s for both, forward and backward scans. An open circuit potential (OCP) of 3 h was monitored prior to performing each CPP test.

The DOS of the sensitized samples was obtained by means of the DL–EPR test. A 0.1 M H$_2$SO$_4$ + 0.01 M KSCN (VWR Chemicals, LLC, Solon, OH, USA) test solution was used at room temperature [22,23,40]. All the samples were polished with sandpaper up to 1200 grit and rinsed with water and ethanol and dried with air. The electrochemical cell set up for the DL–EPR test was the same that for the three-electrodes configuration cell from the CPP. This system avoided the intrusion of air and separated the solution from the sample until the chamber was completely deaerated, avoiding the premature contact of the acid solution and the metal, which could attack the surface. The cell was deaerated with nitrogen for 30 min, then the solution was pumped into the electrochemical cell. Continuous N$_2$ bubbling was kept for the entire test to keep the air from entering the system. The OCP was monitored for 30 min from the moment the solution covered the sample until a stable potential was reached. After recording the OCP, a potentiostatic hold of −1.0 V$_{OCP}$ was applied for 1 min with an imposed current limit of 100 mA/cm^2. Then, the samples were polarized from −500 mV$_{OCP}$ to +500 mV$_{OCP}$, and subsequently reversed for a complete DL–EPR test [22]. The scan rate for the potentiokinetic scans was 0.2 mV/s. The DOS was calculated with the ratio between the current density peak in the activation process (forward scan) (i_a) and the current density peak in the reactivation process (backward scan) (i_r) (see Equation (1)). The DL–EPR tests were done in triplicate for each of the different thermo-mechanically treated samples.

$$\text{DOS} = \frac{i_r}{i_a} \times 100 \ (\%) \tag{1}$$

2.3. Microstructural Characterization

The microstructural study was conducted at the cross-section of the Inconel 718 samples. The samples were cut in half, mounted in epoxy resin and polished with 0.05 µm diamond powder. To reveal the microstructure, an etchant solution containing 17 mL HCl and 1 mL H_2O_2 (VWR Chemicals, LLC, Solon, OH, USA) was used. The optical images were taken with a metallographic microscope Nikon eclipse MA 100 (Nikon Corp., Tokyo, Japan), and a Hitachi TM3030 (Hitachi High-Tech. America Inc., Schaumburg, IL, USA) was used to perform the micrographs analysis with the scanning electron microscopy (SEM) technique, as well as energy dispersed X-ray (EDX). The grain size was calculated based on the optical microscopy images from the metallographic microscope at ×100 magnifications following ASTM E112–13 [41]. The amount of precipitates coverage for each sample was measured with the ImageJ software v.1.8.0_112 (National Institutes of Health, Bethesda, MD, USA).

X-ray diffraction (XRD) analysis was performed using a Rigaku SmartLab 3kW X-ray diffractometer (Rigaku Corp., Tokyo, Japan), with a Cu target (K_α = 1.5406 Å). The scan speed was 2°/min over the 2θ range of 40°–95°. The γ, γ″, δ and $NiFe_2O_4$ phases were elucidated in the XRD patterns.

3. Results and Discussion

3.1. Cyclic Potentiodynamic Polarization (CPP)

The CPP curves for each sample are showed in Figure 1. Sample I3 has the lowest corrosion potential (E_{corr}) from all the samples, having a value of −625 mV_{SCE}. However, its corrosion current density (i_{corr}) is the lowest with a value of 2.37 µA/cm². The i_{corr} remains in the µA/cm² range for all the thermo-mechanical treatments; I2 has an i_{corr} of 3.83 µA/cm², while I4 and I1 4.54 µA /cm² and 5.51 µA/cm², respectively. All the values of the E_{corr} and i_{corr} are presented in Table 3. The most passive sample is I2 with an E_{corr} of −484 mV_{SCE}. Samples I2 and I4 present a peak in the anodic branch around the same current density of 0.36 mA/cm²; this peak is associated with the dissolution of the δ phase and NbC [42]. During the anodic polarization of Sample I2, after the dissolution of the δ phase, the sample shows greater repassivation than was observed in Sample I4. Sample I4 shows higher i_{corr} value and, thus is more active compared to Sample I2, despite having the same thermal treatment.

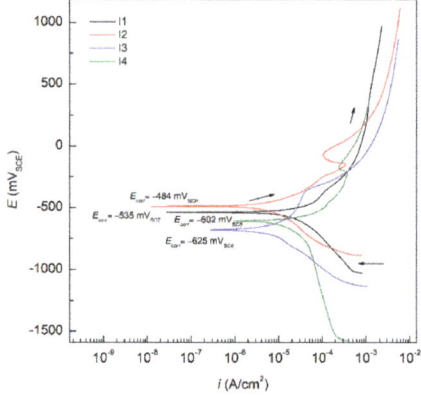

Figure 1. Cyclic potentiodynamic polarization (CPP) curves of each sample in 3.5 wt.% NaCl.

Table 3. E_{corr} and i_{corr} values from the cyclic potentiodinamic polarization (CPP) curves of each sample.

Samples	E_{corr}, mV$_{SCE}$	i_{corr}, µA/cm^2
I1	−535	5.51
I2	−484	3.83
I3	−625	2.37
I4	−602	4.54

3.2. Double Loop Electrochemical Potentiokinetic Eeactivation (DL−EPR)

The effect on the intergranular corrosion susceptibility of the thermo-mechanical process for each of the four Inconel 718 samples was tested with the DL−EPR method. The DL−EPR plot for Sample I1 presents three branches in the backward scan while in the forward resulted in only one; this is because during the backward scan there are three different states (see Figure 2). The upper state is the passive state; here the acid has not broken the passive film that protects the substrate. The second state is the transient state; in this unstable state the passive film is becoming depleted and the acid initiates the surface attack of the Inconel 718. The last state is the active state and it is the only stable one; here the passive film is completely broken and the acid solution can dissolve the metal [21]. The i_r is obtained from the anodic branch which is in between the active and transient states, its value is 75 µA/cm^2 [43]. The DOS for sample I1 is 15.7%, this sensitization is expected for the applied thermal treatment at 675 °C, which is within the range of sensitization temperature for Inconel 718. No intergranular corrosion was produced for Sample I2 after annealing at 1000 °C followed by water quenching (see Figure 3). The forward and backward potentiokinetic scans matched perfectly, meaning that the annealing treatment proved to be an effective way to overcome sensitization. The triple repetition of the annealing treatment at 1000 °C for 10 min followed by water quenching avoid precipitation of γ″ phase. Sample I3 went through three cycles of UNSM treatment drastically increasing the DOS to 59.6% (see Figure 4). This high value of sensitization is the consequence of the increased number of grain boundaries created by the surface plastic deformation produced by the UNSM process. The plastic deformation applied to the top surface of the material increases the intergranular corrosion suffered by the Inconel 718; the current density peaks increased three orders of magnitude from 0.51 mA/cm^2 (i_a Sample I1) to 126.9 mA/cm^2 (i_a Sample I3), proving the higher susceptibility to intergranular corrosion (see Table 4). This plastic deformation is also evidenced in the microstructural analysis where carbides are formed in the newly created twin boundaries. Finally, Sample I4, three cycles of UNSM plus the 1000 °C for 10 min and water quenching, was tested. The DL−EPR plot in Figure 5 suggests an increase of the DOS due to the effect of the UNSM treatment compared to Sample I1. This increase in the DOS up to 40.9% is higher than the 15.7% from Sample I1, however compared to Sample I3 the DOS is almost one-third less. The annealing at 1000 °C for 10 min after each UNSM treatment enables the release of plastic strain and hence reducing the twin boundaries density. In addition to the stress relaxation, due to the annealing at 1000 °C transformation of γ″ occurs, leading to the formation of a more stable and corrosion resistance δ phase. Although the DOS decreases, the current density values of both peaks remain in the same order of magnitude (mA/cm^2). The anodic current density remains in the order of mA/cm^2 for the repassivation of Samples I1, I3 and I4, which is high for a DL−EPR test for stainless steels. However, in the case of Inconel, these current densities are within the expected range (mA/cm^2) as shown by previous authors [44,45].

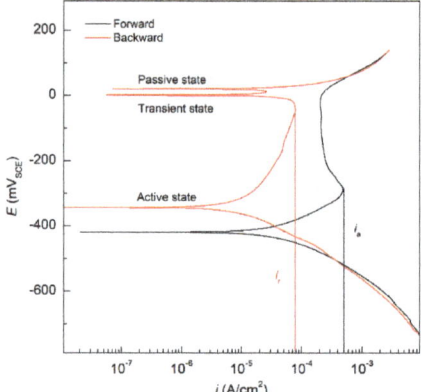

Figure 2. DL−EPR (double loop electrochemical potentiokinetic reactivation) plot for Inconel 718 annealed at 675 °C for 2 h, sample I1.

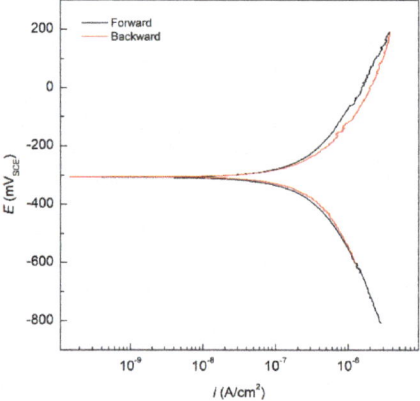

Figure 3. DL−EPR plot for Inconel 718 annealed at 1000 °C for 10 min and water quenched, reapeated three times each, sample I2.

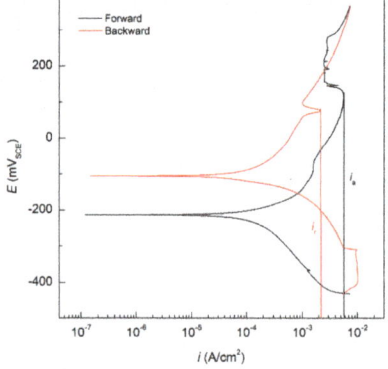

Figure 4. DL−EPR plot for Inconel 718 treated with UNSM × 3, sample I3.

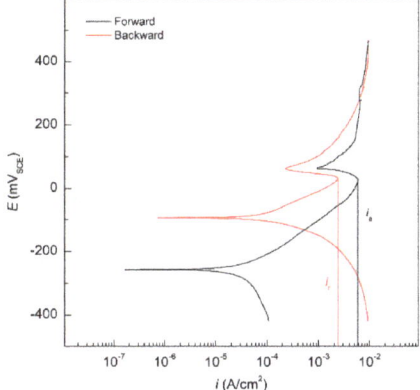

Figure 5. DL–EPR plot of UNSM treated Inconel 718 annealed at 1000 °C for 10 min and water quenched, repeated three times each, Sample I4.

Table 4. Degree of sensitization values for the different thermo-mechanically treated samples of Inconel 718, DOS = ($\frac{i_r}{i_a}$ × 100).

Samples	i_r, mA/cm^2	i_a, mA/cm^2	DOS, %	Standard Deviation
I1	0.08	0.51	15.7	2.1
I2	-	-	-	-
I3	75.7	126.9	59.6	4.5
I4	2.5	6.1	40.9	3.4

3.3. Microstructure Characterization

After the DL–EPR tests were completed, the Inconel 718 samples were cut in half and epoxy mounted to perform the microstructural analysis. The Inconel 718 cross-section was revealed with the etchant containing HCl and H_2O_2, the time to reveal the microstructure was between 15–20 s immersed in the solution.

Figure 6 represents the Inconel 718 after each thermal and/or mechanical treatment. The as-received (AR) Inconel 718 has a microstructure mainly formed by γ phase, having deformation twins scattered throughout the entire microstructure as well as some minor carbide precipitates due to the manufacturing temperatures (see Figure 6a). The Inconel 718 sheets were hot rolled, leaving the top surface with some plastic deformation where the grains were much smaller compared to the bulk. Sample I1 developed more γ″ phase at the grain boundaries because of the thermal treatment at 675 °C for 2 h, as was expected from the TTT (time-temperature-transformation) diagram of Inconel 718 for these temperature and time conditions [46,47] (see Figure 6b). Sample I2 did not show as much γ″ phase on the grain boundaries as Sample I1, but it presented δ phase instead as well as slightly more carbides (see Figure 6c). The annealing treatment at 1000 °C for 10 min and water quenching repeated three times made the γ″ phase transform into its more stable δ phase. In these first three steps, the grain size and the density of deformation twins remained constant as no extra work had been added to the Inconel 718 samples. However, once the UNSM process was applied, the outermost surface layer became heavily deformed, considerably reducing its grain size as well as promoting more twin boundaries. In Sample I3, which suffered three UNSM treatments, three regions were differentiated in the microstructure (UNSM, transition and bulk regions) (see Figure 6d). The closer to the UNSM treated surface, the greater the plastic deformation, reducing the grain size more severely. The transition region between the UNSM treated region and the bulk experienced a higher concentration of twin boundaries; here the plastic deformation was gradually reduced, having a wide average grain size.

Finally, Sample I4 was thermo-mechanically treated with the UNSM, annealed at 1000 °C for 10 min and water quenched, all repeated three times (see Figure 6e). This thermo-mechanical process promotes a higher amount of δ phase in the UNSM region as well as in the transition region. The UNSM thermo-mechanically treated sample now had a higher amount of grain boundaries and hence, a higher probability to promote δ phase and carbides.

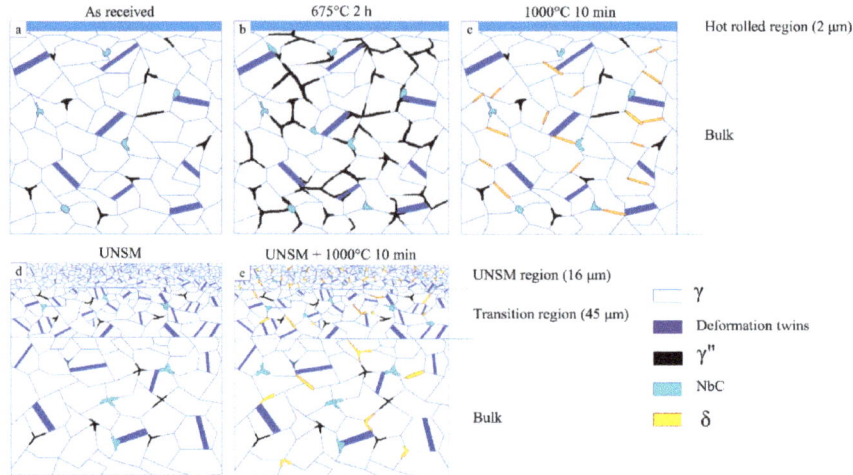

Figure 6. Schematic of the different microstructure as a result of the thermal and UNSM processing of Inconel 718: (**a**) As-received sample, (**b**) I1 sample, (**c**) I2 sample, (**d**) I3 sample and, (**e**) I4 sample.

The optical images in Figure 7 show the microstructure of the Inconel 718 samples, revealing a similar polygonal equiaxed γ phase matrix with scattered twins along the plane [9]. The morphology and deformation twins distribution of the Inconel 718 samples found in this study resemble the one found on the literature [17,48]. Samples AR, I1 and I2 did not experienced UNSM processing, therefore their grain size remained constant with an average size of 11.2 µm, 11.9 µm and 11.4 µm respectively [49] (see Figure 7a–c). While the microstructure of Samples I3 and I4 present three regions: bulk, transition and UNSM (see Figure 7d,e). The bulk region has an average grain size of 11.4 µm, similar to the one for Samples AR, I1 and I2, while in the UNSM region the grain size is reduced to average values of 7.75 µm. It has been found that the DOS value highly increases from Sample I1 to Sample I3 by almost four times, 15.7% and 59.6% respectively. In the transition zone between the bulk and the UNSM region the grain size gradually decreases, having a more dispersed average size, ranging from 9–10 µm. The UNSM processing produced a similar penetration depth of 16.3 µm for surface-modified Samples I3 and I4 (see Figure 7f). Among all the samples, I1 shows the greater amount of precipitates in the grain boundaries; this is due to its thermal treatment temperature being in the sensitization temperature range for Inconel 718, as well as being subjected to it for two hours. In the case of Samples I2 and I4—which were thermally treated at 1000 °C for 10 min and water quenched—the grain boundaries are not affected as much after the annealing treatment, mainly because the time of exposure is minimal compare to the one of Sample I1.

The SEM micrographs of the four treated samples (I1, I2, I3, I4) and the AR sample can be seen in Figure 8, where all the images are at ×600 magnification. The formation of the niobium carbides (NbC) and precipitates are found at the grain boundaries and inside the twins. The AR sample shows the typical microstructure of Inconel 718 that is found in the literature (Figure 8a). The γ matrix has some deformation twins scattered along the microstructure as well as minor niobium carbides, probably formed due to the manufacturing process, which are mainly located at the grain boundaries triple junctions and twins [50]. Sample I1 experiences the highest grain boundary change from among all the samples, drastically increasing the amount of γ″ in the grain boundaries (Figure 8b). This is due to the applied thermal treatment at 675 °C for two hours, which mainly promotes γ″, according to the TTT diagram [46]. In addition to the γ″ formation, higher niobium carbides are seen at the grain boundaries and even inside the grains. By means of EDX, these niobium carbides are characterized, having a 54.52 wt.% Nb, 22.72 wt.% C, 12.68 wt.% Ni, 5.58 wt.% Cr and 4.5 wt.% Fe content. The area analysis also shows the presence of elements that compose the base Inconel 718 alloy. Sample I2 shows δ phase with an equiaxed elongated shape on the grain boundaries due to the annealing treatment at 1000 °C (Figure 8c). However, the exposure time is not enough to promote them as much as Sample I1 did with the γ″ [4]. In the case of Sample I3, as is clearly seen in the three regions from Figure 8d, the bulk shows a similar microstructure to the AR sample, while the transition region becomes more deformed and the formation of twins increases. The high deformation present in the UNSM region produces a very fine grain that is almost undistinguishable at the magnification used (×600). Even if more deformation twins are formed on the transition and UNSM regions, the amount of precipitates does not increase as there is no post-thermal treatment to the UNSM process. Sample I4, annealed at 1000 °C for 10 min and water quenched after USNM process, forms δ phase on the grain boundaries (Figure 8e). The UNSM process produced a smaller grain size on the top layer of the sample, thus increasing grain boundaries surface and hence favoring the formation of precipitates.

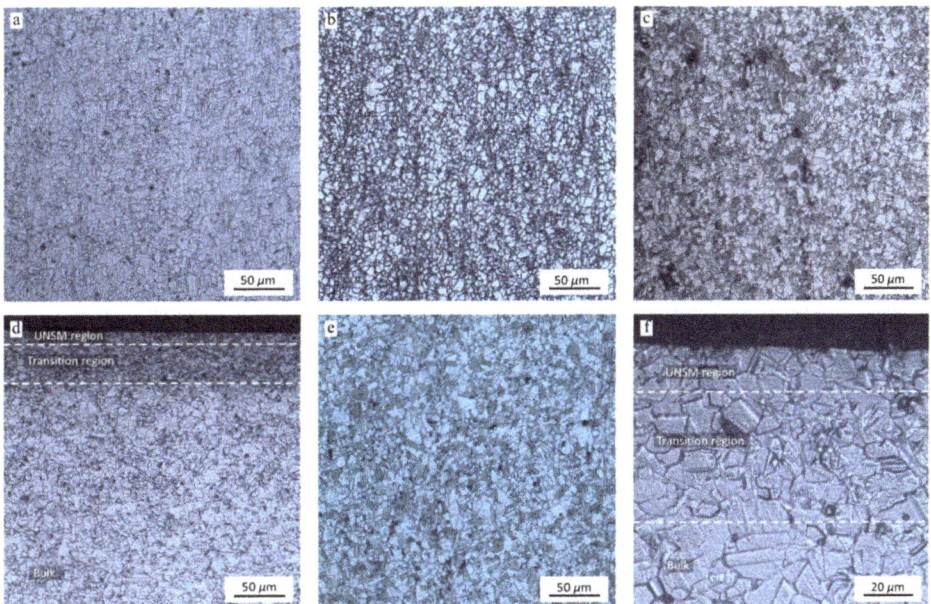

Figure 7. Optical microscope images showing the cross-section microstructure of Inconel 718: (**a**) As-received sample (×10), (**b**) I1 sample (×10), (**c**) I2 sample (×10), (**d**) I3 sample (×10), (**e**) I4 sample (×10) and, (**f**) I4 UNSM region (×50).

In addition to the micrographs of the bulk of each sample, micrographs at ×2000 magnifications were also taken at the top surface. Figure 9a shows the outer most layer of sample AR, where the rolled region can be distinguished from the bulk by the smaller grain size. As seen in Figure 8a, deformation twins and Nb carbides are scattered throughout the microstructure. Sample I1 also shows the numerous formations of γ″ phase, which is increased in the rolled region and nearby due to the density increase of grain boundaries (see Figure 9b). The morphology of Sample I2 is similar to the one seen in Figure 9a with the exception of the increase amount of δ phase in the grain boundaries (see Figure 9c). The top surface of Sample I3, as previously seen in Figure 8d, is heavily deformed to the point at which it is no longer possible to distinguish the grains (see Figure 9d). In the case of Sample I4 (see Figure 9e), the δ phase is more concentrated in the UNSM region, which implies that the volume fraction covered by γ" phase is lower, as shown by previous authors [12,46,51]. Besides the δ phase formation, the microstructure of Sample I4 resembles that previously seen of Sample I3, having higher twin boundaries in the transition region and scattered carbides in the grain boundary triple junctions. Due to the annealing temperature of 1000 °C after each UNSM an oxide scale is formed on top of the UNSM region.

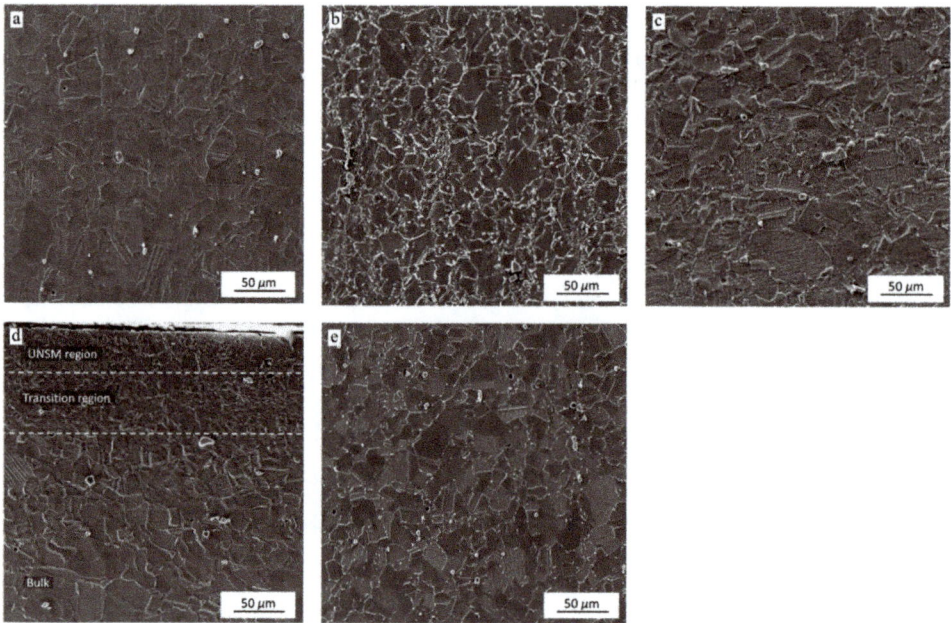

Figure 8. SEM (scanning electron microscopy) images showing the cross-section microstructure of Inconel 718: (**a**) As-received sample (×600), (**b**) I1 sample (×600), (**c**) I2 sample (×600), (**d**) I3 sample (×600) and, (**e**) I4 sample (×600).

In order to quantify the effect of the UNSM process plus the thermal treatment, the area fraction of the grain boundaries covered by δ phase was calculated by ImageJ analysis [49]. Ten different SEM micrographs were used to measure the grain boundaries covered by δ phase; ×1000 magnifications images were used for the bulk region while images at ×2000 magnifications were used for the UNSM region. The image analysis showed that in the bulk region, Sample I2 had a δ phase area fraction of 3.94% while Sample I4 had a δ phase area fraction of 4.09% (see Table 5), while for the UNSM region in Sample I4 the δ phase area fraction increases to 9.87%. In addition to the grain boundary covered by δ phase area fraction study, an analysis of the volume fraction was done using the systematic manual point count by the ASTM E562 [52]. The grid size was set to 100 point grid, and with a relative accuracy of 20% the number of images to measure for Samples I2 and I4 were 10 each. The volume fraction obtained for the bulk of I2 and I4 was 4.15 and 4.25, respectively. The difference between the area and volume fraction is not very significant. The volume fraction for the UNSM region was 10.15. The addition of the annealing treatment at 1000 °C for 10 min reduces the residual stress created by the UNSM process, as well as promoting the formation of δ phase, which also reduces the corrosion susceptibility. Both effects make the DOS decrease from 59.4% to 40.9%, as seen in Figures 4 and 5 for the DL–EPR test.

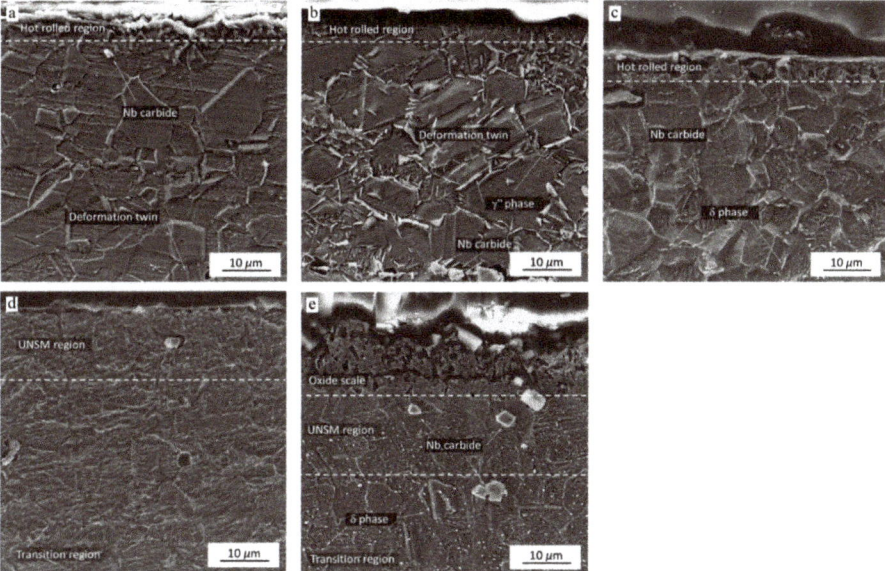

Figure 9. SEM images showing the area close to the treated surface of Inconel 718: (**a**) As-received sample (×2000); (**b**) I1 sample (×2000); (**c**) I2 sample (×2000); (**d**) I3 sample (×2000); (**e**) I4 sample (×2000).

Table 5. δ phase area and volume fraction coverage for Samples I2 and I4.

Region	δ phase Area Fraction, %	Standard Deviation	δ phase Volume Fraction, %	Standard Deviation
		Sample I2		
Bulk	3.94	0.55	4.15	0.23
		Sample I4		
Bulk	4.09	0.47	4.25	0.29
UNSM	9.87	1.53	10.15	1.18

The XRD patterns for the different samples are shown in Figure 10, where the main phases are labeled with their respective peaks in the stick pattern. The AR sample mainly shows γ phase diffraction peaks, as it has neither thermal nor mechanical treatment. These XRD results are in good agreement with the SEM analysis, which did not show precipitates. Sample I1, as previously found in the SEM, shows a higher volume of γ″ in its pattern. The presence of γ″ phase is not directly seen as a single sharp peak; however, it can be confirmed by the distribution of sideband profiles around the γ peak (111) at 2θ = 43° (marked with an arrow in Figure 10), which indicate the presence of an austenite microregion (γ′–FCC) within the γ matrix [53,54]. The γ″ peaks are inside the γ phase peaks, which have a higher intensity. Sample I2 has more Nb precipitates as it was thermally treated at 1000 °C, showing peaks of δ phase. Sample I3 does not show high concentrations of δ phase or γ″ as individual peaks. However, as also seen in Sample I1, sideband profiles are found around 2θ = 43° showing the γ″ inside the γ matrix. Finally, Sample I4, shows the peaks of δ phase with higher intensity, due to the increased amount of δ phase grain boundary coverage because of the UNSM process. It also shows the formation of a new peak around 2θ = 63°, which corresponds to the $NiFe_2O_4$. This spinel phase is a protective oxide that is formed due to the high temperature, which is seen in Figure 9e [55].

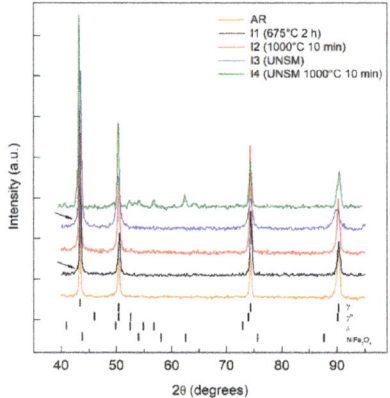

Figure 10. XRD (X-ray diffraction) patterns for Inconel 718 samples with different thermal and/or UNSM treatments.

The effect that the UNSM treatment has on the DOS of Inconel 718 has been studied by different techniques. In this study, it has been shown that the plastic deformation generated during the UNSM treatment reduces the grain size, as seen in both the optical and SEM microstructures images. The grain size reduction increments the concentration of grain boundaries close to the UNSM treated surface, making Inconel 718 more susceptible to intergranular corrosion. The DOS of the UNSM treated Sample I3 is 59.6% while Sample I1 treated at 675 °C for two hours only has a DOS of 15.7%. Nevertheless, the annealing treatment of 1000 °C promotes the formation of δ phase, stable form of the γ″ phase; which reduces the intergranular corrosion susceptibility. This annealing treatment after UNSM process reduces the DOS to 40.9% (Sample I4), compare with 59.6% DOS value obtained for UNSM treated Sample I3 with no post-thermal treatment. The δ phase obtained from the annealing treatment (γ″ → δ transformation) is seen in the diffractograms as a sharp peak at 2θ = 57°.

4. Conclusions

In this work the influence of the UNSM process and post-thermal treatments on the degree of sensitization of Inconel 718 was studied. The main conclusions can be drawn as follows:

The UNSM process depletes the corrosion properties of the Inconel 718 compared to the AR sample. Nevertheless, the application of the annealing treatment at 1000 °C releases stresses and lowers the DOS down to 40.9%, almost one-third the value of non-annealed UNSM sample (59.6%).

The decrease in the grain size due to the UNSM treatment—from 11.3 μm to 7.75 μm—increases the grain boundary density, consequently enhancing the formation of niobium carbides and δ phase due to thermal treatments.

The area fraction of grain boundary covered by δ phase increases from 4.09% in the bulk region to 9.87% in the UNSM region after the annealing treatment is applied to the Inconel 718.

The increased amount of δ phase in the grain boundaries reduces the intergranular corrosion susceptibility as the δ phase is more stable than the γ″ phase. The δ phase preferentially forms in the grain boundaries' triple junctions and the twin boundaries.

Formation of the δ phase in the top surface after the UNSM processing and the annealing thermal treatment of 1000 °C was elucidated with XRD patterns. The thermo-mechanical treated Sample I4, developed a higher intensity δ phase diffraction peak, thus producing a lower DOS value compared to UNSM treated Sample I3, and conferring on Inconel 718 a better performance against intergranular corrosion.

This work enhances the current knowledge of UNSM treatment of Inconel 718 by assessing the resulting DOS and corrosion performance. Therefore, further research is necessary to evaluate the changes in mechanical properties such as wear, friction, and micro-hardness caused by sensitization.

Author Contributions: Conceptualization, D.M.B. and U.M.; methodology, D.M.B. and U.M.; formal analysis, U.M., J.R., J.B. and D.M.B.; investigation, U.M., J.R., J.B. and D.M.B.; resources, D.M.B.; writing—original draft preparation, U.M., J.R., J.B. and D.M.B.; writing—review and editing, U.M. and D.M.B.; supervision, D.M.B.; project administration, D.M.B.; funding acquisition, D.M.B. All authors have read and agreed to the published version of the manuscript.

Funding: This research was funded by The University of Akron, Fellowship Program FRC-207367.

Acknowledgments: The authors acknowledge funding support from The University of Akron, Fellowship Program FRC-207367.

Conflicts of Interest: The authors declare no conflict of interest.

References

1. Hosseini, E.; Popovich, V.A. A review of mechanical properties of additively manufactured Inconel 718. *Addit. Manuf.* **2019**, *30*, 100877. [CrossRef]
2. Kang, Y.J.; Yang, S.; Kim, Y.K.; AlMangour, B.; Lee, K.A. Effect of post-treatment on the microstructure and high-temperature oxidation behaviour of additively manufactured Inconel 718 alloy. *Corros. Sci.* **2019**, *158*, 108082. [CrossRef]
3. Chamanfar, A.; Monajati, H.; Rosenbaum, A.; Jahazi, M.; Bonakdar, A.; Morin, E. Microstructure and mechanical properties of surface and subsurface layers in broached and shot-peened Inconel-718 gas turbine disc fir-trees. *Mater. Charact.* **2017**, *132*, 53–68. [CrossRef]
4. Azadian, S.; Wei, L.Y.; Warren, R. Delta phase precipitation in Inconel 718. *Mater. Charact.* **2004**, *53*, 7–16. [CrossRef]
5. Li, R.B.; Yao, M.; Liu, W.C.; He, X.C. Isolation and determination for δ, γ′ and γ″ phases in Inconel 718 alloy. *Scr. Mater.* **2002**, *46*, 635–638. [CrossRef]
6. Zhang, H.; Li, C.; Guo, Q.; Ma, Z.; Huang, Y.; Li, H.; Liu, Y. Hot tensile behavior of cold-rolled Inconel 718 alloy at 650 °C: The role of δ phase. *Mater. Sci. Eng. A* **2018**, *722*, 136–146. [CrossRef]
7. Amanov, A.; Pyun, Y.S.; Kim, J.H.; Suh, C.M.; Cho, I.S.; Kim, H.D.; Wang, Q.; Khan, M.K. Ultrasonic fatigue performance of high temperature structural material Inconel 718 alloys at high temperature after UNSM treatment. *Fatigue Fract. Eng. Mater. Struct.* **2015**, *38*, 1266–1273. [CrossRef]
8. Campos-Silva, I.; Contla-Pacheco, A.D.; Figueroa-López, U.; Martínez-Trinidad, J.; Garduño-Alva, A.; Ortega-Avilés, M. Sliding wear resistance of nickel boride layers on an Inconel 718 superalloy. *Surf. Coat. Technol.* **2019**, 124862. [CrossRef]
9. Zhang, H.; Li, C.; Liu, Y.; Guo, Q.; Huang, Y.; Li, H.; Yu, J. Effect of hot deformation on γ″ and δ phase precipitation of Inconel 718 alloy during deformation&isothermal treatment. *J. Alloys Compd.* **2017**, *716*, 65–72.
10. Jelvani, S.; Shoja Razavi, R.; Barekat, M.; Dehnavi, M.R.; Erfanmanesh, M. Evaluation of solidification and microstructure in laser cladding Inconel 718 superalloy. *Opt. Laser Technol.* **2019**, *120*, 105761. [CrossRef]
11. Kulkarni, A.; Dwivedi, D.K.; Vasudevan, M. Dissimilar metal welding of P91 steel-AISI 316L SS with Incoloy 800 and Inconel 600 interlayers by using activated TIG welding process and its effect on the microstructure and mechanical properties. *J. Mater. Process. Technol.* **2019**, *274*, 116280. [CrossRef]
12. Kañetas, P.J.P.; Osorio, L.A.R.; Mata, M.P.G.; La Garza, M.D.; López, V.P. Influence of the delta phase in the microstructure of the Inconel 718 subjected to "Delta-processing" heat treatment and hot deformed. *Procedia Mater. Sci.* **2015**, *8*, 1160–1165. [CrossRef]
13. Rafiei, M.; Mirzadeh, H.; Malekan, M. Micro-mechanisms and precipitation kinetics of delta (δ) phase in Inconel 718 superalloy during aging. *J. Alloys Compd.* **2019**, *795*, 207–212. [CrossRef]
14. Anderson, M.; Thielin, A.L.; Bridier, F.; Bocher, P.; Savoie, J. δ Phase precipitation in Inconel 718 and associated mechanical properties. *Mater. Sci. Eng. A* **2017**, *679*, 48–55. [CrossRef]
15. Gao, Y.; Zhang, D.; Cao, M.; Chen, R.; Feng, Z.; Poprawe, R.; Schleifenbaum, J.H.; Ziegler, S. Effect of δ phase on high temperature mechanical performances of Inconel 718 fabricated with SLM process. *Mater. Sci. Eng. A* **2019**, *767*, 138327. [CrossRef]

16. Luo, S.; Huang, W.; Yang, H.; Yang, J.; Wang, Z.; Zeng, X. Microstructural evolution and corrosion behaviors of Inconel 718 alloy produced by selective laser melting following different heat treatments. *Addit. Manuf.* **2019**, *30*, 100875. [CrossRef]
17. Gill, A.; Telang, A.; Mannava, S.R.; Qian, D.; Pyoun, Y.S.; Soyama, H.; Vasudevan, V.K. Comparison of mechanisms of advanced mechanical surface treatments in nickel-based superalloy. *Mater. Sci. Eng. A* **2013**, *576*, 346–355. [CrossRef]
18. Bazarbayev, Y.; Kattoura, M.; Mao, K.S.; Song, J.; Vasudevan, V.K.; Wharry, J.P. Effects of corrosion-inhibiting surface treatments on irradiated microstructure development in Ni-base alloy 718. *J. Nucl. Mater.* **2018**, *512*, 276–287. [CrossRef]
19. Amanov, A.; Pyun, Y.S. Local heat treatment with and without ultrasonic nanocrystal surface modification of Ti-6Al-4V alloy: Mechanical and tribological properties. *Surf. Coat. Technol.* **2017**, *326*, 343–354. [CrossRef]
20. Kondavalasa, S.R.; Prakash, A.; Jagtap, R.; Shanmugam, S.; Samajdar, I.; Vasudevan, V.K.; Wilde, G. On the comparison of graded microstructures developed through high reduction (per pass) cold rolling (HRCR) and ultrasonic nanocrystal surface modification (UNSM) in nickel-base alloy 602CA. *Mater. Charact.* **2019**, *153*, 328–338. [CrossRef]
21. Momeni, M.; Moayed, M.H.; Davoodi, A. Tuning DOS measuring parameters based on double-loop EPR in H_2SO_4 containing KSCN by Taguchi method. *Corros. Sci.* **2010**, *52*, 2653–2660. [CrossRef]
22. Maday, M.F.; Mignone, A.; Vittori, M. The application of the electrochemical potentiokinetic reactivation method for detecting sensitization in Inconel 600. The influence of some testing parameters. *Corros. Sci.* **1988**, *28*, 887–900. [CrossRef]
23. Aydoğdu, G.H.; Aydinol, M.K. Determination of susceptibility to intergranular corrosion and electrochemical reactivation behaviour of AISI 316L type stainless steel. *Corros. Sci.* **2006**, *48*, 3565–3583. [CrossRef]
24. Borello, A.; Mignone, A. Intergranular corrosion in alloy 800: Intercomparison between the Strauss test, the EPR method and magnetic permeability measurements. *Br. Corros. J.* **1982**, *17*, 176–183. [CrossRef]
25. Majidi, A.P.; Streicher, M.A. Potentiodynamic reactivation method for detecting sensitization in AISI 304 and 304L stainless steels. *Corrosion* **1984**, *40*, 393–408. [CrossRef]
26. Roelandt, A.; Vereecken, J. A modified electrochemical technique (electrochemical potentiokinetic reactivation) for evaluating the susceptibility of Inconel 600 to intergranular corrosion. *Corrosion* **1986**, *42*, 289–298. [CrossRef]
27. Abe, S.; Kojima, M.; Hosoi, Y. Stress corrosion cracking susceptibility index, ISCC, of austenitic stainless steels in constant strain-rate test. In *Stress Corrosion Cracking; Slow Strain-Rate Technique*; Ugiansky, G.M., Payer, J.H., Eds.; ASTM International: West Conshohocken, PA, USA, 1979; pp. 294–304.
28. Lo, K.H.; Kwok, C.T.; Chan, W.K. Characterisation of duplex stainless steel subjected to long-term annealing in the sigma phase formation temperature range by the DLEPR test. *Corros. Sci.* **2011**, *53*, 3697–3703. [CrossRef]
29. Taiwade, R.; Shukla, R.; Vashishtha, H.; Ingle, A.; Dayal, R. Effect of grain size on degree of sensitization of chrome-manganese stainless steel. *ISIJ Int.* **2013**, *53*, 2206–2212. [CrossRef]
30. *ASTM G108-94, Standard Test Method for Electrochemical Reactivation (EPR) for Detecting Sensitization of AISI Type 304 and 304L Stainless Steels*; ASTM International: West Conshohocken, PA, USA, 2015.
31. Wu, T.F.; Cheng, T.P.; Tsai, W.T. Effect of electrolyte composition on the electrochemical potentiokinetic reactivation behavior of alloy 600. *J. Nucl. Mater.* **2001**, *295*, 233–243. [CrossRef]
32. Lim, Y.S.; Kim, H.P.; Han, J.H.; Kim, J.S.; Kwon, H.S. Influence of laser surface melting on the susceptibility to intergranular corrosion of sensitized alloy 600. *Corros. Sci.* **2001**, *43*, 1321–1335. [CrossRef]
33. Telang, A.; Gill, A.S.; Kumar, M.; Teysseyre, S.; Qian, D.; Mannava, S.R.; Vasudevan, V.K. Iterative thermomechanical processing of alloy 600 for improved resistance to corrosion and stress corrosion cracking. *Acta Mater.* **2016**, *113*, 180–193. [CrossRef]
34. Telang, A.; Gill, A.S.; Zweiacker, K.; Liu, C.; Wiezorek, J.M.K.; Vasudevan, V.K. Effect of thermo-mechanical processing on sensitization and corrosion in alloy 600 studied by SEM- and TEM-Based diffraction and orientation imaging techniques. *J. Nucl. Mater.* **2018**, *505*, 276–288. [CrossRef]
35. Fang, X.Y.; Li, H.Q.; Wang, M.; Li, C.; Guo, Y.B. Characterization of texture and grain boundary character distributions of selective laser melted Inconel 625 alloy. *Mater. Charact.* **2018**, *143*, 182–190. [CrossRef]
36. Shukla, A.; Patil, A.P. Effect of strain induced martensite reversal on the degree of sensitization of metastable austenitic stainless steel. *Procedia Struct. Integr.* **2019**, *14*, 259–264. [CrossRef]

37. Shukla, S.; Patil, A.P.; Bansod, A.V.; Tandon, V. Effect of cold work and thermal ageing on corrosion and mechanical behavior of Cr-Mn ASS. *Mater. Today Proc.* **2018**, *5*, 17769–17777. [CrossRef]
38. Liu, T.; Xia, S.; Du, D.; Bai, Q.; Zhang, L.; Lu, Y. Grain boundary engineering of large-size 316 stainless steel via warm-rolling for improving resistance to intergranular attack. *Mater. Lett.* **2019**, *234*, 201–204. [CrossRef]
39. Amanov, A.; Umarov, R. The effects of ultrasonic nanocrystal surface modification temperature on the mechanical properties and fretting wear resistance of Inconel 690 alloy. *Appl. Surf. Sci.* **2018**, *441*, 515–529. [CrossRef]
40. Loto, R.T. Comparative study of the pitting corrosion resistance, passivation behavior and metastable pitting activity of N07718, N07208 and 439L super alloys in chloride/sulphate media. *J. Mater. Res. Technol.* **2019**, *8*, 623–629. [CrossRef]
41. *ASTM E112-13, Standard Test Methods for Determining Average Grain Size*; ASTM International: West Conshohocken, PA, USA, 2013.
42. Kurzynowski, T.; Smolina, I.; Kobiela, K.; Kuźnicka, B.; Chlebus, E. Wear and corrosion behaviour of Inconel 718 laser surface alloyed with rhenium. *Mater. Des.* **2017**, *132*, 349–359. [CrossRef]
43. Stansbury, E.E.; Buchanan, R.A. Relationship of individual anodic and cathodic polarization curves to experimentally measured curves. In *Fundamentals Electrochemical Corrosion*, 1st ed.; Stansbury, E.E., Buchanan, R.A., Eds.; ASM international: Materials Park, OH, USA, 2000; pp. 199–201.
44. Ahn, M.K.; Kwon, H.S.; Lee, J.H. Predicting susceptibility of alloy 600 to intergranular stress corrosion cracking using a modified electrochemical potentiokinetic reactivation test. *Corros. Sci.* **1995**, *51*, 441–449. [CrossRef]
45. Yu, G.P.; Yao, H.C. The relation between the resistance of IGA and IGSCC and the chromium depletion of alloy 690. *Corrosion* **1990**, *46*, 391–402. [CrossRef]
46. Jambor, M.; Bokuvka, O.; Novy, F.; Trško, L.; Belan, J. Phase transformations in nickel base superalloy Inconel 718 during cyclic loading at high temperature. *Prod. Eng. Arch.* **2017**, *15*, 15–18. [CrossRef]
47. Chandler, H. *Heat Treater's Guide: Practices and Procedures for Nonferrous alloys*, 1st ed.; Chandler, H., Ed.; ASM international: Materials Park, OH, USA, 1996; p. 48.
48. Kattoura, M.; Telang, A.; Mannava, S.R.; Qian, D.; Vasudevan, V.K. Effect of ultrasonic nanocrystal surface modification on residual stress, microstructure and fatigue behavior of ATI 718Plus alloy. *Mater. Sci. Eng. A* **2018**, *711*, 364–377. [CrossRef]
49. Vanderesse, N.; Anderson, M.; Bridier, F.; Bocher, P. Inter- and intragranular delta phase quantitative characterization in Inconel 718 by means of image analysis. *J. Microsc.* **2016**, *261*, 79–87. [CrossRef]
50. Nabavi, B.; Goodarzi, M.; Khan, A.K. Metallurgical effects of nitrogen on the microstructure and hot corrosion behavior of alloy 718 weldment. *Mater. Charact.* **2019**, *157*, 109916. [CrossRef]
51. Niang, A.; Viguier, B.; Lacaze, J. Some features of anisothermal solid-state transformations in alloy 718. *Mater. Charact.* **2010**, *61*, 525–534. [CrossRef]
52. *ASTM E562-19, Standard Test Method for Determining Volume Fraction by Systematic Manual Point Count*; ASTM International: West Conshohocken, PA, USA, 2019.
53. Liu, W.C.; Yao, M.; Chen, Z.L.; Wang, S.G. Niobium segregation in Inconel 718. *J. Mater. Sci.* **1999**, *34*, 2583–2586. [CrossRef]
54. Anbarasan, N.; Gupta, B.K.; Prakash, S.; Muthukumar, P.; Oyyaravelu, R.; Kumar, R.J.F.; Jerome, S. Effect of heat treatment on the microstructure and mechanical properties of Inconel 718. *Mater. Today Proc.* **2018**, *5*, 7716–7724. [CrossRef]
55. Ramkumar, K.D.; Abraham, W.S.; Viyash, V.; Arivazhagan, N.; Rabel, A.M. Investigations on the microstructure, tensile strength and high temperature corrosion behaviour of Inconel 625 and Inconel 718 dissimilar joints. *J. Manuf. Process.* **2017**, *25*, 306–322. [CrossRef]

© 2020 by the authors. Licensee MDPI, Basel, Switzerland. This article is an open access article distributed under the terms and conditions of the Creative Commons Attribution (CC BY) license (http://creativecommons.org/licenses/by/4.0/).

Article

The Influence of Test-Panel Orientation and Exposure Angle on the Corrosion Rate of Carbon Steel. Mathematical Modelling

Juan J. Santana [1], Víctor Cano [2], Helena C. Vasconcelos [3,4,5] and Ricardo M. Souto [6,7,*]

1. Department of Process Engineering, University of Las Palmas de Gran Canaria, 35017 Las Palmas de Gran Canaria, Gran Canaria, Canary Islands, Spain; juan.santana@ulpgc.es
2. Department of Applied Economics and Quantitative Methods, University of La Laguna, 38071 La Laguna, Tenerife, Spain; vcano@ull.edu.es
3. Faculty of Sciences and Technology, Azores University, 9500-321 Ponta Delgada, São Miguel, Azores Islands, Portugal; helena.cs.vasconcelos@uac.pt
4. Centre of Physics and Technological Research (CEFITEC), Faculty of Sciences and Technology, Universidade Nova de Lisboa, 2829-516 Caparica, Portugal
5. Centre of Biotechnology of Azores (CBA), 9500-321 Ponta Delgada, São Miguel, Azores Islands, Portugal
6. Department of Chemistry, University of La Laguna, 38200 La Laguna, Tenerife, Canary Islands, Spain
7. Institute of Material Science and Nanotechnology, University of La Laguna, 38200 La Laguna, Tenerife, Canary Islands, Spain
* Correspondence: rsouto@ull.es; Tel.: +34-922-318-067

Received: 29 December 2019; Accepted: 27 January 2020; Published: 29 January 2020

Abstract: The effects of both test-panel orientation and exposure angle on the atmospheric corrosion rates of carbon steel probes exposed to a marine atmosphere were investigated. Test samples were exposed in a tree-shape metallic frame with either three exposure angles of 30°, 45° and 60° and orientation north-northeast (N-NE), or eight different orientation angles around a circumference. It was found that the experimental corrosion rates of carbon steel decreased for the specimens exposed with greater exposure angles, whereas the highest corrosion rates were found for those oriented to N-NE due to the influence of the prevailing winds. The obtained data obtained were fitted using the bi-logarithmic law and its variations as to take in account the amounts of pollutants and the time of wetness (TOW) for each particular case with somewhat good agreement, although these models failed when all the effects were considered simultaneously. In this work, we propose a new mathematical model including qualitative variables to account for the effects of both exposure and orientation angles while producing the highest quality fits. The goodness of the fit was used to determine the performance of the mathematical models.

Keywords: Atmospheric corrosion; corrosion rates; exposure angle; orientation angle; predictive models; carbon steel

1. Introduction

Corrosion prevention is an essential task in many areas of society, especially in engineering applications where metals or metal alloys are used [1,2]. Many industries are often faced with serious economic consequences due to unexpected component failures when regular maintenance was not foreseen. In particular, the damage caused by atmospheric corrosion accounts for more than half of the total cost caused by the corrosion phenomena [1,3,4]. Metals are consumed by electrochemical reactions which rates depend on the exposure time ($TEXP$) but the phenomenon itself is very complex since is also highly dependent on numerous damage factors [4–7], each of which are extremely variable. These factors include natural air pollutants and anthropogenic sources [4,5], mainly sulfur dioxide (SO_2), salinity (chlorides, CL) and other pollutants, such as *'particulate matter'* or 'PM', as well as climatic factors such as relative humidity, time of wetness (TOW), temperature, rainfall and wind speed, but also physical characteristics such as shape and type of metal (ferrous or non-ferrous) [4,8], exposure angle [9,10], orientation [8–12] and geographic location [4].

The effects of the atmosphere on the corrosion rate are generally studied by exposing metallic samples to the environment. Then, when both corrosion rate and the environmental parameters are properly measured, relationships between the various damage factors are established through mathematical models [13–21]. The goodness of a fit has been employed to compare the theoretical model to observed data. These tools have become essential for corrosion prevention because they allow forecasting metal behavior in potentially corrosive real operating situations. For example, carbon steel undergoes a less severe attack in urban environments (C3) than in marine environments (C5) [10,22,23].

Many of these models address the combination of climate and air pollutant variables and their influence on the corrosion rate in order to estimate the loss of thickness or the loss of mass per unit area of metallic material. A very popular approach to estimate corrosion rates is the use of linear logarithmic or bi-logarithmic laws (i.e., Equations (1) and (2), respectively) to describe the damage due to atmospheric corrosion versus time in mathematical terms, because the atmospheric corrosion rate is generally non-linear with time [19], and the surface accumulation of corrosion products (e.g. rust layer) strongly influences the subsequent corrosion behavior of the material and tends to reduce the corrosion rate over time [24–26].

$$\ln(CR) = k_i + k_f(TEXP) \tag{1}$$

$$\ln(CR) = k_i + k_f \ln(TEXP) \tag{2}$$

where CR is the corrosion rate. According to these laws, the corrosive behavior of a metal exposed to specific atmosphere can be defined by the two parameters k_i and k_f. The initial corrosion rate, observed during the first year of exposure [19], is described by k_i, while k_f is a measure of the long-term decrease in corrosion rate or passivation that depends directly on the characteristics of the atmosphere and the exposure conditions. The improvement of such equations is a key issue in the effort to fight corrosion. So, these equations can be eventually generalized to account for a vast variety of situations by adequately defining k_i and k_f values as a function of relevant atmospheric variables (AV), which may include TOW, CL, SO_2, etc.

$$\ln(CR) = \underbrace{k_0 + \sum_{n=1}^{m} k_n(AV)}_{k_i} + k_{n+1}(TEXP) \tag{3}$$

$$\ln(CR) = \underbrace{k_0 + \sum_{n=1}^{m} k_n(AV)}_{k_i} + \sum_{n}^{n+1} k_{n+1}\ln(AV) + k_{n+2}\ln(TEXP) \tag{4}$$

In a previous work [13], we developed models to predict atmospheric corrosion rates for carbon steel using statistical regression, "power-law" and other approaches that resulted in forecasts adapted to the wide variety of microclimates found in the Canary Islands (Spain). However, none of these models considered the effects of either the exposure angle (with respect to the horizontal) or the orientation of the tested panel samples. It has been reported in the literature that the orientation of the metal surface and its exposure angle influence the corrosion process thus introducing a further complexity for the development of forecast models [8–12]. Indeed, changes in the time of sun exposure, time of wetness (TOW), dust accumulation, cleaning action of rainfall, etc., occur in these cases. It is usually accepted that the rate of corrosion decreases as the angle of inclination increases from 0° (horizontal) to 90° (vertical) [10]. However, this dependence is not well understood so far. There are two competing phenomena likely to affect the corrosion rate: fast, dry, and wet accumulation of corrosion products. The latter situation often happens in urban and industrial environments where horizontal samples will be more severely corroded that vertical ones due to the accumulation of dirt on the horizontal surface, which increases the TOW and accelerates corrosion rates [27]. Yet, the reverse situation is possible if surfaces are quickly dried. Indeed, metal surfaces exposed to the South (S) exhibited smaller corrosion rates than those oriented to the north (N) because solar radiation from the south resulted in a lower TOW [12].

From the above, it can be observed that there is still a gap in ascertaining and quantifying the phenomena that regulate the dependence of atmospheric corrosion damages on the exposure angle and orientation the metals experience in service. In this work we propose new mathematical models with the objective of evaluating the influence of exposure angle and orientation on the corrosion rate of carbon steel exposed in a site with marine environment located in the Grand Canary Island. Moreover, additional qualitative variables were included in Equations (3) and (4) to yield changes in the independent coefficient k0 that account for the different initial corrosion characteristics associated with variables introduced in the field that are related to their exposure angle (I) and orientation (O):

$$(3) \text{ or } (4) + \sum_{n=1}^{8} \delta_n O_n + \sum_{n=1}^{3} \gamma_n I_n \tag{5}$$

where O_n = N (north), NE (northeast), E (east), SE (southeast), S (south), SW (southwest), W (west), NW (northwest); I_n = 30°, 45° and 60°; and δ_n, γ_n are constants. These new models explicitly include a wider variability of damage factors, and so they are able to forecast more accurate corrosion rates. In summary, we report here an analysis on the effect of orientation and inclination of carbon steel probes exposed in a power station located in Gran Canaria (Canary Islands, Spain) is carried out. The data were fitted to a novel mathematical model using qualitative variables, including both the exposure angle and orientation effects.

2. Materials and Methods

The test site was located in the power station of Jinámar (Gran Canaria, Canary Islands, Spain) and it was selected among those employed to elaborate the Corrosion Maps of the Canary Islands [28,29]. A summary of the location coordinates and atmosphere conditions is given in Table 1.

Table 1. Location and characteristic environment type of the test site.

Test Site	Elevation (m)	Geographic Coordinate		Atmosphere
		North Latitude	West Longitude	
Power station of Jinámar	30	28° 02′ 30″ N	15° 24′ 39″ W	Industrial marine

The test site consisted of two metallic frames on which the metal samples were attached using a nylon screw to avoid the formation of galvanic couples. For the orientation analysis, a tree-shaped metallic frame was built as shown in Figure 1A. The probes were located with an exposure angle of 45° and with eight different orientation angles (N, NE, E, SE, S, SW, W and NW). The distribution of the test panels in the different levels of this metallic frame prevented the downwards drainage of liquid or solid materials on the exposed panels. In a second metallic frame, carbon steel probes were placed with different exposure angles with respect to the horizontal (namely, 30°, 45° and 60°; see Figure 1B). The sensors for pollutants were located at the rear side of this frame, and they were collected on a monthly basis. The determination of sulphur dioxide pollution was made by the candle lead dioxide method according to the ASTM D 2010-85 norm [30]. Chloride measurements were performed using the wet candle method following the specifications of ISO 9225:1992 (E) [31].

The composition of the carbon steel samples is given in Table 2. Plates of approximate dimensions 100 mm × 40 mm × 20 mm were employed. The evaluation of the corrosion rates was made by weight loss of the samples according to the ASTM G1-90 norm [32]. Before being placed in the frames, the samples were marked for identification, cleaned according to the ASTM G1-90 norm [32], subsequently measured and weighed. Samples were collected from the test sites every two months for one year. Corrosion products were removed by chemical operation as described by the ASTM G1-90 standard [32]. After the samples were cleaned and dried, they were weighed again.

The time of wetness (TOW) was determined from the data collected using relative humidity hygrometers placed in a small cabinet at the rear of the station frames, and they were complemented with data supplied by the National Meteorological Institute of Spain (AEMET, Madrid, Spain). The latter were cumulative values taken over 8 h periods in a systematic way, whereas the autonomous hygrometers produced a continuous recording with autonomy for about one month. Data on the speed and direction of the winds were kindly supplied by AEMET.

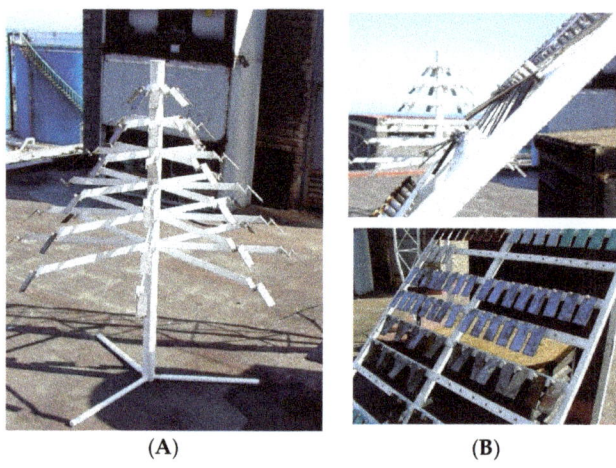

(A) (B)

Figure 1. Metallic frames employed to expose the carbon steel panels for the investigation of the influence of (**A**) the orientation and (**B**) the exposure angles on the atmospheric corrosion rate.

Table 2. Composition of the carbon steel.

Composition (wt. %)										
Si	Fe	C	Mn	Zn	Ti	Cu	Mg	Al	Others	Fe
0.08	99.47	0.06	0.37	-	-	-	-	-	0.02	balance

3. Results and Discussion

3.1. Concentrations of Pollutants and Measurement of Corrosion Rates

The test site selected for this study was one of the used in the elaboration of the Corrosion Map of the Canary Islands where the carbon steel was one of the metals analysed [28]. In a recent review of the Corrosion Maps performed after the modification of the ISO 9223 norm [33], the previous results were confirmed [29]. Table 3 shows the amounts of pollutants as well as the TOW recorded bimonthly along the full period of study. The chloride levels obtained during the exposure time were constants with an average value of 74.98 ± 7.86 mg/(m^2·day), whereas the sulphur dioxide (SO$_2$) levels obtained show greater variability (5.14 ± 2.24 mg/(m^2·day)). TOW increases linearly until 4312 h/year with a rate of increase 10 h/day. The main component of the winds is N-NE that corresponds with the Alisios Trade Winds affecting the Archipelago during most of the year.

Table 3. Chloride and sulphur dioxide deposition rates and time of wetness recorded bimonthly at the test station over one year.

Exposure Time (day)	SO_2 (mg/(m²·day))	Cl^- (mg/(m²·day))	TOW (h/year)
60	2.99	85.43	862
123	2.17	78.10	1552
182	4.47	80.86	1897
243	6.74	68.45	2673
305	7.12	64.68	3553
396	7.33	72.36	4312

3.2. Corrosion Rates

Table 4 shows the corrosion rates for the carbon steel samples exposed with different inclination and orientation angles. From an analysis of these data, it is observed that the corrosion rate varied with the exposure and orientation angles for each period of time under consideration. As for the exposure angle, the corrosion rates showed a maximum value of 71.39 µm/year and a minimum of 25.86 µm/year for all the periods. For one year of exposure, the higher corrosion rate corresponded to the carbon steel panels exposed with an angle of 30°, and the lowest corrosion rates for those exposed with 60°. For the sake of comparison, the 45° inclination as taken as reference because this is the typical exposure angle previously used all the studies of atmospheric corrosion that are carried in the North Hemisphere [6], and it was the exposure angles employed for the elaboration of the Corrosion Map of the Canary Islands [28,29].

Table 4. Variation of the corrosion rates measured with different exposure angles and orientations of the carbon steel panels. N (north), NE (northeast), E (east), SE (southeast), S (south), SW (southwest), W (west), NW (northwest).

Time (day)	r_{corr} (µm/year)										
	Exposure Angle			Orientation							
	30°	45°	60°	N	NE	E	SE	S	SW	W	NW
60	71.1	56.1	55.1	61.38	62.60	56.15	54.40	64.22	56.91	57.07	58.53
123	52.4	51.7	46.9	52.52	51.72	46.29	51.82	47.19	52.03	52.50	51.26
182	44.6	39.5	38.8	41.77	39.46	39.70	45.86	40.88	45.86	31.08	39.60
243	39.3	37.5	33.4	39.78	37.51	36.22	36.91	34.87	37.57	27.48	35.10
305	35.7	31.4	29.6	32.25	31.44	30.19	28.63	27.81	29.78	25.89	32.96
396	31.9	24.7	26.0	25.18	28.76	23.50	22.92	22.72	23.33	24.91	26.48

In this way, it can be observed the corrosion rate was 29.1% greater for 30° and −5.6% for 60° exposure angles after 1-year exposure. The changes with the exposure angles determined for each time were not constant, and they showed a quasi-linear dependence. According to the data obtained in this work, the corrosion rate decreased with increasing exposure angles of the carbon steel panels, in good agreement with previous observations by other authors for carbon steel probes exposed to different atmospheres [10,12,34–38].

On the other hand, in regards to the corrosion rates determined for different orientation angles, a maximum corrosion rate of 64.24 µm/year and a minimum of 22.72 µm/year were obtained for all the periods. The corrosion rates for one year of exposure ranged between 22.70 (NW) and 24.65 (NE) µm/year (ca. 26.6% higher than those oriented to the south). The highest corrosion rates after one year of exposure were found for those steel panels oriented N-NE that were facing the prevailing winds. Figure 2 shows the evolution of the weight loss for each exposure angle (see Figure 2A) and orientation angle (Figure 2B). These results are in good agreement with those obtained for Vera et al. [10] in Chile in a similar test site.

In summary, according to the ISO 9223 norm [33], this test site shows a corrosivity category of C4, a S_2 pollution level by airborne salinity ($60 < S_d \leq 300$ mg/(m^2·day)) and a P_1 pollution level for by sulphur-containing substances represented by SO_2 ($4 < P_d \leq 24$ mg/(m^2·day)), and a time of wetness (TOW) level of τ_4 ($2500 < \tau \leq 5500$ h/year).

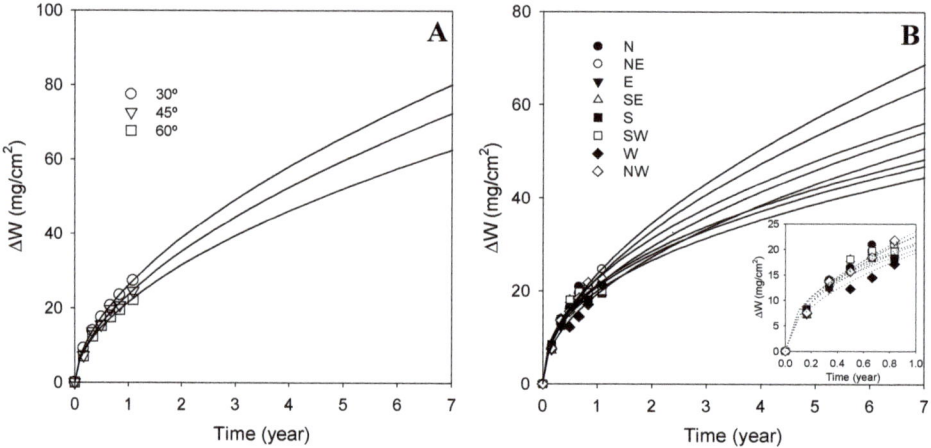

Figure 2. Weight loss of carbon steel as a function of time for panels exposed with varying: (**A**) exposure angles with respect to the horizontal, and (**B**) orientations.

3.3. Mathematical Models

The mathematical models considered in this work were derived from the generic model showed in Equation (5), and they are listed next:

$$\ln(CR) = k_0 + k_1 TEXP \tag{6}$$

$$\ln(CR) = k_0 + k_1 TEXP + \delta_2 O2 + \delta_3 O3 + \delta_4 O4 + \delta_5 O5 + \delta_6 O6 + \delta_7 O7 + \delta_8 O8 + \gamma_1 I1 + \gamma_2 I2 \tag{7}$$

$$\ln(CR) = k_0 + k_1 \ln(TEXP) \tag{8}$$

$$\ln(CR) = k_0 + k_1 \ln(TEXP) + \delta_2 O2 + \delta_3 O3 + \delta_4 O4 + \delta_5 O5 + \delta_6 O6 + \delta_7 O7 + \delta_8 O8 + \gamma_1 I1 + \gamma_2 I2 \tag{9}$$

$$\ln(CR) = k_0 + k_1 SO2 + k_2 CL + k_3 TOW + k_4 \ln(TEXP) \tag{10}$$

$$\ln(CR) = k_0 + k_1 SO2 + k_2 CL + k_3 TOW + k_4 \ln(TEXP) + \delta_2 O2 + \delta_3 O3 + \delta_4 O4 + \delta_5 O5 + \delta_6 O6 + \delta_7 O7 + \delta_8 O8 + \gamma_1 I1 + \gamma_2 I2 \tag{11}$$

$$\ln(CR) = k_0 + k_1 SO2 + k_2 CL + k_3 \ln(TOW) + k_4 \ln(TEXP) \tag{12}$$

$$\ln(CR) = k_0 + k_1 SO2 + k_2 CL + k_3 \ln(TOW) + k_4 \ln(TEXP) + \delta_2 O2 + \delta_3 O3 + \delta_4 O4 + \delta_5 O5 + \delta_6 O6 + \delta_7 O7 + \delta_8 O8 + \gamma_1 I1 + \gamma_2 I2 \tag{13}$$

$$\ln(CR) = k_0 + k_1 \ln(SO2) + k_2 \ln(CL) + k_3 \ln(TOW) + k_4 \ln(TEXP) \tag{14}$$

$$\ln(CR) = k_0 + k_1 \ln(SO2) + k_2 \ln(CL) + k_3 \ln(TOW) + k_4 \ln(TEXP) + \delta_2 O2 + \delta_3 O3 + \delta_4 O4 + \delta_5 O5 + \delta_6 O6 + \delta_7 O7 + \delta_8 O8 + \gamma_1 I1 + \gamma_2 I2 \tag{15}$$

$$\ln(CR) = k_0 + k_1 \ln(SO2) + k_2 \ln(CL) + k_3 TOW + k_4 \ln(TEXP) \tag{16}$$

$$\ln(CR) = k_0 + k_1 \ln(SO2) + k_2 \ln(CL) + k_3 TOW + k_4 \ln(TEXP) + \delta_2 O2 + \delta_3 O3 + \delta_4 O4 + \delta_5 O5 + \delta_6 O6 + \delta_7 O7 + \delta_8 O8 + \gamma_1 I1 + \gamma_2 I2 \tag{17}$$

where CR is the corrosion rate expressed in µm/year; TEXP, the exposure time (year); TOW, time of wetness (year); CL, concentration of chlorides (g/(m^2·year)); and SO2, concentration of SO$_2$ (g/(m^2·year)). The qualitative variables (O2 to O8) were included in Equations (7), (9), (11), (13) and (15) to produce changes in the independent coefficient k_0 accounting for the different initial corrosion characteristics associated with the orientation of the probes. Analogously, I1 and I2 are qualitative variables included in Equations (7), (9), (11), (13) and (15) to produce changes in the independent coefficient k_0 accounting for the different initial corrosion characteristics associated to the exposure angle of the probes.

The analysis of the different models was first performed considering the separate effects of the orientation and the exposure angle of the samples, and later evaluating their combined effect as they would operate together.

Figure 3 shows the corrosion rates measured for carbon steel panels exposed with different exposure and orientation angles as a function of elapsed time (cf. Figure 3A,B, respectively). It is observed that the data conformed to the potential law with a very good fit quality (R^2 = 0.9999 for 30°, 0.8902 for 45°, and 0.9629 for 60° exposure angles), as shown in Table 5. When the evolution of the corrosion rate was analyzed exclusively in terms of the orientation angle, the bi-logarithmic model also fitted very well in most cases except for the SE orientation (namely, R^2 = 0.7935; cf. Table 5). The results of the fists are given in the Supplementary Information, where Table S-1 shows the results using only the data related with the exposure angle, whereas Table S-2 shows those taking into account only the data according to the orientation angle.

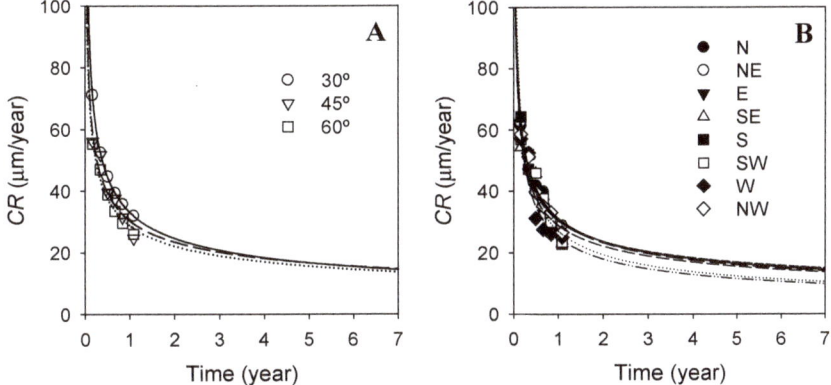

Figure 3. Data fitting to the bi-logarithmic law for the corrosion rates of carbon steel panels exposed with varying: (**A**) exposure angles with respect to the horizontal, and (**B**) orientations.

Table 5. Values of the k and n constants for the potential law model (CR in µm/year) and fit quality.

Constants	Exposure Angle			Orientation						
	30°	45°	60°	N	NE	E	SE	S	SW	W	NW
k	33.0920	30.3580	28.7564	31.1281	30.6990	28.8354	30.4171	27.0422	30.7166	25.0282	30.3778
	(0.049)	(2.3768)	(1.2736)	(2.1182)	(1.3036)	(1.6732)	(3.4883)	(1.3639)	(3.1321)	(3.0355)	(1.7038)
n	−0.4236	−0.3714	−0.3789	−0.3993	−0.4079	−0.3877	−0.3687	−0.4901	−0.3810	−0.4847	−0.3848
	(0.0012)	(0.06505)	(0.03663)	(0.0556)	(0.0345)	(0.04774)	(0.0954)	(0.0392)	(0.0842)	(0.0944)	(0.0462)
$\overline{R}^2$	0.9999	0.8902	0.9629	0.9273	0.9710	0.9419	0.7935	0.9795	0.8392	0.8664	0.9441

Notes: Standard errors are given within brackets below the corresponding estimated coefficient.

3.3.1. Exposure Angle

When considering the models (6) and (8) taking TEXP as the only explanatory variable, it was found that both models exhibited similar fit qualities ($\overline{R}^2$ = 0.8667 and 0.8974, respectively). In these models it was assumed that the exponent k_1 depended on the variables corresponding to the environment [39], and they were further developed to account for the levels of chlorides, SO_2 and TOW [40], producing the models (10), (12), (14) and (16). But the application of these new models did not produce any significant improvement as compared to the results obtained from the application of models (6) and (8).

From a more detailed analysis of the results, TOW was found to be the most significant variable among them. Since the different probes were exposed to the same levels of pollutants in the atmosphere, it was deduced that the lifetime of the water film over the surface of the probe must be an important factor influencing the corrosion rate. The actual levels of pollutants deposited on the metallic surface would depend on the exposure angle of the sample, and their effect should also depend on the kind of particle involved. On the other hand, the exposure angle can affect deposition of pollutants in two ways, namely gravitational settling of particles and condensation of water. According to Spence et al. [8], the deposition rate would be expected to be proportional to the projected horizontal area, which is given by the cosine of the exposure angle. Therefore, a bigger angle should facilitate the run-off of particles from the surface leading to a decrease in the corrosion rate of the metal. The new models described by Equations (7), (9), (11), (13), (15) and (17) were obtained by introducing a new qualitative variable related to the exposure angle, I. These models showed better fit qualities than the corresponding ones without the qualitative variable, although major differences between them was not observed ($\overline{R}^2 \approx 0.98$ for all of them) except for model (7) (with $\overline{R}^2 = 0.9482$). The estimation of the parameter of the qualitative variable related to the inclination angle of 30°, $I1$, indicates a greater effect than for 45°, with an increase of 9.41%. The estimation of the parameter related to $I2$ shows a negative value in relation to the reference angle, with a fall of 8.24%. This is in good agreement with the corrosion rates observed so to biggest angles lowest corrosion rates.

3.3.2. Orientation Angle

The data were fitted with the models (6)–(17) where the orientation angle is introduced as explanatory variable (O) together with the levels of pollutants, TOW and TEXP. In this case, the inclination angle was fixed at 45° because it is the typical procedure employed for this geographical zone.

From an analysis of the models (6) to (9), models (6) and (7) exhibited better fit qualities (i.e., $\overline{R}^2 = 0.9194$ and 0.9363, respectively) than models (8) and (9). In the latter, although model (9) includes the qualitative variable O, the obtained fit did not improve compared to that with model (8). In models (10) to (17), the variables related with the levels of pollutants and the TOW are included, being the models with the qualitative variables those delivering the best fits. Indeed, the best regression indexes (namely, $\overline{R}^2 = 0.9401$) were obtained for the models (15) and (17). In this case, all the coefficients related with the qualitative variables were negative. This feature implies that lower corrosion rates will be obtained with regards to the panels oriented to the North, with the biggest reductions occurring for the orientations East ($O3$), South ($O5$) and West ($O7$), with ratios of 7.91%, 7.47% and 14.91%, respectively.

3.3.3. Global Mode

In the previous two sections, the fits were obtained by working solely on either the exposure or the orientation angles. With the aim of obtaining a global model including all the variables, models (6)–(17) were next applied to the whole set of data. The results are shown in Table 6. It can be concluded that the models including the qualitative variables showed very good fit qualities, being difficult to make a clear distinction among them based on their performance (that is, $\overline{R}^2 > 0.91$ in all the cases). Anyway, those models including the environmental variables (i.e., models (11), (13), (15) and (17)) exhibited the best fits (with $\overline{R}^2 > 0.94$) in comparison with models (7) and (9).

Table 6. Values of the constants k_i, δ_i and γ_i obtained from the application of models (6)–(15).

Variables	Model (6)	Model (7)	Model (8)	Model (9)	Model (10)	Model (11)	Model (12)	Model (13)	Model (14)	Model (15)	Model (16)	Model (17)
Constant	4.2192 * (0.0258)	4.2484 * (0.0353)	3.3580 * (0.0190)	3.3943 * (0.0379)	4.2499 * (0.2587)	4.2862 * (0.1851)	3.8302 * (0.2637)	3.8666 * (0.1887)	5.2157 * (0.7226)	5.2521 * (0.5119)	4.7733 * (0.6890)	4.8097 * (0.4880)
TEXP	−0.9193 * (0.0385)	−0.9193 * (0.0289)	-	-	-	-	-	-	-	-	-	-
SO2	-	-	-	-	−0.0726 ** (0.0383)	−0.0726 * (0.0271)	−0.1092 * (0.0345)	−0.1092 * (0.0245)	-	-	-	-
CL	-	-	-	-	−0.0133 (0.0082)	−0.0133 ** (0.0058)	−0.0287 * (0.0093)	−0.0287 * (0.0066)	-	-	-	-
TOW	-	-	-	-	−0.8768 * (0.2996)	−0.8768 * (0.2121)	-	-	-	-	-	-
LTEXP	-	-	−0.4463 * (0.0206)	−0.4463 * (0.0167)	−0.2363 * (0.0616)	−0.2363 * (0.0435)	0.1289 (0.1760)	0.1289 (0.1247)	0.2260 (0.1800)	0.2260 *** (0.1274)	−0.9763 * (0.2796)	−0.9763 * (0.1978)
LSO2	-	-	-	-	-	-	-	-	−0.1547 * (0.0486)	−0.1547 * (0.0332)	−0.2211 * (0.0612)	−0.2211 * (0.0395)
LCL	-	-	-	-	-	-	-	-	−0.7264 * (0.2455)	−0.7264 * (0.01737)	−0.0963 *** (0.0489)	−0.0963 * (0.0346)
LTOW	-	-	-	-	-	-	−0.6523 * (0.2236)	−0.6523 * (0.1584)	−0.7762 * (0.2225)	−0.7762 * (0.1574)	−0.2828 (0.2066)	−0.2828 *** (0.1461)
O2	-	−0.0006 (0.0434)	-	−0.0006 (0.0511)	-	−0.0006 (0.0396)	-	−0.0006 (0.0396)	-	−0.0006 (0.0439)	-	−0.0006 (0.0395)
O3	-	−0.0825 ** (0.0434)	-	−0.0824 (0.0511)	-	−0.0824 ** (0.0396)	-	−0.0824 ** (0.0396)	-	−0.0824 ** (0.0439)	-	−0.0824 * (0.0395)
O4	-	−0.0548 (0.0434)	-	−0.0548 (0.0511)	-	−0.0548 (0.0396)	-	−0.0548 (0.0396)	-	−0.0548 (0.0441)	-	−0.0548 (0.0395)
O5	-	−0.0777 *** (0.0434)	-	−0.0776 (0.0511)	-	−0.0776 *** (0.0396)	-	−0.0776 *** (0.0396)	-	−0.0776 *** (0.0439)	-	−0.0776 *** (0.0395)
O6	-	−0.0341 (0.0434)	-	−0.0341 (0.0511)	-	−0.0341 (0.0396)	-	−0.0341 (0.0396)	-	−0.0341 (0.0439)	-	−0.0341 (0.0395)
O7	-	−0.1615 * (0.0434)	-	−0.1615 * (0.0511)	-	−0.1615 * (0.0396)	-	−0.1615 * (0.0396)	-	−0.1615 * (0.0439)	-	−0.1615 * (0.0395)
O8	-	−0.0297 (0.0434)	-	−0.0297 (0.0511)	-	−0.0297 (0.0396)	-	−0.0297 (0.0396)	-	−0.0297 (0.0439)	-	−0.0297 (0.0395)
I1	-	0.0899 * (0.0377)	-	0.0899 * (0.0443)	-	0.0899 ** (0.0343)	-	0.0899 ** (0.0343)	-	0.0899 ** (0.0343)	-	0.0899 ** (0.0342)
I2	-	−0.0864 ** (0.0377)	-	−0.0864 * (0.0443)	-	−0.0864 ** (0.0343)	-	−0.0864 ** (0.0343)	-	−0.0864 ** (0.0343)	-	−0.0864 ** (0.0342)
$\bar{R}^2$	0.8908	0.9373	0.8684	0.9133	0.8964	0.9480	0.8964	0.9480	0.8966	0.9482	0.8966	0.9483
N	72	72	72	72	72	72	72	72	72	72	72	72

Notes: Standard errors are given within brackets below the corresponding estimated coefficient; * Significant at 1% level. ** Significant at 5% level. *** Significant at 10% level.

It can be concluded that the models including the qualitative variables showed very good fit qualities, being difficult to make a clear distinction among them based on their performance (that is, $\overline{R}^2 > 0.91$ in all the cases). Anyway, those models including the environmental variables (i.e., models (11), (13), (15) and (17)) exhibited the best fits (with $\overline{R}^2 > 0.94$) in comparison with models (7) and (9).

From the foregoing, it can be concluded that model (18) would be a good mathematical model to describe all the effects participating in an atmospheric corrosion process as it accounts for the effect of the pollutants, TOW, time of exposure, as well as the exposure angle and the orientation of the test panels.

$$\begin{aligned}\ln(CR) = &\ 4.8097 - 0.0963\ln(SO2) - 0.2828\ln(CL) - 0.9763 TOW - 0.2211\ln(TEXP) - 0.0006 O2 \\ &- 0.0824 O3 - 0.0548 O4 - 0.0776 O5 - 0.0341 O6 - 0.1615 O7 - 0.0297 O8 + 0.0899 I1 - 0.0864 I2\end{aligned} \quad (18)$$

4. Conclusions

The corrosive process occurring on carbon steel specimens exposed in a marine test site has been characterized taking into account the exposure angle and the orientation of the probes. It was found that the corrosion rates would diminish with greater exposure angles, in good accordance with previous reports in the literature. Next, the carbon steel probes with N-NE orientation were those with the highest corrosion rates because they were directly facing the prevailing winds.

Although data could be adjusted to the bi-logarithmic law with good fits when either the exposure angle or the orientation were considered separately, the fits became worst when all the data were taken into account simultaneously. The quality of the fits could be improved with respect to the bi-logarithmic law models when the variable concentration of pollutants and TOW were introduced in the model, but not to full satisfaction. A definite improvement of the model was attained by incorporating qualitative variables, thus leading to obtaining a global model with a high degree of adjustment ($\overline{R}^2 = 0.99$), being the first mathematical model to incorporate the effect of inclination and orientation together.

Supplementary Materials: The following are available online at http://www.mdpi.com/2075-4701/10/2/196/s1, Table S-1: Values of the constants k_i and δ_i in the models (6) to (15) when considering solely the effect of the exposure angle (i.e., $\delta_i = 0$ in all cases); and Table S-2: Values of the constants k_i and δ_i in the models (6) to (15) when considering solely the effect of the orientation (i.e., $\gamma_n = 0$ in all cases).

Author Contributions: Conceptualization, J.J.S and R.M.S.; methodology, J.J.S., R.M.S. and H.C.V.; software, J.J.S. and V.C.; validation, J.J.S., V.C., R.M.S. and H.C.V.; formal analysis, J.J.S., R.M.S. and V.C.; investigation, J.J.S., V.C., H.C.V. and R.M.S.; resources, J.J.S. and R.M.S.; data curation, J.J.S. and V.C.; writing—original draft preparation, J.J.S. and H.C.V.; writing—review and editing, J.J.S. and R.M.S.; visualization, J.J.S. and R.M.S.; supervision, J.J.S. and R.M.S.; project administration, J.J.S. and R.M.S.; funding acquisition, J.J.S. and R.M.S. All authors have read and agreed to the published version of the manuscript.

Funding: This research was funded by the Canarian Agency for Research, Innovation and Information Society (Las Palmas de Gran Canaria, Spain) and the European Social Fund (Brussels, Belgium) under grant ProID2017010042.

Conflicts of Interest: The authors declare no conflict of interest.

References

1. Ghali, E.; Sastri, V.S.; Elboujdaini, M. *Corrosion Prevention and Protection. Practical Solutions*; John Wiley & Sons, Ltd.: Chichester, UK, 2007; ISBN 9780470024546.
2. Cramer, S.D.; Covino, B.S., Jr. *Corrosion, Fundamentals, Testing and Applications, ASM Handbook Series*, 1st ed.; American Society for Testing Materials Int.: Columbus, OH, USA, 2003; Volume 13A, ISBN 978-1-62708-182-5.
3. Schweitzer, P.A. *Fundamentals of Metallic Corrosion*; CRC Press: Boca Raton, FL, USA, 2006; ISBN 9780429127137.
4. Leygraf, C.; Wallinder, I.O.; Tidblad, J.; Graedel, T. *Atmospheric Corrosion*; John Wiley & Sons, Inc.: Hoboken, NJ, USA, 2016; ISBN 9781118762134.
5. Schweitzer, P.A. *Atmospheric Degradation and Corrosion Control*; CRC Press: New York, NY, USA, 1999; ISBN 9780824777098.

6. Veleva, L.; Kane, R.D. Atmospheric corrosion. In *ASM Handbook Volume 13A: Corrosion: Fundamentals, Testing, and Protection*; Cramer, S.D., Covino, B.S., Jr., Eds.; American Society for Testing Materials Int.: Columbus, OH, USA, 2003; Volume 13A, pp. 196–209. ISBN 978-0-87170-705-5.
7. Giardina, M.; Buffa, P. A new approach for modeling dry deposition velocity of particles. *Atmos. Environ.* **2018**, *180*, 11–22. [CrossRef]
8. Spence, J.W.; Lipfert, F.W.; Katz, S. The effect of specimen size, shape, and orientation on dry deposition to galvanized steel surfaces. *Atmos. Environ. Part A. Gen. Top.* **1993**, *27*, 2327–2336. [CrossRef]
9. Morcillo, M.; Chico, B.; Díaz, I.; Cano, H.; de la Fuente, D. Atmospheric corrosion data of weathering steels. A review. *Corros. Sci.* **2013**, *77*, 6–24. [CrossRef]
10. Vera, R.; Rosales, B.M.; Tapia, C. Effect of the exposure angle in the corrosion rate of plain carbon steel in a marine atmosphere. *Corros. Sci.* **2003**, *45*, 321–337. [CrossRef]
11. Edney, E.O.; Cheek, S.F.; Stiles, D.C.; Corse, E.W. Field study investigations of the impact of shape, size and orientation on dry deposition induced corrosion of galvanized steel. *Atmos. Environ. Part A. Gen. Top.* **1992**, *26*, 2353–2363. [CrossRef]
12. Coburn, S.K.; Komp, M.E.; Lore, S.C. Atmospheric corrosion rates of weathering steels at test sites in the Eastern United States—Effect of environment and test-panel orientation. In *Atmospheric Corrosion*; Kirk, W., Lawson, H., Eds.; ASTM STP 1239; American Society for Testing Materials Int.: West Conshocken, PA, USA, 1995; pp. 100–113.
13. Vasconcelos, H.C.; Fernández-Pérez, B.M.; Morales, J.; Souto, R.M.; González, S.; Cano, V.; Santana, J.J. Development of mathematical models to predict the atmospheric corrosion rate of carbon steel in fragmented subtropical environments. *Int. J. Electrochem. Sci.* **2014**, *9*, 6514–6528.
14. Spence, J.W.; McHenry, J.N. Development of regional corrosion maps for galvanized steel by linking the RADM engineering model with an atmospheric corrosion model. *Atmos. Environ.* **1994**, *28*, 3033–3046. [CrossRef]
15. Feliu, S.; Morcillo, M.; Feliu, S. The prediction of atmospheric corrosion from meteorological and pollution parameters—II. Long-term forecasts. *Corros. Sci.* **1993**, *34*, 415–422. [CrossRef]
16. Tidblad, J.; Mikhailov, A.; Kucera, V. Application of a model for prediction of atmospheric corrosion in tropical environments. In *Marine Corrosion in Tropical Environments*; Dean, S., Delgadillo, G.-D., Bushman, J., Eds.; ASTM International: West Conshohocken, PA, USA, 2000; pp. 250–285. ISBN 978-0-8031-2873-6.
17. Tidblad, J.; Kucera, V.; Mikhailov, A.; Knotkova, D. Improvement of the ISO classification system based on dose-response functions describing the corrosivity of outdoor atmospheres. In *Outdoor Atmospheric Corrosion*; Townsend, H.E., Ed.; ASTM International: West Conshohocken, PA, USA, 2002; pp. 73–87. ISBN 978-0-8031-5467-4.
18. Spence, J.W.; Haynie, F.H.; Lipfert, F.W.; Cramer, S.D.; McDonald, L.G. Atmospheric corrosion model for galvanized steel structures. *Corrosion* **1992**, *48*, 1009–1019. [CrossRef]
19. Pourbaix, M. The linear bilogarithmic law for atmospheric corrosion. In *Atmospheric Corrosion*; Ailor, W.H., Ed.; J. Wiley & Sons: New York, NY, USA, 1982; pp. 107–121.
20. Panchenko, Y.M.; Strekalov, P.V. Correlation between corrosion mass losses and corrosion product quantity retained on metals in a cold, moderate and tropical climate. *Prot. Met.* **1994**, *30*, 459–467.
21. Morcillo, M.; Feliu, S.; Simancas, J. Deviation from bilogarithmic law for atmospheric corrosion of steel. *Br. Corros. J.* **1993**, *28*, 50–52. [CrossRef]
22. Almeida, E.; Morcillo, M.; Rosales, B.; Marrocos, M. Atmospheric corrosion of mild steel. Part I—Rural and urban atmospheres. *Mater. Corros.* **2000**, *51*, 859–864. [CrossRef]
23. Almeida, E.; Morcillo, M.; Rosales, B. Atmospheric corrosion of mild steel. Part II—Marine atmospheres. *Mater. Corros.* **2000**, *51*, 865–874. [CrossRef]
24. Graedel, T.E. Corrosion mechanisms for iron and low alloy steels exposed to the atmosphere. *J. Electrochem. Soc.* **1990**, *137*, 2385. [CrossRef]
25. Kamimura, T.; Hara, S.; Miyuki, H.; Yamashita, M.; Uchida, H. Composition and protective ability of rust layer formed on weathering steel exposed to various environments. *Corros. Sci.* **2006**, *48*, 2799–2812. [CrossRef]
26. Santana Rodríguez, J.J.; Santana Hernández, F.J.; González González, J.E. Mathematical and electro-chemical characterisation of the layer of corrosion products on carbon steel in various environments. *Corros. Sci.* **2002**, *44*, 2597–2610. [CrossRef]

27. Zoccola, J.C.; Permoda, A.J.; Oehler, L.T.; Horton, J.B. Performance of Mayari R Weathering Steel (ASTM A242) in Bridges at the Eight Mile Road and John Lodge Expressway in Detroit, Michigan. Bethlehem Steel Corporation: Bethlehem, PA, USA. Available online: https://www.michigan.gov/documents/mdot/RR213CON_67_534507_7.pdf (accessed on 28 January 2020).
28. Santana, J.J.; Santana, F.J.; González, J.E.; de la Fuente, D.; Chico, B.; Morcillo, M. Atmospheric corrosivity map for steel in Canary Isles. *Br. Corros. J.* **2001**, *36*, 266–271. [CrossRef]
29. Santana, J.J.; Ramos, A.; Rodriguez-Gonzalez, A.; Vasconcelos, H.C.; Mena, V.; Fernández-Pérez, B.M.; Souto, R.M. Shortcomings of International Standard ISO 9223 for the classification, determination, and estimation of atmosphere corrosivities in subtropical archipelagic conditions—The case of the Canary Islands (Spain). *Metals* **2019**, *9*, 1105. [CrossRef]
30. *ASTM D2010/D2010M-98(2017), Standard Test Methods for Evaluation of Total Sulfation Activity in the Atmosphere by the Lead Dioxide Technique*; ASTM International: West Conshohocken, PA, USA, 2017.
31. *ISO 9225:2012, Corrosion of Metals and Alloys—Corrosivity of Atmospheres—Measurement of Environmental Parameters Affecting Corrosivity of Atmospheres*, 2nd ed.; International Organization for Standardization: Geneva, The Switzerland, 2012.
32. *ASTM G1-90, Standard Practice for Preparing, Cleaning, and Evaluating Corrosion Test Specimens*; American Society for Testing Materials: Philadelphia, PA, USA, 1990.
33. *ISO 9223:2012, Corrosion of Metals and Alloys—Corrosivity of Atmospheres—Clasification, Determination and Estimation*, 2nd ed.; International Organization for Standardization: Geneva, The Switzerland, 2012.
34. Larrabee, C.P. Corrosion resistance of high-strength low-alloy steels as influenced by composition and environment. *Corrosion* **1953**, *9*, 259–271. [CrossRef]
35. LaQue, F.L. Corrosion testing. In *ASTM Proceeding 1951—Volume 51*; American Society for Testing and Materials: Philadelphia, PA, USA, 1951; Volume 51, pp. 495–582.
36. Binh, D.T.; Strekalov, P.V.; Van Khuong, N. Effects of seasonal conditions and the slope of a specimen on the atmospheric corrosion of steel in the tropics of Viet-Nam. *Prot. Met.* **2003**, *39*, 278–287. [CrossRef]
37. Veleva, L.; Maldonado, L. Classification of atmospheric corrosivity in humid tropical climates. *Br. Corros. J.* **1998**, *33*, 53–58. [CrossRef]
38. Rostron, P.; Belbarak, C. Atmospheric corrosion issues in Abu Dhabi. *Mater. Perform.* **2015**, *54*, 1–7.
39. Porro, A.; Otero, T.F.; Elola, A.S. Gravimetric corrosion monitoring and mathematical fitting of atmospheric corrosion data for carbon steel. *Br. Corros. J.* **1992**, *27*, 231–235. [CrossRef]
40. Santana Rodríguez, J.J.; Santana Hernández, F.J.; González González, J.E. XRD and SEM studies of the layer of corrosion products for carbon steel in various different environments in the province of Las Palmas (The Canary Islands, Spain). *Corros. Sci.* **2002**, *44*, 2425–2438. [CrossRef]

© 2020 by the authors. Licensee MDPI, Basel, Switzerland. This article is an open access article distributed under the terms and conditions of the Creative Commons Attribution (CC BY) license (http://creativecommons.org/licenses/by/4.0/).

Article

Corrosion Study of Pipeline Steel under Stress at Different Cathodic Potentials by EIS

Ricardo Galván-Martínez [1], Ricardo Orozco-Cruz [1,*], Andrés Carmona-Hernández [1], Edgar Mejía-Sánchez [1], Miguel A. Morales-Cabrera [2] and Antonio Contreras [3]

[1] CA-245- Ingeniería de Corrosión y Protección, Instituto de Ingeniería, Universidad Veracruzana, S. S. Juan Pablo II, Zona Universitaria, Boca del Río, Veracruz 94294, Mexico; rigalvan@uv.mx (R.G.-M.); andres_carmona_hernandez@hotmail.com (A.C.-H.); edmejia@uv.mx (E.M.-S.)

[2] Facultad de Ciencias Químicas, Universidad Veracruzana, Circuito Gonzalo Aguirre Beltrán S/N, Zona Universitaria, Xalapa, Veracruz 91000, Mexico; miguelmorales.uv@gmail.com

[3] Instituto Mexicano del Petróleo, Eje Central Lázaro Cárdenas Norte 152, Col. San Bartolo Atepehuacan, Alcaldia Gustavo A. Madero, Ciudad de México 07730, Mexico; acontrer@imp.mx

* Correspondence: rorozco@uv.mx

Received: 13 November 2019; Accepted: 11 December 2019; Published: 16 December 2019

Abstract: The effect of different cathodic potentials applied to the X70 pipeline steel immersed in acidified and aerated synthetic soil solution under stress using a slow strain rate test (SSRT) and electrochemical impedance spectroscopy (EIS) was studied. According to SSRT results and the fracture surface analysis by scanning electron microscopy (SEM), the steel susceptibility to stress corrosion cracking (SCC) increased as the cathodic polarization increased (E_{cp}). This behavior is attributed to the anodic dissolution at the tip of the crack and the increment of the cathodic reaction (hydrogen evolution) producing hydrogen embrittlement. Nevertheless, when the E_{cp} was subjected to the maximum cathodic potential applied (−970 mV), the susceptibility decreased; this behavior is attributed to the fact that the anodic dissolution was suppressed and the process of the SCC was dominated only by hydrogen embrittlement (HE). The EIS results showed that the cathodic process was influenced by the mass transport (hydrogen diffusion) due to the steel undergoing so many changes in the metallic surface as a result of the applied strain that it generated active sites at the surface.

Keywords: X70 steel; stress corrosion cracking (SCC); slow strain rate tests (SSRT); electrochemical impedance spectroscopy (EIS); cathodic potentials

1. Introduction

Stress corrosion cracking (SCC) is one of the most important causes of failures in buried pipelines, mainly in the transport of hydrocarbons [1,2]. The environments that generate SCC in the pipelines are generally electrolytes caught between the detached coating and the surface of the pipeline steel. According the characteristics of electrolytes developed under disbonded coatings, pipelines can suffer two different types of SCC: high-pH SCC, and near-neutral pH SCC [3,4]. The SCC at high pH generally occurs as a result of the generation of a concentrated solution of carbonate-bicarbonate (pH > 9) with an intergranular crack morphology. This mechanism is attributed to selective anodic dissolution of the grain border and to the repeated rupture of the passive film that forms in the tip of the crack. On the other hand, the SCC at almost neutral pH (pH 5–8.5) is associated with diluted solutions generated by the underground water with a transgranular crack morphology; [5–7] however, there is no consensus about its mechanism. Some research [8–10] proposes that the crack can be induced by the synergic effect between the anodic dissolution and the diffusion of the H in the steel. In addition, some SCC types have been found in acid environments (pH 3–6), where the surface of the pipeline is in direct contact with the soil in coating defects [11]. Nevertheless, the SCC behavior in acid soils is not clear, and investigations into it have not been well defined [3].

Cathodic protection (CP) is one of the most important methods of protection against the corrosion in buried pipelines; however, this CP can contribute to the SCC process through hydrogen generation. The current in the CP can fluctuate as a result of the permeability of the coating and variations in soil resistivity, provoking potential fluctuations in some zones of the pipeline steel and generating the formation of localized corrosion, such as pitting and crevice corrosion. Pitting corrosion in combination with the stress (residual or operational) in the pipeline can nucleate and generate cracks [12].

Some researchers [13–16] have studied the influence of CP in the SCC susceptibility of steels, and determined that CP could influence in the mechanism of cracking induced by SCC. Liu et al. [16] investigated the SCC mechanism of X70 steel under different cathodic potentials (E_{cp}) in a synthetic soil solution with an almost neutral pH; they found a critical potential interval (from −730 to −920 mV vs. saturated calomel electrode); if the E_{cp} was more positive than above interval, the SCC mechanism would be based on the anodic dissolution of the steel, but if the E_{cp} were more negative, the SCC mechanism would be hydrogen embrittlement. Finally, if the E_{cp} had a value in between, the SCC would present both mechanisms.

In this work, electrochemical polarization measurements were combined with slow strain rate tensile (SSRT) tests and surface characterization to investigate the relationship between the SCC characteristics and the electrochemical corrosion properties of an API X70 steel in a near-neutral pH solution.

This research was focused on characterizing the SCC process of X70 pipeline steel immersed in synthetic soil solution under the application of different cathodic potentials, using a slow strain rate test (SSRT) and electrochemical impedance spectroscopy (EIS).

2. Materials and Methods

2.1. Working Electrode

The material used in the present study as working electrode was API 5L X70 pipeline steel with a chemical composition as shown in Table 1. Steel samples were obtained from a pipeline that had an external diameter of 36 in (914.4 mm) and a wall thickness of 0.902 in. (22.91 mm). The samples used in the slow strain rate test were machined according to NACE TM 0198 [17].

Table 1. Chemical composition of X70 pipeline steel (wt.%).

C	Mn	Si	P	S	Al	Nb	Cu	Cr	Ni	V	Ti	Fe
0.031	1.48	0.13	0.012	0.002	0.033	0.1	0.29	0.27	0.16	0.004	0.012	Balanced

The surface of the gauge section was polished up to 1200 SiC grit paper in an orientation parallel to the subsequent loading direction of the SSRT in order to ensure similar surface conditions for all tests.

2.2. Test Solution

The corrosive environment used in all of the electrochemical tests was a simulated groundwater solution (called NS4) with a pH of 3. NS4 synthetic solution has been used widely to simulate soil solution in the study of near neutral pH-SCC behavior. However, other synthetic soil solutions such as NS1, NS2, NS3, NS4 and NOVA have been used in similar studies. Table 2 shows the chemical composition of the NS4 solution used. The pH of the NS4 solution was around 8.0; however, for this study, the pH solution was adjusted with hydrochloric acid to obtain a pH of 3.

Table 2. Chemical composition of NS4 solution (g/L).

NaHCO$_3$	CaCl$_2$. 2H$_2$O	MgSO$_4$. 7H$_2$O	KCl
0.483	0.181	0.131	0.122

2.3. Slow Strain Rate Tests (SSRT)

SSRT were carried out on smooth cylindrical tensile samples, using an MCERT machine (Mobile Constant Extension Rate Tests) at a strain rate of 1×10^{-6} s^{-1} in air and synthetic soil solution (NS4 solution). All SSRT were conducted at room temperature and atmospheric pressure. The length direction of the sample was parallel to the circumferential direction of the pipeline steel, with the goal that, if a crack appeared, it would grow in the longitudinal direction of the pipeline as is typically observed in underground pipelines. Cylindrical samples with a reduced length of 1 inch and 0.150 inches in diameter were machined with a total exposed area of 3.04 cm^2. These cylindrical samples were machined according to the NACE TM 0198 standard [17]. After the SSRT was completed, the fractured sample was removed, and cleaned using inhibited acid and acetone for SEM examination.

2.4. Electrochemical Impedance Spectroscopy (EIS)

The EIS measurements were carried out simultaneously with the SSRT using a potentiostat/galvanostat and an electrochemical cell with a three-electrode arrangement where the working electrode was the X70 pipeline steel, sintered graphite bar was used as the auxiliary electrode, and saturated calomel electrode, SCE, was used as the reference electrode. In all EIS tests, the frequency range from 0.01 to 10 kHz with the amplitude of 0.01 V against to E_{corr} was used. Seven points per decade of frequency were recorded. In order to apply the overvoltage to the working electrode, an external direct current source and a second auxiliary electrode were used to close the circuit in the cathodic protection system by impressed current.

EIS tests and SSRT were carried out at corrosion potential (E_{corr} = −650 mV vs. SCE) and at three different cathodic potentials (E_{cp}): −770 mV (cathodic protection potential, CPP), −870 mV (100 mV of overvoltage as a function of CPP), and −970 mV (200 mV of overvoltage as a function of CPP) and compared to the saturated calomel electrode.

2.5. Potentiodynamic Polarization Curves (PC)

Two polarization curves were recorded at two different sweep rates, at 5 mV and 50 mV per second. All PC were carried out at room temperature atmospheric pressure and without stress conditions. The focus of this experiment is to predict a potential range where the steel is susceptible to SCC. Polarization range used in the carry out the PC was from −1400 mV to 0 vs. SCE.

3. Results and Discussion

3.1. SSRT at Different Cathodic Potentials (E_{cp})

Figure 1 shows the stress (σ) vs. strain (ε) curve of X70 pipeline steel immersed in acidified NS4 solution (pH 3) at the E_{corr} (−650 mV vs. SCE) and at different applied cathodic potentials (−770, −870 and −970 mV vs. SCE) during the SSRT.

According to Figure 1, the profiles at the different cathodic potentials show an increase in the yielding strength (σ_{YS}) and ultimate tensile strength (σ_{UTS}) compared to the test in air. This increment is attributed to the diffusion of atomic hydrogen (H) into the crystal lattice, because the E_{cp} thermodynamically favors the evolution reaction of the H [18]. According to Liu et al. [16], there is a critical concentration of H ($C_{H,crit}$) that determines the increase or decrease of σ_{YS}. When the H concentration (C_H) on metallic surface is below the $C_{H,crit}$, the H impedes the sliding dislocations that generate the increment of the σ_{YS}. However, if the C_H exceeds the $C_{H,crit}$ value, the dislocations activity is facilitated, and the σ_{YS} decreases.

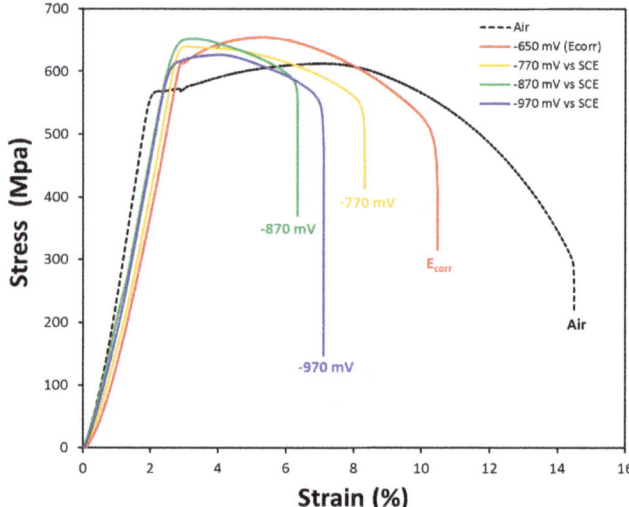

Figure 1. Stress-strain (σ vs. ε) curves of X70 pipeline steel immersed in NS4 solution (pH 3) at the E_{corr} (−650 mV vs. SCE) under the application of different cathodic potentials.

In accordance with this fact, the σ_{YS} increased between the E_{corr} and −870 mV vs. SCE, because the C_H also increased. It is important to point out that the C_H did not exceed the $C_{H,crit}$. At −970 mV vs. SCE, the σ_{YS} decreased because the C_H exceeded the $C_{H,crit}$.

The SCC susceptibility of X70 steel was calculated on the basis of the reduction area percent (%RA) and the plastic elongation (%PE), using the following equations:

$$\%RA = \frac{Di^2 - Df^2}{Di^2} \times 100, \qquad (1)$$

where D_i and D_f are the initial and final diameter of the fracture surface.

$$\%PE = \left\{ \frac{E_f}{L_I} - \left[\frac{\sigma_F}{\sigma_{PL}} \right] \times \frac{E_{PL}}{L_I} \right\} \times 100, \qquad (2)$$

where E_f is the elongation at failure, L_I the initial gauge length, σ_F is the stress at failure; σ_{PL} is the stress at the proportional limit, E_{PL} is the elongation at the proportional limit.

The SCC index considering %RA and %PE were calculated according to the following equations:

$$I_{RA} = RA_{sol}/RA_{Air}, \qquad (3)$$

$$I_{PE} = PE_{sol}/PE_{Air}, \qquad (4)$$

where the suffix Sol and Air correspond to the values obtained in the NS4 solution and air, respectively. The I_{RA} and I_{PE} are the SCC susceptibility indexes.

According to the indexes (I_{RA} and I_{PE}) shown in Figure 2, and to the classification proposed by McIntyre et al. [19], the X70 steel at E_{corr} had moderate susceptibility to SCC, but when the X70 steel was analyzed with different E_{cp}, the steel exhibited great susceptibility to SCC.

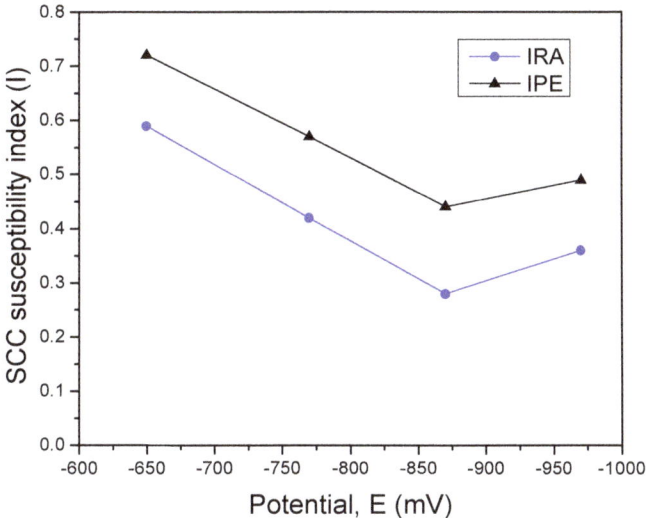

Figure 2. SCC susceptibility indexes of X70 pipeline steel as a function of potential.

The I_{RA} and I_{PE} values decreased as the E_{cp} became more negative; however at the E_{cp} of −970 mV vs. SCE, the indexes increased, indicating a slight decrease in the susceptibility to SCC. The SCC susceptibility of X70 steel shows a complex dependence on cathodic potential.

As it is showed in the Figure 2, the SCC susceptibility tends to increase as the potential reaches more negative values. With the different applied cathodic potentials, the SCC process changes. At less negative potential (−650 and −770 mV), the SCC is based primarily on the anodic dissolution mechanism. When the applied potential reaches more negative values, hydrogen is involved in the cracking process (−870 and −970 mV), resulting in a transgranular cracking mode with brittle feature of the fracture surface.

Generally, more negative cathodic potentials enhance the hydrogen evolution reaction. Therefore, it is commonly assumed that there will be more hydrogen atoms penetrating into the steel, contributing to the SCC process. According to the Pourbaix diagram, it can be observed that over potential applied to steel specimens (−870 and −970 mV) results in hydrogen evolution. The concentration of hydrogen atoms depends on cathodic potential; as cathodic potential decreases, a greater concentration of hydrogen atoms is generated. Hydrogen can diffuse into the steel specimen around the crack tip during the SCC process, resulting in hydrogen embrittlement, leading to increased SCC susceptibility. Cheng [8] developed a thermodynamic model to illustrate the interaction of hydrogen, stress and anodic dissolution at the crack tip. He suggested that crack growth rate was dependent on the synergistic effect of hydrogen and stress, and the concentration of hydrogen atoms.

3.2. Surface Fracture after SSRT

Figure 3 shows SEM micrographs of the fractured surface of X70 pipeline steel at two different magnifications (100× and 1000×) after performing SSRT in air and NS4 solution (pH 3) at the E_{corr} (−650 mV) and when different cathodic potentials (−770, −870 and −970 mV vs. SCE) were applied.

The X70 steel tested in air exhibited a ductile facture type, Figure 3a,a1. The same behavior was observed for samples tested at −650, Figure 3b,b1 and −770 mV, Figure 3c,c1. In these figures, the dimples produced by micro-plastic deformation can be observed, as well as the presence of micro-pores that can act as stress concentrators, with these sites being the typical places where cracking is preferentially generated. These micropores produce metal cracking by the coalescence mechanism [13,20].

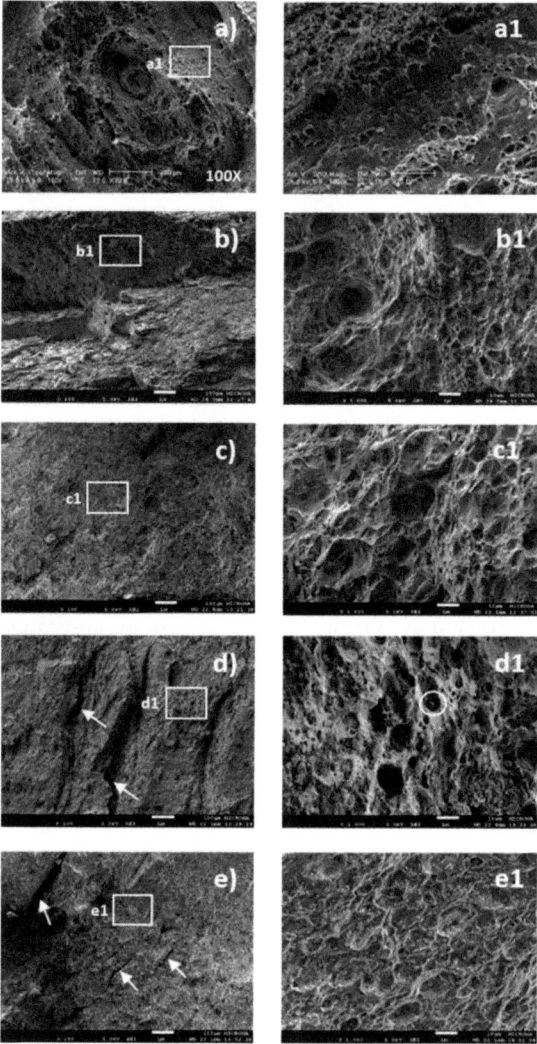

Figure 3. SEM micrographs of the fracture surface of X70 steel after performing SSRT at two different magnifications (100× and 1000×): (**a, a1**) in air; and (**b, b1**) in NS4 solution with E_{corr} = −650 mV, and in NS4 solution with different cathodic potentials, (**c, c1**) −770 mV, (**d, d1**) −870 mV, and (**e, e1**) −970 mV.

Figure 3d,d1,e,e1 shows the fracture surface for samples tested at −870 and −970mV, respectively. In these figures, a mix fracture types—ductile and brittle—can be observed. The presence of some brittle regions and the presence of some microcracks can be clearly observed. Most of the samples with cathodic potentials of −870 and −970 mV show the presence of some internal cracks, which is correlated with the more brittle fracture type observed in these samples.

It is important to point out that Lynch et al. [21] found that the hydrogen embrittlement induced by the medium could produce ductile fracture through the Adsorption Induced Dislocation Emission (AIDE) mechanism. The AIDE mechanism includes both nucleation and subsequent movement of dislocations away from the crack tip. This mechanism consists of the weakening of the interatomic bond at the tip of the crack due to adsorbed hydrogen; this behavior promotes a change in the fracture to a relatively lower deformation due to the local reduction of ductility.

In Figure 3d,d1,e,e1, it is possible to observe internal cracks, which can be indicative of the diffusion of hydrogen into the steel; the hydrogen is preferentially trapped in some defects such as microcavities and inclusions, the crystalline lattice is dilated, and the interatomic cohesion decreases, leading to the appearance of internal cracks. It is generally known that pitting and pre-existing defects initiate SCC; the lattice defects—vacancies, dislocations, grain boundaries and precipitates—provide a variety of trapping sites for hydrogen diffusion. Hydrogen diffusion promotes both intergranular and transgranular cracking.

The hydrogen atoms diffuse into the metal and have a preference for accumulating in internal defects like inclusions, precipitates, cavities, and dislocations, among others. Hydrogen atoms in these regions result in embrittlement, and when the stress increases to a critical value, cracking is initiated. It is important to note that the level of stress or strain depends on the steel grade. High tensile strength steels with elasticity limits exceeding 650 MPa are highly susceptible to SCC. Martensitic structures are considerably more susceptible than bainitic and ferrite structures. In addition, coarse-grained materials are more susceptible to embrittlement than fine microstructures.

Crack propagation can be either intergranular or transgranular; sometimes, both types are observed on the same fracture surface. In the case of near neutral pH-SCC at corrosion potential (−680 to −710 mV SCE), a transgranular crack morphology is generally observed. Meanwhile, with high pH SCC, intergranular cracking is commonly observed in the potential range of −520 to −670 mV SCE.

It is important to point out that in Figure 3d,d1 (−870 mV), it is possible to observe some inclusions that could have contributed to the cracking. Liu et al. [22] concluded that the inclusions in the X70 steel are rich in brittle oxides that are incoherent with the matrix of the steel; thus, a great deformation of the crystalline lattice is formed. This deformation is the main factor for the preferential generation of microcracks between the borders of these inclusions and the steel.

Finally, at −970 mV, Figure 3e,e1 it is possible to observe that the fracture morphology is different from the morphologies found at −770 and −870 mV, because the surface of the facture shows a smooth scission of brittle character, corresponding to the SCC process dominated by the hydrogen embrittlement. This fact indicates that the concentration of the atomic hydrogen adsorbed on the surface of the X70 steel was greater when the E_{cp} was applied, resulting in an increase in yield strength (YS) and ultimate tensile strength (UTS), but decreases in strain and elongation (Figure 1) due to the embrittlement of the microstructure. According to this behavior, it is possible to say that the mechanism in the SCC process was hydrogen-enhanced localized plasticity (HELP), which improves the movement of dislocations because they are surrounded by atomic hydrogens, facilitating plastic deformation in a localized way. According to Martin et al. [23], the hydrogen-enhanced localized plasticity (HELP) mechanism as a viable mechanism of hydrogen embrittlement.

Hydrogen embrittlement (HE) has been studied for many years; however, it remains a consistent problem in the SCC process. In addition, there is no precise mechanism for HE. Various mechanisms proposed in the literature include hydrogen-enhanced decohesion and hydrogen-enhanced local plasticity (HELP) [23].

The HELP mechanism is supported by experimental observations of enhanced dislocation motion and localized slip bands in the vicinity of the crack tip in hydrogen-charged test specimens. The most accepted mechanism for enhancing HELP is that hydrogen increases dislocation mobility, leading to the material being more ductile.

3.3. Cracks in the Gauge Section after SSRT

The secondary cracking commonly presented in the gauge section of the specimen corresponded to typical cracking in a medium that promotes the SCC process. Figure 4 shows some images obtained by SEM of cracks located in the gauge section of the SSR specimens. These cracks were sharp and without ramifications, which is characteristic of a hydrogen-induced cracking (HIC) process [3]. On the other hand, the cracks observed in the steel at the E_{corr} (Figure 4a) show ramifications and aggressive attacks on the surface with no cracks, produced by the anodic activity that prevails under these conditions.

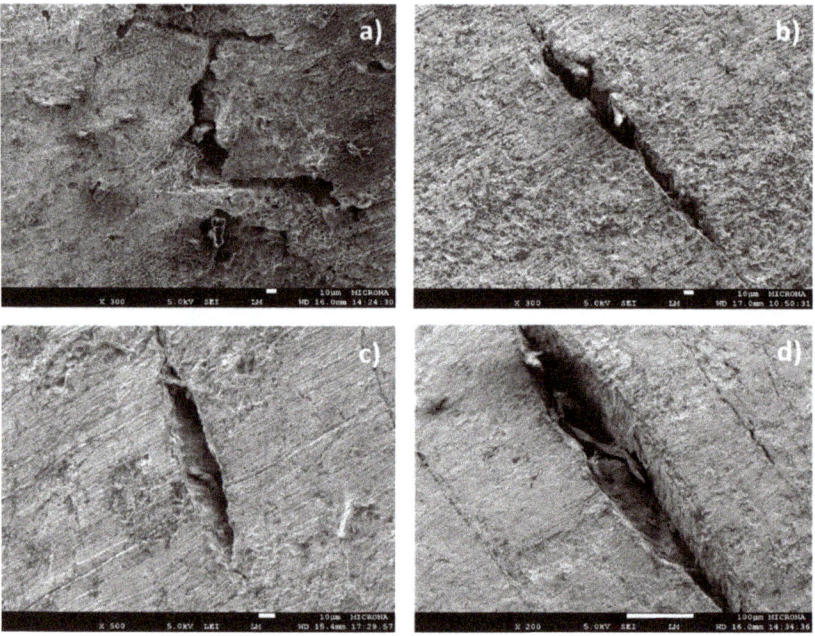

Figure 4. Secondary cracking of X70 pipeline steel after SSRT: (**a**) −650 mV (E_{corr}), (**b**) −770 mV (**c**) −870 mV (**d**) −970 mV vs. SCE.

3.4. SCC Analysis under different E_{cp} by Polarization Curves

To gain a better understanding of the SCC mechanism of X70 steel under different cathodic potentials, a conceptual model developed by Liu et al. [16] was used. This model is based on the polarization curves (PC) of the X70 steel without stress at fast and slow sweep rates. Figure 5 shows three zones or potential intervals between the two polarization curves (at fast and slow sweep rate) of the X70 steel immersed in the NS4 solution (pH 3), limited by the potential values where the current density trends to zero. Zone I corresponds to potential values lower than −700 mV vs. SCE, where the PCs at both sweep rates are located in the anodic polarization region. Zone II is localized to the interval between −700 and −900 mV vs. SCE. In this zone, the PC obtained with the slow sweep rate was found to be in the cathodic polarization region, while the PC obtained with the fast sweep rate was found to be in the anodic polarization region; finally, zone III encompasses the more negative potential at −970 mV vs. SCE, where the two PCs, at both sweep rates, were located in the cathodic polarization region [24]. According the behavior of the three zones, if the X70 steel is polarized with a potential that is in zone I, this fact indicates that in the tip of the crack, in the wall of the crack and in the uncracked metallic surface, the anodic reaction mainly could be carried out, and the cracking mechanism would depend only on the anodic dissolution. If the applied potential is found in zone II, this fact indicates that the anodic dissolution can be carried out in the tip of the crack, contributing to the acceleration of its propagation, while in the uncracked metallic surface, hydrogen reduction is carried out on other species like O_2 and H_2CO_3; therefore, the SCC process can be controlled by the combination of two processes, the mechanism of the hydrogen embrittlement and the anodic dissolution.

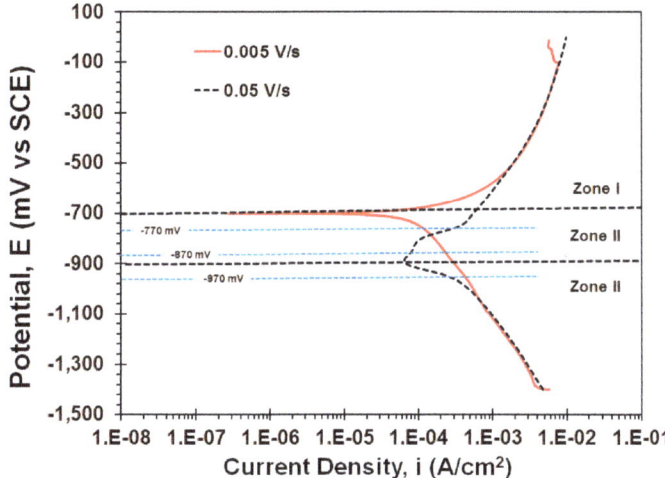

Figure 5. Polarization curves of the X70 steel in NS4 solution at two different sweep rates: 5 mV/s and 50 mV/s.

On the other hand, if the applied potential is found in zone III, it means that in the tip and wall of the crack could benefit the cathodic reaction, specifically the H reduction, generating H atoms with this reaction that can then be actively involved in the SCC process. These results are in agreement with the research of Javidi and Galvan-Martinez [24,25]. Accordingly, the SCC mechanism would be dominated by hydrogen embrittlement.

In addition, in Figure 5, it is possible to observe that the E_{cp} of −0.77 and −0.87 V vs. SCE are within zone II, while the E_{cp} of −0.97 V vs. SCE is located in zone III. Therefore, the behavior of the SCC susceptibility with respect to E_{cp}, as shown in Figure 2, can be attributed to the contribution of the anodic dissolution on the SCC mechanism. That is, when the potential is sufficiently negative, the anodic dissolution of the steel will be negligible and will not contribute to the SCC process, and this process will be dominated only by hydrogen embrittlement; on the other hand, at more positive potentials, even when the steel is under cathodic protection, in the area of tip of the crack, the steel will be in a non-stationary state and the anodic dissolution will contribute to the cracking process.

3.5. Qualitative analysis of the EIS Spectra

EIS measurements of the X70 pipeline steel immersed in NS4 solution (pH 3) at the E_{corr} and at under different E_{cp} (−0.77, −0.87 and −0.97 V vs. SCE) are shown in Figure 6. It is important to point out that five different points were measured in the stress strain curve at each E_{cp}: at beginning of the test (T0), in the elastic zone (EZ), at yield strength (YS), at ultimate tensile strength (UTS), and before fracture (BF). In Figure 6, the experimental data (dots) and the fit line (continuous line) can be observed.

The EIS spectra of the X70 steel at the five points recorded and at the E_{corr} (Figure 6a) show a characteristic semicircle of a capacitive process related t the electrochemical double layer capacitance (C_{dl}) and the charge transfer resistance (R_{ct}) associated with the redox reactions that occur at the electrolyte/metal interface. On the basis of this behavior, the proposal for fitting the EIS spectra is a simple equivalent electric circuit (EEC) including only the charge transfer resistance process, the most common of which is the Randles circuit. It is important to point out that all EIS spectra only show one time constant, which is attributed to the activational process or, as it is also called, charge transfer resistance.

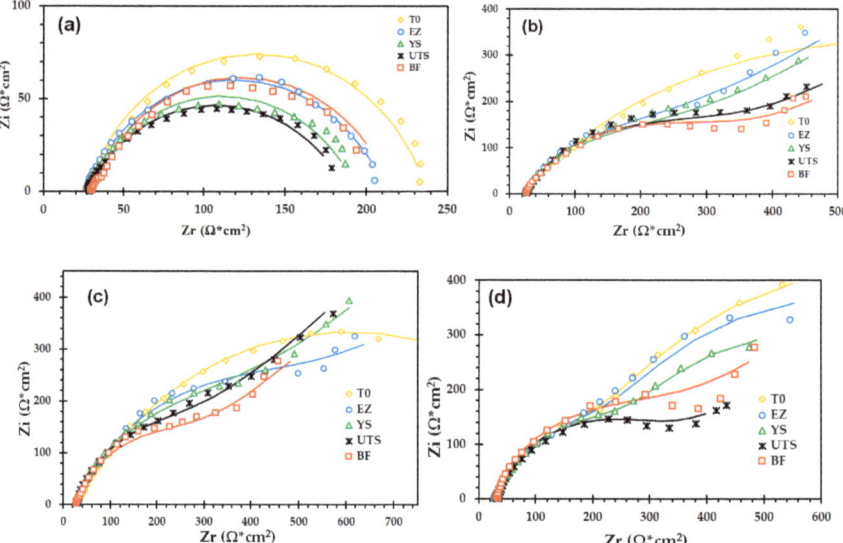

Figure 6. EIS spectra in Nyquist diagram of the X70 steel immersed in NS4 solution (with pH 3) during the SSRT test. At the E_{corr} (**a**), 0.77 (**b**), 0.87 (**c**) and 0.97 (**d**) V vs. SCE at different times. Dots correspond to experimental data and continuous lines correspond to fitting.

In addition, the R_{ct} decreased from T0 to UTS, and increased again at the BF point. The decrease of the R_{ct} should be attributed to the gradual increase of the stress that prompting the generation and concentration of dislocations in the metal and the formation of active sites close to tip of the crack [23,26], and to the hydrogen (H) diffusion into the metal, whereby H comes from the cathodic reaction. Cheng [8] proposed that the combination of the H and stress causes the anodic dissolution to be more thermodynamically favorable. Finally, the increase of the R_{ct} at the BF point is attributed to effects of corrosion film products on the cracked surface.

At E_{cp} of −0.77 and −0.87 V vs. SCE (Figure 6b,c) and at the T0 point, the EIS spectrum shows a capacitive semicircle that can be attributed to the charge transfer process. However, at the EZ, YS, UTS and BF points, it is possible to observe two time constants (τ): the first τ, located between the high and middle frequencies, is attributed to the charge transfer process; while the second τ is attributed to the mass transfer process. Some researchers, like Liu [24,27], have reported Nyquist plots with this behavior. Taking into consideration the fact that the SCC is a dynamic process (this is because in the Slow Strain Rate Tests the machine that carries out the extension of the metal sample applies a constant rate, 1×10^{-6} s^{-1}) and the metallic surface during the SSRT can undergo small changes that affect the kinetics of the electrochemical reactions such that the transition from the activational to the mixing process is observed in the Nyquist plots of the EIS measurements between T0 and EZ. This behavior can be attributed to the model of local additional potential (LAP) proposed by Liu et al. [25,28]. When the steel subjected to stress is in the macroscopic elastic zone, the local concentration of the stress can result in defects such as microcracks and inclusions, which can generate active sites through the movement of the dislocations. Due to the steel being under cathodic polarization and the supplied flow of electrons being preferentially concentrated in these sites, a negative LAP is generated, which can benefit the cathodic process, and which can be diffused by the electroactive species because its consumption is faster than the supply at the interface. The mass transport from the bulk to the interface can thus be the limiting factor of the corrosion kinetics, specifically the cathodic process.

Finally, at 0.97 V vs. SCE, Figure 6d shows two τ when the steel was subjected to elastic stress (points T0 to LE). The τ corresponded to two overlapping capacitive semicircles. The τ was located in interval between high and medium frequencies was attributed to the charge transfer (activational process) from the interfaces to the metallic surface. Meanwhile, the τ located at low frequencies was attributed to the response of the adsorbed species, like hydrogen atoms and bubbles of molecular H_2, on the surface, because when the voltage was applied to the steel, the hydrogen evolution (hydrogen reduction) increased, strongly provoking another interface with these hydrogens and generating a pseudo-capacitance phenomenon by adsorption [26,29].

Pseudo-capacitance is the differential capacitance of an interface between the metal and the electrolyte, which is caused by the concentrated change of an adsorbed electroactive species and which is shown in electrochemical reactions where the charge transfer process precedes the determinant stage of the speed. It is important to point out that the electrolytic reaction of the hydrogen evolution (cathodic reaction) is an example of these electrochemical reactions. This is because in the first stage, the speed of the charge transfer could be fast, while in the second stage, the reaction between the two H atoms in order to form a H_2 molecule could be slow [27,30]. At the UTS and BF points, and in the low frequency zone, the τ changed to a straight line, which was attributed to the diffusion process.

3.6. Quantitative Analysis of the EIS Spectra

Figure 7 shows the equivalent electric circuit (EEC) scheme and its physical explanation. According to Figure 7, it is possible to explain the EEC used in the fitting of the EIS spectra; that is to say, Figure 7a shows when the charge transfer process was the only process limiting the corrosion. As stated in the Section 3.5, on Qualitative analysis, only one time constant can be observed in the EIS spectra, and this is attributed to the corrosion process. In the case of Figure 7b, the charge transfer process is influenced by a mass transfer of hydrogen ions or oxygen through the diffusion layer, in this case, the first time constant is attributed to the corrosion process and the second time constant to the diffusion process. Finally, Figure 7c shows the EEC with two time constants, but in this case the second constant is attributed to the layer of hydrogen reduced adsorbed on the metallic surface. It is important to point out that R_s is the solution resistance, and CPE_{dl} is the constant phase element of the double layer, which substitutes the double layer capacitor, because in real systems, the capacitive behavior of the electrode is not ideal. Due to corrosion products, film is adsorbed onto the metallic surface, and this generates a non-homogeneous surface [28,31].

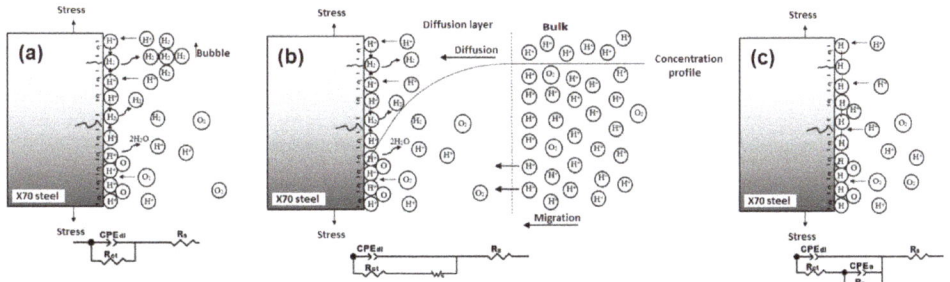

Figure 7. Scheme of EEC employed to fit the EIS spectra shown in Figure 6. (**a**) charge transfer process at T0, (**b**) diffusion process at EZ, YS, UTS and RS points, and (**c**) formation of a non-homogeneous surface at T0, EZ and YS points.

Figure 7 shows the EEC used to fit all of the EIS spectra in this research. All EIS spectra of the X70 steel obtained at the beginning of the test (T0) at E_{corr} and at different E_{cp}, −0.77 and −0.87 V vs. SCE, were fitted with a simple equivalent electric circuit (Figure 7a), the so-called Randles circuit.

Figure 7b shows the EEC used to fit the EIS spectra obtained at the EZ, YS, UTS and RS points. In this figure, (W_z) is the semi-infinite diffusion Warburg impedance (Z_w); in this case, the diffusion of the electroactive specie occurs when the mass transport is carried out from the bulk solution to the metallic surface in the absence of corrosion products [29,32]. When corrosion products are able to partially isolate the metallic surface, the study by Bastidas [33] defines the diffusion process for a layer of finite thickness using the transmission line model. Despite this model not being used in this research, it is important to note that this model is capable of providing a good analysis of corrosion. At E_{cp} of −0.97 V vs. SCE and at T0, EZ and YS, the EEC used is shown in Figure 7c [30]. In this EEC, it is possible to observe that an interface was adsorbed onto the metallic surface, and this can generate an electric response [30,34]. The parameter R_a is the adsorbed resistance and CPE_a is the pseudo-capacitance associated with the adsorbed H. At the UTS and RS points, the EIS spectra were fitted with the Randles circuit.

To obtain a better analysis, the parameters calculated with the EEC fit of the EIS spectra shown in Figure 6 are presented in Table 3, where C_{dl} and C_a are the double layer and the adsorbed capacitance, respectively, calculated using the G. J. Brug expression [35]. It is important to point out that the CPE and R_{ct} values obtained in the EIS spectra fitting were used to calculate the capacitor by the Brug expression.

Table 3. Calculated parameters obtained for the EIS spectra of Figure 6 fitted by EEC in Figure 7.

Point	R_s		C_{dl}	R_{ct}		C_a	R_a		σ_w	
	(Ωcm^2)	Error (%)	($\mu F/cm^2$)	(Ωcm^2)	Error (%)	($\mu F/cm^2$)	(Ωcm^2)	Error (%)	$\Omega cm^2 s^{-0.5}$	Error (%)
At the E_{corr} (−650 mV vs. SCE), Using the EEC Shown in Figure 7a										
T0	26.62	2.88	49.23	211	4.28	—	—	—	—	—
EZ	26.05	1.58	60.975	185.9	2.56	—	—	—	—	—
YS	27.24	1.73	62.22	168	3.01	—	—	—	—	—
UTS	26.43	1.63	72.45	162.1	3.08	—	—	—	—	—
BF	30.22	1.71	135.34	186	3.60	—	—	—	—	—
At −770 mV vs. SCE, using the EEC shown in Figure 7a (for T0) and Figure 7b										
T0	23.24	3.23	91.52	1135	14.20	—	—	—	—	—
EZ	24.09	2.99	75.73	305.82	7.26	—	—	—	278.90	11.93
YS	24.71	3.36	65.21	314.25	5.89	—	—	—	264.22	7.15
UTS	26.32	2.90	62.84	304.72	7.26	—	—	—	193.61	11.93
BF	26.00	2.56	57	295.41	4.55	—	—	—	179.64	5.49
At −870 mV vs. SCE, using the EEC show in Figure 7a (for T0) and Figure 7b										
T0	27.92	4.42	39.71	1056.00	11.20				—	—
EZ	28.11	3.43	55.19	496.04	8.31				203.99	4.70
YS	27.02	3.11	45.18	398.39	8.34				331.41	2.40
UTS	26.482	8.41	43.60	344.23	2.21				319.53	2.33
BF	27.505	10.86	50.15	268.13	2.10				241.60	3.53
At the E_{corr} (−970 mV vs. SCE) using the EEC show in Figure 7b (for UTS and BF) and Figure 7c										
T0	27.1	9.09	67.76	486.69	3.00	536.67	744.2	6.67	—	
EZ	26.74	4.00	73.84	392.76	4.32	637.78	615.93	8.65	—	
YS	28.04	7.06	83.04	332.79	2.15	625.46	642.78	7.10	—	
UTS	36.08	4.46	104.99	302.02	2.66	—	—	—	110	3.76
BF	34.50	11.10	99.75	318.80	4.37	—	—	—	190.75	5.77

In this table, is possible to see that the values of charge transfer resistance (R_{ct}), corresponding to the EIS spectra measured at the E_{corr}, have the lowest values. This fact indicates that the contribution of the anodic process tends to lead to a decrease in R_{ct} value and limits the SCC process. To obtain the influence of the E_{cp} on the corrosion phenomenon of X70 steel under stress conditions, an analysis with the R_{ct} as a function of the E_{cp} was carried out.

3.7. R_{ct} Behavior as a Function of E_{cp}

Figure 8 shows the R_{ct} behavior obtained from the EEC shown in Figure 7. These R_{ct} were obtained from the corrosion phenomenon of the X70 steel immersed in NS4 solution at the E_{corr} and at different

E_{cp} during the SSRT test. In a general way, it is possible to point out that at the E_{corr}, all R_{ct} values are lower than the R_{ct} values obtained at any E_{cp}. According to the mix potential theory, at the E_{corr}, the R_{ct} is the equilibrium result of the anodic and cathodic reactions occurring at the interface.

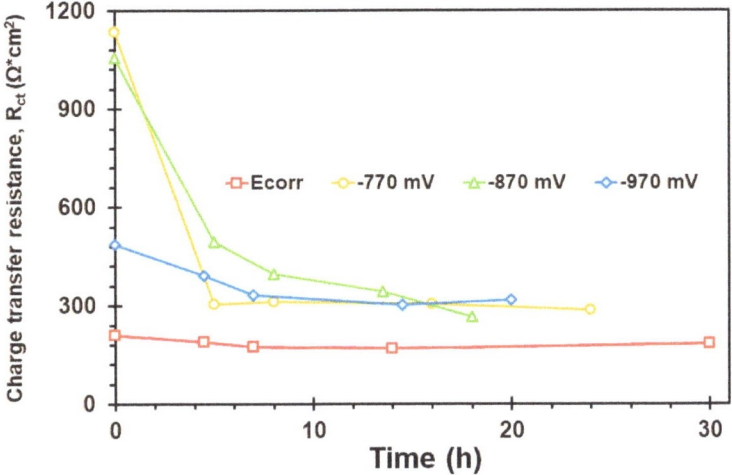

Figure 8. R_{ct} as a function of time. X70 steel immersed in NS4 solution (pH3) under different E_{cp}.

Some researchers, like Liu et al., have indicated that the R_{ct} can be expressed as follows [31,36]:

$$\frac{1}{R_{ct}} = \frac{1-\theta_c}{R_{ct,a}} + \frac{\theta_c}{R_{ct,c}}, \qquad (5)$$

where $R_{ct,a}$ and $R_{ct,c}$ are the charge transfer resistance for the anodic and cathodic reactions, respectively, θ_c is the fraction of the active area, where the cathodic reaction is carried out. When an E_{cp} is applied to the X70 steel, the $R_{ct,a}$ and θ_c will increase and $R_{ct,c}$ will decrease, with the R_{ct} value contributing to the cathodic process to a great extent. Therefore, at the E_{corr}, the anodic process contributes to a reduction in the R_{ct} value.

On the other hand, at the E_{cp} of −770 and −870 mV vs. SCE, the R_{ct} decreased throughout the entire time that the test was carried out. This behavior can be attributed to the propagation of the SCC in the tip of the crack, generating a new active surface, which is subject to a non-stationary stage of polarization. Liu et al. [24,27] proposed that the tip of the crack resembled an active specimen recently exposed to electrochemical medium; for this reason, the anodic dissolution in the main crack increases, contributing to the R_{ct} reduction.

It is important to point out that at the beginning of the SSRT test, the R_{ct} was lower, as the E_{cp} was more negative due to the major supply of electrons to the interface, improving the cathodic reaction (Hydrogen evolution); however, when the stress levels increased, the R_{ct} became independent of the E_{cp}. This behavior can be attributed to the fact that as the SSRT test is carried out, the R_{ct} is a function of two parameters, the level stress and the applied potential, that can improve the cathodic process through the negative LAP generated at the active sites, and produced by the dislocation movement on the surface [25,28,31,34].

4. Conclusions

Characterization of the SCC process of X70 pipeline steel immersed in synthetic soil solution under the application of different cathodic potentials, using slow strain rate test (SSRT) and electrochemical impedance spectroscopy (EIS) allowed the following conclusions to be drawn:

- According to the SCC indexes obtained (I_{RA} and I_{PE}), it is clear that SCC susceptibility increases with the increase in cathodic potential. These indexes indicate that X70 steel could be susceptible to SCC.
- SEM observations revealed the presence of some internal cracks on the fracture surface (which is indicative of hydrogen diffusion); additionally, some secondary cracks in the gauge section of the SSRT specimens were observed. These cracks grew perpendicular to applied stress.
- The application of cathodic potentials (E_{cp}) decreased the corrosive attack on the metal surface; however, they increased the SCC susceptibility of the steel, which is attributed to the H^+ reduction process inducing hydrogen embrittlement by H diffusion into the steel.
- The influence of E_{cp} on the SCC susceptibility of X70 steel meant that by decreasing the E_{cp} from −770 to −870 mV, the SCC susceptibility increased through the improvement of the cathodic process and the contribution of the anodic dissolution at the tip of the crack. However, when the E_{cp} reached −970 mV, the susceptibility decreased, because the anodic dissolution of the steel became negligible and the mechanism was dominated solely by hydrogen embrittlement. The above was verified with the analysis of the fracture surface using SEM.
- The results of EIS at −770 and −870 mV showed an active behavior at the beginning of the test, which changed to a mixed process when the steel was subjected to strains higher than YS (elastic region), which generated active sites, thus improving the cathodic process, meaning that mass transport became the speed limiting step in the cathodic process.
- At the E_{corr}, all R_{ct} values are lower than the R_{ct} values obtained at any E_{cp}. This behavior indicates that the trend of the corrosion rate (CR) was affected by the applied cathodic potential, which prompted a decrease in CR; however, it is important to point out that the E_{cp} could improve the cathodic reaction, generating atomic H and increasing the susceptibility to SCC.

Author Contributions: Conceptualization, R.G.-M. and A.C.-H.; methodology, R.G.-M.; validation, R.G.-M., R.O.-C. and A.C.-H.; formal analysis, R.G.-M. and A.C.; investigation, R.O.-C. and E.M.-S.; resources, R.G.-M.; writing—original draft preparation, A.C. and R.G.-M.; writing—review and editing, R.O.-C. and M.A.M.-C.; visualization, E.M.-S.; supervision, R.G.-M. and A.C.; project administration, A.C.; funding acquisition, R.O.-C.

Funding: This research received no external funding.

Conflicts of Interest: The authors declare no conflicts of interest.

References

1. Wang, L.W.; Du, C.W.; Liu, Z.Y.; Wang, X.H.; Li, X.G. Influence of carbon on stress corrosion cracking of high strength pipeline steel. *Corros. Sci.* **2013**, *76*, 486–493. [CrossRef]
2. Torres-Islas, A.; Serna, S.; Campillo, B.; Colin, J.; Molina, A. Hydrogen embrittlement behavior on microalloyed pipeline steel in NS-4 solution. *Int. J. Electrochem. Sci.* **2013**, *8*, 7608–7624.
3. Cheng, Y. *Stress Corrosion Cracking of Pipelines*; John Wiley & Sons: Hoboken, NJ, USA, 2013.
4. Vasil'ev, V.Y.; Sergeeva, T.K.; Baldokhin, Y.V.; Ivanov, E.S.; Novosadov, V.V.; Bayankin, V.Y. Internal stresses, corrosion and electrochemical behavior in soils, and stress corrosion of pipe steels. *Prot. Met.* **2002**, *38*, 166–171. [CrossRef]
5. Fang, B.Y.; Atrens, A.; Wang, J.Q.; Han, E.H.; Zhu, Z.Y.; Ke, W. Review of stress corrosion cracking of pipeline steels in "low" and "high" pH solutions. *J. Mater. Sci.* **2003**, *8*, 127–132. [CrossRef]
6. Liang, P.; Li, X.; Du, C.; Chen, X. Stress corrosion cracking of X80 pipeline steel in simulated alkaline soil solution. *Mater. Des.* **2009**, *30*, 1712–1717. [CrossRef]
7. Cui, Z.Y.; Liu, Z.Y.; Wang, L.W.; Ma, H.C.; Du, C.W.; Li, X.G.; Wang, X. Effect of pH value on the electrochemical and stress corrosion cracking behavior of X70 pipeline steel in the dilute bicarbonate solutions. *JMEP* **2015**, *24*, 4400–4408. [CrossRef]
8. Cheng, Y.F. Thermodynamically modeling the interactions of hydrogen, stress and anodic dissolution at crack-tip during near-neutral pH SCC in pipelines. *J. Mater. Sci.* **2007**, *42*, 2701–2705. [CrossRef]
9. Niu, L.; Cheng, Y.F. Corrosion behavior of X-70 pipe steel in near-neutral pH solution. *Appl. Surf. Sci.* **2007**, *253*, 8626–8631. [CrossRef]

10. Tang, X.; Cheng, Y.F. Quantitative characterization by micro-electrochemical measurements of the synergism of hydrogen, stress and dissolution on near-neutral pH stress corrosion cracking of pipelines. *Corros. Sci.* **2011**, *53*, 2927–2933. [CrossRef]
11. Liu, Z.Y.; Li, X.G.; Du, C.W.; Zhai, G.L.; Cheng, Y.F. Stress corrosion cracking behavior of X70 pipe steel in an acidic soil environment. *Corros. Sci.* **2008**, *50*, 2251–2257. [CrossRef]
12. Fu, A.Q.; Cheng, Y.F. Electrochemical polarization behavior of X70 steel in thin carbonate/bicarbonate solution layers trapped under a disbonded coating and its implication on pipeline SCC. *Corros. Sci.* **2010**, *52*, 2511–2518. [CrossRef]
13. Contreras, A.; Hernández, S.L.; Orozco-Cruz, R.; Galvan-Martínez, R. Mechanical and environmental effects on stress corrosion cracking of low carbon pipeline steel in a soil solution. *Mater. Des.* **2012**, *35*, 281–289. [CrossRef]
14. Ohaeri, E.; Eduok, U.; Szpunar, J. Hydrogen related degradation in pipeline steel: A review. *Int. J. Hydrog. Energy* **2018**, *43*, 14584–14617. [CrossRef]
15. Shahriari, A.; Shahrabi, T.; Oskuie, A.A. Effects of Cathodic Potential, Bicarbonate, and Chloride Ions on SCC of X70 Pipeline Steel. *JMEP* **2013**, *22*, 1421–1429. [CrossRef]
16. Liu, Z.Y.; Li, X.G.; Cheng, Y.F. Mechanistic aspect of near-neutral pH stress corrosion cracking of pipelines under cathodic polarization. *Corros. Sci.* **2012**, *55*, 54–60. [CrossRef]
17. NACE TM 0198. *Slow Strain Rate Test Method for Screening Corrosion-Resistant Alloys for Stress Corrosion Cracking in Sour Oilfield Service*; NACE International: Houston, TX, USA, 2016.
18. Zhang, G.A.; Cheng, Y.F. Micro-electrochemical characterization of corrosion of welded X70 pipeline steel in near-neutral pH solution. *Corros. Sci.* **2009**, *51*, 1714–1724. [CrossRef]
19. McIntyre, D.R.; Kane, R.D.; Wilhelm, S.M. Slow strain rate testing for materials evaluation in high-pressure H_2S environments. *Corrosion* **1988**, *44*, 920–926. [CrossRef]
20. Contreras, A.; Salazar, M.; Carmona, A.; Galván-Martínez, R. Electrochemical Noise for Detection of Stress Corrosion Cracking of Low Carbon Steel Exposed to Synthetic Soil Solution. *Mater. Res.* **2017**, *20*, 1201–1210. [CrossRef]
21. Lynch, S.P.; Raja, V.S.; Shoji, T. Stress corrosion cracking: Theory and practice. Cambric mechanistic and fractographic aspects of stress-corrosion cracking (SCC). In *Stress Corrosion Cracking: Theory and Practice*; Raja, V.S., Shoji, T., Eds.; Woodhead Publishing: Cambridge, UK, 2011; pp. 1–88.
22. Liu, Z.Y.; Li, X.G.; Du, C.W.; Lu, L.; Zhang, Y.R.; Cheng, Y.F. Effect of inclusions on initiation of stress corrosion cracks in X70 pipeline steel in an acidic soil environment. *Corros. Sci.* **2009**, *51*, 895–900. [CrossRef]
23. Martin, M.L.; Dadfarnia, M.; Nagao, A.; Wang, S.; Sofronis, P. Enumeration of the hydrogen-enhanced localized plasticity mechanism for hydrogen embrittlement in structural materials. *Acta Mater.* **2019**, *165*, 734–750. [CrossRef]
24. Javidi, S.; Horeh, B. Investigating the mechanism of stress corrosion cracking in near-neutral and high pH environments for API 5L X52 steel. *Corros. Sci.* **2014**, *80*, 213–220. [CrossRef]
25. Galván-Martínez, R.; Carmona, A.; Baltazar, M.; Contreras, A.; Orozco-Cruz, R. Stress Corrosion Cracking of X70 Pipeline Steel immersed in Synthetic Soil Solution. *Afinidad* **2018**, *76*, 52–62.
26. Marvasti, M.H. Crack Growth Behavior of Pipeline Steels in near Neutral pH Soil Environment. Master's Thesis, Department of Chemical and Materials Engineering, University of Alberta, Edmonton, AB, Canada, 2010.
27. Liu, Z.Y.; Lu, L.; Huang, Y.Z.; Du, C.W.; Li, X.G. Mechanistic aspect of non-steady electrochemical characteristic during stress corrosion cracking of an X70 pipeline steel in simulated underground water. *Corrosion* **2014**, *70*, 678–685. [CrossRef]
28. Liu, Z.Y.; Li, X.G.; Du, C.W.; Cheng, Y.F. Local additional potential model for effect of strain rate on SCC of pipeline steel in an acidic soil solution. *Corros. Sci.* **2009**, *51*, 2863–2871. [CrossRef]
29. Herraiz-Cardona, I.; Ortega, E.; Vázquez-Gómez, L.; Pérez-Herranz, V. Double-template fabrication of three-dimensional porous nickel electrodes for hydrogen evolution reaction. *Int. J. Hydrog. Energy* **2012**, *37*, 2147–2156. [CrossRef]
30. Srinivasan, S. *Fuel Cells: From Fundamentals to Applications*; Springer Science & Business: New York, NY, USA, 2006.

31. Lebrini, M.; Lagrenée, M.; Vezin, H.; Traisnel, M.; Bentiss, F. Experimental and theoretical study for corrosion inhibition of mild steel in normal hydrochloric acid solution by some new macrocyclic polyether compounds. *Corros. Sci.* **2007**, *49*, 2254–2269. [CrossRef]
32. Taylor, S.R.; Gileadi, E. Physical interpretation of the Warburg impedance. *Corrosion* **1995**, *51*, 664–671. [CrossRef]
33. Bastidas, D.M. Interpretation of impedance data for porous electrodes and diffusion processes. *Corrosion* **2007**, *63*, 515–521. [CrossRef]
34. Lasia, A. *Electrochemical Impedance Spectroscopy and its Applications*; Springer: New York, NY, USA, 2014.
35. Brug, G.J.; Van den Eeden, A.L.G.; Sluyters-Rehbach, M.; Sluyters, J.N.H. The analysis of electrode impedances complicated by the presence of a constant phase element. *J. Electroanal. Chem. Interfacial Electrochem.* **1984**, *176*, 275–295. [CrossRef]
36. Liu, Z.Y.; Li, X.G.; Cheng, Y.F. Effect of strain rate on cathodic reaction during stress corrosion cracking of X70 pipeline steel in a near-neutral pH solution. *JMEP* **2011**, *20*, 1242–1246. [CrossRef]

 © 2019 by the authors. Licensee MDPI, Basel, Switzerland. This article is an open access article distributed under the terms and conditions of the Creative Commons Attribution (CC BY) license (http://creativecommons.org/licenses/by/4.0/).

Article

Shortcomings of International Standard ISO 9223 for the Classification, Determination, and Estimation of Atmosphere Corrosivities in Subtropical Archipelagic Conditions—The Case of the Canary Islands (Spain)

Juan J. Santana [1,*], Alejandro Ramos [1], Alejandro Rodriguez-Gonzalez [2], Helena C. Vasconcelos [3,4,5], Vicente Mena [6], Bibiana M. Fernández-Pérez [6] and Ricardo M. Souto [6,7,*]

1. Department of Process Engineering, University of Las Palmas de Gran Canaria, 35017 Las Palmas de Gran Canaria, Spain; alejandro.ramos@ulpgc.es
2. Instituto de Estudios Ambientales y Recursos Naturales (i-UNAT), Universidad de Las Palmas de Gran Canaria (ULPGC), 35017 Las Palmas de Gran Canaria, Spain; alejandro.rodriguezgonzalez@ulpgc.es
3. Faculty of Sciences and Technology, Azores University, 9500-321 Ponta Delgada, Portugal; helena.cs.vasconcelos@uac.pt
4. Faculty of Sciences and Technology, Centre of Physics and Technological Research (CEFITEC), Universidade Nova de Lisboa, 2829-516 Caparica, Portugal
5. Biotechnology Centre of Azores (CBA), Azores University, 9500-321 Ponta Delgada, Portugal
6. Department of Chemistry, University of La Laguna, 38200 La Laguna, Spain; vmenagon@ull.es (V.M.); bfernand@ull.edu.es (B.M.F.-P.)
7. Institute of Materials Science and Nanotechnology, University of La Laguna, 38200 La Laguna, Spain
* Correspondence: juan.santana@ulpgc.es (J.J.S.); rsouto@ull.es (R.M.S.); Tel.: +34-928-451-945 (J.J.S.); +34-922-318-067 (R.M.S.)

Received: 23 September 2019; Accepted: 11 October 2019; Published: 15 October 2019

Abstract: The classification, assessment, and estimation of the atmospheric corrosivity are fixed by the ISO 9223 standard. Its recent second edition introduced a new corrosivity category for extreme environments CX, and defined mathematical models that contain dose–response functions for normative corrosivity estimations. It is shown here that application of the ISO 9223 standard to archipelagic subtropical areas exhibits major shortcomings. Firstly, the corrosion rates of zinc and copper exceed the range employed to define the CX category. Secondly, normative corrosivity estimation would require the mathematical models to be redefined introducing the time of wetness and a new set of operation constants.

Keywords: atmospheric corrosion; ISO 9223; corrosivity categories; predictive models; archipelagic regions; Canary Islands

1. Introduction

Atmospheric corrosion is a process caused by the interaction of metals with the atmosphere causing their degradation. The relevance of atmospheric corrosion is often quantified in terms of the high costs caused by its action, because repairs and replacements due to corrosion amount ca. 5% of the gross domestic product (GDP) in Western countries, China, and India per year [1]. Even if this process was not producing the costs of material replacement, it would also account for production losses, energy-based costs, and the release of toxic substances to the environment. Given its impact, numerous studies on atmospheric corrosion are available in the scientific and technical literature [2–10], often directed to the acquisition of atmospheric corrosivity maps for a given geographic region. The methodology required to perform these studies is established by a series of international standards (namely ISO 9223 to 9226) that were first published in 1992, and subjected to technical revision in 2012 [11–14]. In particular, the revised ISO 9223 contains substantial changes from its first edition [15]. Among them, the procedure to assign the corrosivity categories based on environmental data (i.e., SO_2 and chloride dry depositions, temperature and time of wetness) was removed from the international standard, whereas dose–response functions were introduced for the normative corrosivity estimation based on environmental data. In addition, a new corrosivity category, CX, corresponding to extreme environments, was included for classifying the corrosion rates of standard metals (i.e., carbon steel, zinc, copper, and aluminum). Indeed, several reports in the literature had previously shown the shortcomings of the first edition of ISO 9223 to rank, determine, and estimate the corrosion of metals and alloys in subtropical and tropical regions [3,16–22], because category C5 failed to determine their actual aggressiveness, thus requiring higher corrosivity categories.

The Canary Islands are subject to the climatic dynamics of the subtropical latitudes, which, together with its proximity to the African continent and its abrupt orography, originate very specific climate conditions. The action of the trade winds determines the climate of the islands [23]. They are very humid winds of Northeast (NE) component with an annual frequency higher than 80% that bestow a very stable weather to the archipelago. The most eastern islands (i.e., Lanzarote and Fuerteventura) have desert-like climates, associated to a slightly rugged terrain with low mountains that are not able to retain the moisture of the trade winds. The remaining islands have a Mediterranean-type climate [23]. As they are more abrupt islands, the moisture of the trade winds is effectively sustained. The complex combination of climate and orography conditions may originate various climatic zones (e.g., microclimates) to develop on the same island, which supports the popular topic that the Canary Islands are a continent in miniature. These climatic zones have been determined and characterized in a recent project named CLIMCAN-010 [24]. The main objective of this project was to perform a complete climatological characterization of the Canary archipelago aimed for inclusion into the Technical Building Catalogue of Spain [25]. A major outcome of that work was probing several distinct climate zones to be present in all the islands, highlighting the islands of Tenerife and Gran Canaria with 6 and 5 zones, respectively. As a result, there is a complex distribution of Canarian atmospheres of varying aggressiveness, given that in a small geographical area there are large climatic variations. Another major project performed in the Canaries aimed to obtain the corrosion map of the Canary Islands by measuring the weight losses of carbon steel, copper, zinc, and aluminum from a large number of corrosion stations distributed along the seven main islands of the archipelago [21,26]. It was found that the ISO 9223:1992 failed to characterize the atmospheric corrosivities because the weight losses measured for the standard metals in a large number of the stations exceeded by far the highest C5 corrosivity category [21,26].

In this work, the atmospheric corrosivities of the atmospheres occurring in the Canary Islands have been reassessed in order to classify them using the revised ISO 9223:2012 standard [11], as well as to verify the validity of the proposed dose–response functions for the estimation of normative corrosivities from corrosion losses.

2. Materials and Methods

The Canary Islands are located near the Northwest (NW) African coast, between 27°37′ and 29°27′ North (N) and 13°20′ and 18°20′ West (W) (see Figure 1). The main orographic characteristics of the archipelago together with the geographical coordinates of each island are shown in Figure 2. In addition, Figure 2 depicts the climatic zones defined by the CLIMCAN-010 project [24] as well as the distribution of the 74 corrosion exposure sites through the 7 islands. Table S1 in the Supplementary Material gives the localization and the elevation of the test sites, together with the type of atmosphere on the basis of classification criteria other than corrosivity according to ISO 9223:1992(E) [15] and ISO 9223:2012 standards [11].

Metal samples of carbon steel, zinc, and copper of dimensions 10 cm × 4 cm × 2 mm were exposed at the corrosion exposure sites, and their chemical composition is given in Table 1. The specimens were cleaned according to ASTM G1-90 standard [27], weighed, and duly codified for identification. Subsequently, they were placed in a metal frame oriented towards the North-Northeast (NNE), with an inclination of 45 degrees with respect to the horizontal. Samples were collected every six months during the first year for copper and zinc, and with quarterly periodicity for carbon steel. In each collection, four specimens of each metal were taken. Three of these samples were cleaned according to ASTM G1-90 standard [27], and corrosivity categories were assigned from first-year weight losses according to ISO 9223:2012 standard [11].

The relative humidity level was quantified using a thermohygrometer, whereas chloride and SO_2 dry deposition rates were determined monthly according to standard procedures. Namely, two methods were employed to determine SO_2 pollution, namely the Husy method according to the ISO/TC 156 N 250 standard [28], and the lead dioxide candle according to the ASTM D 2010-85 standard [29]. The concentration of chloride was monitored by the wet candle method according to the ISO 9225 standard [13]. Finally, the ISO 9223:1992 standard [15] was employed to characterize the atmosphere of the localities in terms of pollution categories based on airborne salinity contamination (S_d) and with sulfur compounds based on sulfur dioxide (P_d), and of time of wetness (τ).

The effect of the environmental parameters on the average corrosion rates of metals for the first year of atmospheric exposure, r_{corr}, was analyzed using a multivariate variance analysis (ANOVA).

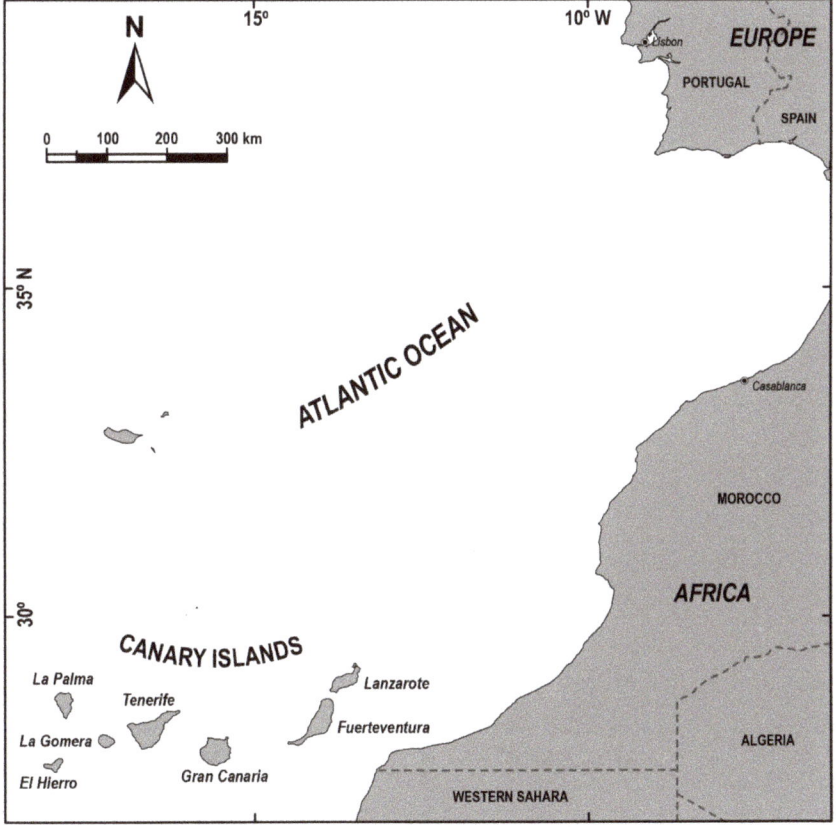

Figure 1. Location of the Canary Islands.

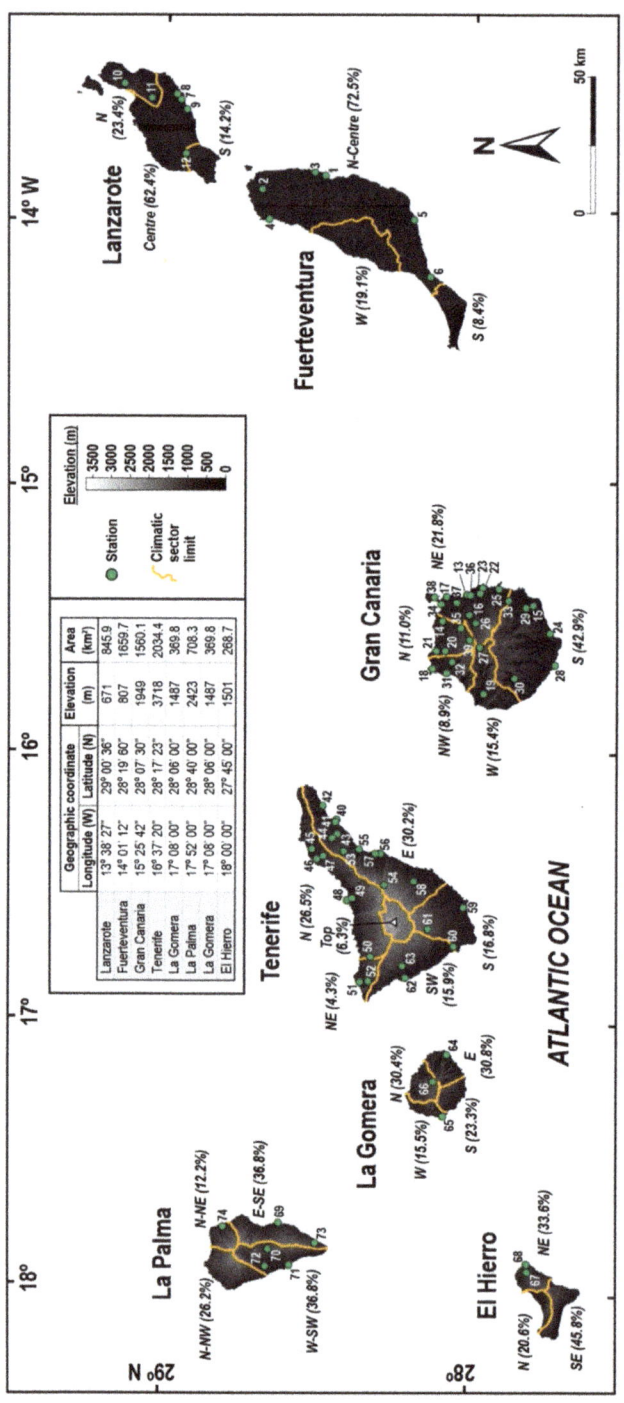

Figure 2. Subdivision of Canary Islands into local climate sectors (adapted from [24]), and location of the 74 corrosion exposure sites (see Table S1 in Supplementary Material for codes and local orography details).

Table 1. Chemical composition of the carbon steel, zinc, and copper test samples.

Metal	Element (wt.%)									
	Si	Fe	C	Mn	Zn	Ti	Cu	Mg	Al	Others
Carbon steel	0.08	99.47	0.06	0.37	-	-	-	-	-	0.023
Copper	0.28	0.9	-	0.05	0.09	0.05	98.5	0.05	-	0.09
Zinc	-	-	-	-	99.77	0.02	0.2	-	0.006	-

X-Ray Diffractometry (XRD) was performed using a Siemens D-5000 instrument (Bruker-Siemens, Billerica, MA, USA) provided with a copper anode (Cu Kα 5406 Å) and a scintillation detector.

3. Results

3.1. Classification of Corrosivity of the Atmosphere and General Corrosivity Estimation

The great variability of local environmental conditions occurring along the Canary archipelago is readily observable by inspecting Figure 3. This graph depicts time of wetness, and SO_2 and Cl^- deposition distributions measured during 3 years at the 74 corrosion exposure sites. Based on the local environmental conditions occurring at each location, corrosivity categories were assigned according to the ISO 9223:2012 standard [11], and they are listed in Table S1 in the Supplementary Material.

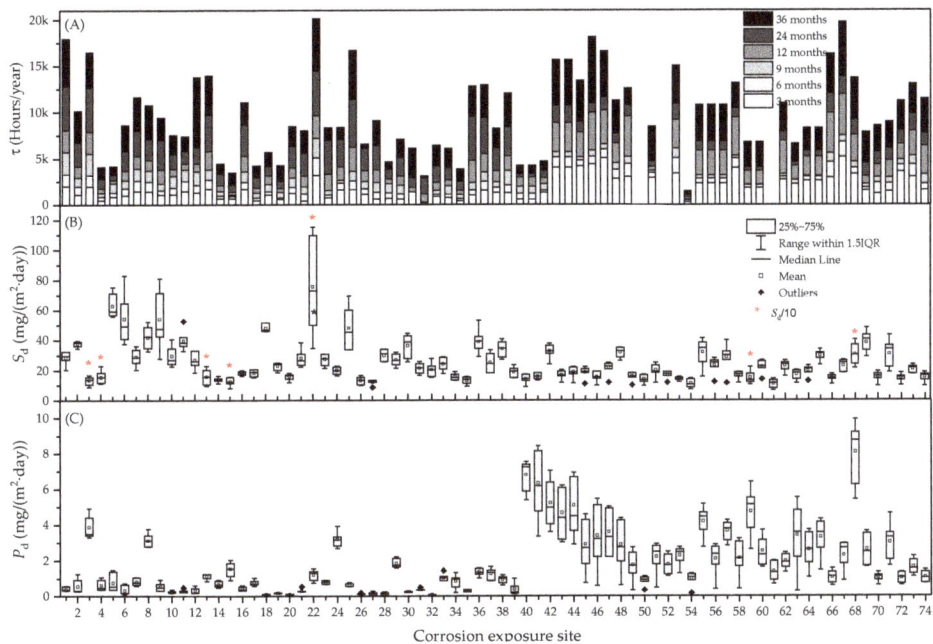

Figure 3. Local environmental conditions at the 74 corrosion exposure sites determined during 3 years. (A) Time of wetness, (B) average Cl^- deposition, and (C) average SO_2 deposition.

Weight losses were measured for carbon steel, zinc, and copper after 1-year exposure, and they are given as first-year corrosion rates in Table S1 (Supplementary Material). Local atmosphere corrosivities were assigned for the three metals according to the ISO 9223:2012 standard [11], and they are also included in Table S1 (Supplementary Material). In addition, atmosphere corrosivity maps were drawn in Figure 4 together with the local climate sectors.

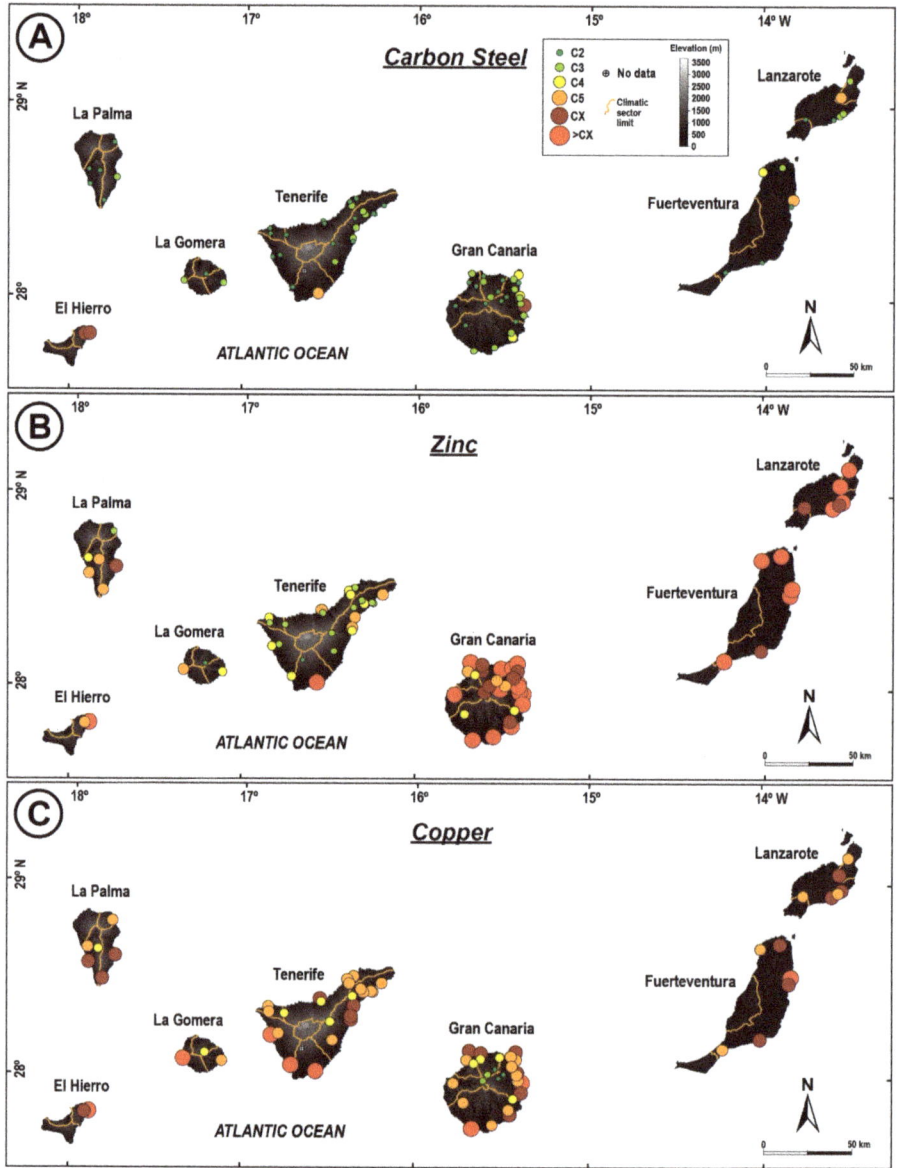

Figure 4. Corrosion maps for (**A**) carbon steel, (**B**) zinc, (**C**) copper according to ISO 9223:2012 [11].

Since the local environmental conditions influence the corrosion rates of metals, multivariate variance analysis (ANOVA) was performed in order to evaluate the effect of these parameters on the first-year corrosion rates. The following environmental parameters were considered: annual average air temperature (T), SO_2 deposition rate (P_d), Cl^- deposition rate (S_d), time of wetness (τ), and relative humidity (RH). To carry out this analysis, the values adopted by these environmental conditions were grouped into levels according to ISO 9223:2012 [11], and they are given in Table 2 together with the number of corrosion exposure sites included in each level. Levels were assigned to T and RH by establishing intervals of 2 °C and 10% allowance, respectively.

Table 2. Environmental parameters and levels employed in the ANOVA analysis.

Environmental Parameter	Level/Number of Samples per Level					
T	$T_1/10$	$T_2/14$	$T_3/40$	-	-	-
P_d	$P_0/49$	$P_1/15$	$P_2/0$	$P_3/0$	-	-
S_d	$S_0/0$	$S_1/4$	$S_2/60$	$S_3/0$	-	-
τ	$\tau_1/0$	$\tau_2/0$	$\tau_3/25$	$\tau_4/31$	$\tau_5/8$	-
RH	$RH_1/3$	$RH_2/0$	$RH_3/0$	$RH_4/2$	$RH_5/46$	$RH_6/13$

Next, the ANOVA variance analysis was performed to determine the corrosivity for the three metals based on the corrosion rates measured at each exposure site, and the results for carbon steel, zinc, and copper, are respectively listed in Tables 3–5. The analysis was done on the data from 64 corrosion exposure sites, because incomplete or not reproducible data were found at sites 13, 15, 18, 22, 50, 52, 54, 61, 67, and 68, and they were discarded for the rest of the study.

Table 3. Statistical analysis of results for carbon steel.

Environmental Parameter	Sum sq.	Freedom Degrees	Mean sq.	f-Value	p-Value
RH	392.5535	3	130.8512	0.7204	0.5442
P_d	680.9685	1	680.9685	3.7489	0.0581
S_d	**4995.7843**	1	**4995.7843**	**27.5031**	**0.0000**
T	236.6076	2	118.3038	0.6513	0.5254
τ	1071.0287	2	535.5143	2.9482	0.0609
Error	9808.7825	54	181.6441	-	-
Total	20,246.1375	63	-	-	-

Table 4. Statistical analysis of results for zinc.

Environmental Parameter	Sum sq.	Freedom Degrees	Mean sq.	f-Value	p-Value
RH	**8635.9756**	3	**2878.6585**	**5.1955**	**0.0032**
P_d	1302.4139	1	1302.4139	2.3506	0.1311
S_d	**7581.5157**	1	**7581.5157**	**13.6833**	**0.0005**
T	2580.5849	2	1290.2925	2.3288	0.1071
τ	**5884.669**	2	**2942.3345**	**5.3104**	**0.0078**
Error	29,919.7938	54	554.0703	-	-
Total	60,270.1119	63	-	-	-

Table 5. Statistical analysis of results for copper.

Environmental Parameter	Sum sq.	Freedom Degrees	Mean sq.	f-Value	p-Value
RH	18.194	3	6.0647	0.6768	0.57
P_d	3.7699	1	3.7699	0.4207	0.5193
S_d	**77.0076**	1	**77.0076**	**8.5939**	**0.0049**
T	45.2614	2	22.6307	2.5255	0.0894
τ	7.9729	2	3.9865	0.4449	0.6432
Error	483.8814	54	8.9608	-	-
Total	650.7247	63	-	-	-

Next, the ANOVA variance analysis was performed to determine the corrosivity for the three metals based on the corrosion rates measured at each exposure site, and the results for carbon steel, zinc, and copper, are respectively listed in Tables 3–5. The analysis was done on the data from 64 corrosion exposure sites, because incomplete or not reproducible data were found at sites 13, 15, 18, 22, 50, 52, 54, 61, 67, and 68, and they were discarded for the rest of the study.

3.2. Normative Corrosivity Estimation

Corrosivity estimation was first attempted employing the dose–response functions of exposure proposed in ISO 9223:2012 (Section 8.2) [11]. Namely, the norm establishes both the function (given by Equation (1)) and the corresponding set of constants (see Table 6):

$$r_{corr} = a \cdot (P_d)^b \cdot e^{(c \cdot RH - d \cdot (T-10))} + e \cdot (S_d)^f \cdot e^{(g \cdot RH + h \cdot T)} \tag{1}$$

Table 6. Normative corrosivity estimation based on calculated first-year corrosion losses. Set of constants and sum of the quadratic error for carbon steel, zinc, and copper using the dose–response functions given by Equations (1) and (3).

Equation	Metal	Constants								SSE
		a	b	c	d	e	f	g	h	
(1)	CS	1.77	0.52	0.02	0.054	0.102	0.62	0.033	0.04	116.7368
(1)	Zn	0.0129	0.44	0.046	0.071	0.0175	0.57	0.008	0.085	295.5821
(1)	Cu	0.0053	0.26	0.059	0.08	0.01	0.27	0.036	0.049	41.4758
(1) *	CS	4.855	0.6	0.01589	0.1089	0.871	0.9	2.2×10^{-4}	2.0×10^{-4}	85.1416
(1) *	Zn	0.0015	0.44	0.0359	0.371	0.4435	0.86	0.0096	0.0189	187.3936
(1) *	Cu	2.6528	0.25	0.00011	0.0048	0.09	0.9	0.0001	0.0162	22.7040
(3)	CS	0.3592	0.6	0.0005	1×10^{-6}	0.8403	0.9	4×10^{-5}	0.00163	86.7857
(3)	Zn	6.1796	0.44	0.0013	0.095	1.0156	0.86	0.0001	0.1×10^{-6}	197.68
(3)	Cu	2.6528	0.25	1×10^{-6}	21×10^{-6}	0.09	0.9	5.7×10^{-6}	0.0003	21.4148

* Equation (1) modified with a new set of constants adapted to the environmental conditions observed in the corrosion exposure sites.

The criteria for establishing the quality of the dose–response function for the estimation of corrosion rates was made in terms of the sum of squared errors (SSE) between the experimental observations and those predicted by the model under consideration, as shown in Equation (2):

$$SSE = \sum_{i=1}^{n} \left(r_{corr_i} - \hat{r}_{corr_i} \right)^2 \tag{2}$$

In this way, the applicability of the dose response function of exposure and the set of constants defined in ISO 9223:2012 for each metal is described by the SSE values included in the first set of rows of Table 6. Values in excess of 100 were found for carbon steel and zinc, whereas the errors for copper amounted ca. 41. A new attempt to improve the estimation of corrosion rates for carbon steel and zinc using the dose–response function given by Equation (1) consisted in the modification of the set of constants for each metal given by the norm as to better fit the experimental observations. The procedure consisted in introducing these constants as fitting parameters in the function, and using the algorithm of the simplex method of Nelder–Mead [30] to obtain the best set of parameters [31,32]. Accordingly, the new sets of constants for the dose–response function and the resulting fit qualities, expressed in terms of SSE values, are included in the second set of rows in Table 6 (i.e., labeled as Equation (1) *). It is observed that corrosivity estimation based on environmental information using the dose–response function defined in the ISO 9223:2012 [11] requires obtaining a new set of constants based on the first-year corrosion rates of the corresponding metal. The improvement of the fit quality was significant even for copper, even though the corrosion rates observed in this fragmented subtropical territory could still be assigned to the corrosivity categories included in the norm.

An alternate method for the estimation of corrosivity based on metal corrosion losses consisted in defining new dose–response functions for each metal using the time of wetness instead of the relative humidity. The resulting dose–response function is Equation (3):

$$r_{\text{corr}} = a \cdot (P_d)^b \cdot e^{(c \cdot \tau - d \cdot (T-10))} + e \cdot (S_d)^f \cdot e^{(g \cdot \tau + h \cdot T)} \tag{3}$$

The third set of rows in Table 6 gives the new set of constants that fit Equation (3), corresponding to the new proposal, as well as the values of the sums of quadratic errors SSE. A better agreement between the estimated corrosion rates and the experimental observations is also observed in this case.

Fit quality analysis was also performed by considering the residual error for each corrosion exposure site that was determined using Equation (4):

$$Residue_i = r_{\text{corr}_i} - \hat{r}_{\text{corr}_i} \tag{4}$$

The residues are plotted in Figure 5 for each metal by comparing the estimations done using either the new set of constants (i.e., Equation (1) *, see Figure 5A–C) or the new dose–response function (i.e., Equation (3), cf. Figure 5D–F) with the estimates from ISO 9223:2012 [11]. In all cases, the worst results were obtained using the norm. This section may be divided by subheadings. It should provide a concise and precise description of the experimental results, their interpretation as well as the experimental conclusions that can be drawn.

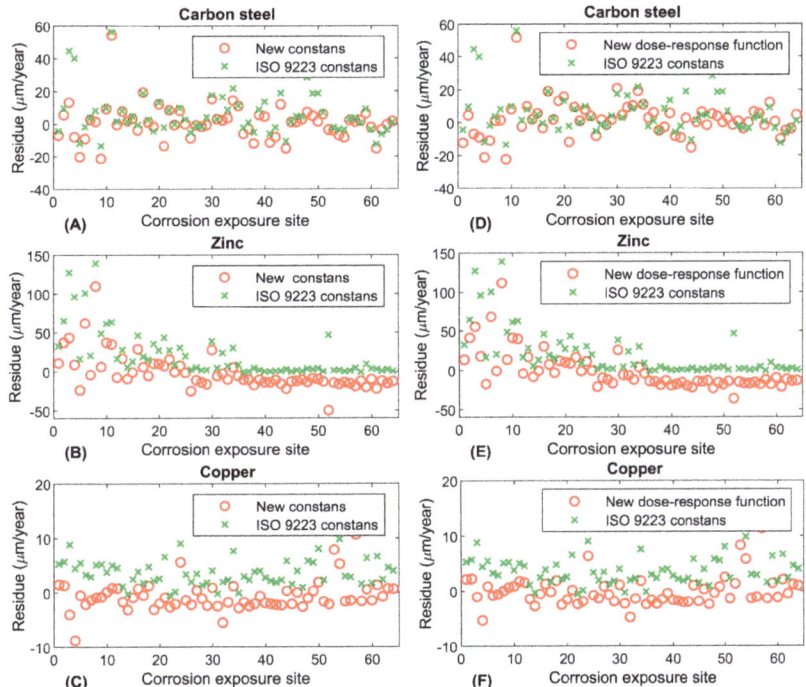

Figure 5. Residue distributions from the application of dose–response functions for normative corrosivity estimation based on calculated first-year corrosion losses. (**A**–**C**) Application of Equation (1) using either the set of constants given by ISO 9223:2012 [11] or a new set of constants that best fit the results from corrosion exposure sites considered in this work; (**D**–**F**) application of the new dose–response function given by Equation (3). Metals: (**A,D**) carbon steel; (**B,E**) zinc; and (**C,F**) copper.

4. Discussion

4.1. Corrosivity of the Atmospheres

Table 7 lists the changes introduced in the second edition of the ISO 9223 standard with respect to the corrosivity categories based on the deposition rate of SO_2, as well as the definition of a new corrosivity category CX. Only the ranges that affect the levels of SO_2 have been modified, effectively decreasing the amounts of pollutant assigned to the categories of rural atmosphere (namely down from 10 to 4 mg/(m² day)), urban atmosphere (the ranges are modified from 10–35 to 4–24 mg/(m² day), and industrial atmosphere (modifying only the lower limit that changes from 35 to 24 mg/(m² day). These changes have a very small effect on the classification of the Canary atmospheres in terms of SO_2 deposition. That is, they only affect the corrosion exposure sites 23, 25, and 38 in the island of Gran Canaria, as well as sites 40 to 46, 55, 57, 59, and 63 in the island of Tenerife, and site 68 in the island of El Hierro. In general, they are moved from category P0 to category P1, although they all remain near the lower limit of the interval. As result, the ratio of corrosion exposure sites with level P0 changes from 94.4% to 74% by applying the modifications in the new edition of the standard with respect to the reports made using its first edition [18,19]. In this way, most of the territory in the archipelago apparently would exhibit urban atmosphere corrosivity, although it must be noticed that SO_2 deposition in most corrosion exposure sites originates from marine contributions as sulphate ion, and therefore most of the atmospheres in the Canaries are predominantly rural. In addition, the ISO 9223:1992 [15] specifies that SO_2 deposition rates corresponding to the P_0 category must be considered as background concentration and would not affect the corrosion process.

Table 7. Changes made between the first and the second editions of the ISO 9223 norm.

SO_2 deposition rate in mg/(m² day)	9223:1992 [15]	9223:2012 [11]	Level
	$P_d \leq 10$	$P_d \leq 4$	P_0—Rural atmosphere
	$10 < P_d \leq 35$	$4 < P_d \leq 24$	P_1—Urban atmosphere
	$35 < P_d \leq 80$	$24 < P_d \leq 80$	P_2—Industrial atmosphere
	$80 < P_d \leq 200$	$80 < P_d \leq 200$	P_3—Highly polluted industrial atmosphere
New corrosivity category CX (according to ISO 9223:2012).			
Carbon steel	$200 < r_{corr} \leq 700$ (µm/year)		$1500 < r_{corr} \leq 5500$ (g/(m² year))
Zinc	$8.4 < r_{corr} \leq 25$ (µm/year)		$60 < r_{corr} \leq 180$ (g/(m² year))
Copper	$5.6 < r_{corr} \leq 10$ (µm/year)		$50 < r_{corr} \leq 90$ (g/(m² year))

Conversely, the classification based on the rate of chloride deposition did not undergo any change by applying the revised norm. About 87.8% of the stations belong to category S_1, 9.5% to category S_2, and 2.7% to category S_3, a fact that reveals the relevance of this pollutant in the atmospheres of the archipelago. Indeed, category S_0 could not be assigned anywhere, not even for corrosion exposure sites located either far from the coast or in high elevation. Although the second edition of the ISO 9223 standard states that atmospheres with high levels of chloride pollutant are outside its scope, this should not be the case of the Canary Islands, where even two corrosion exposure sites are classified into category S_3 (namely, sites 22 and 68), with 933.7 and 334.7 mg/(m² day), thus being far from the upper range of the interval that is established at 1500 mg/(m² day).

Regarding the time of wetness (TOW), the classification established in the first edition experienced no changes by applying the revised standard. Therefore, for the Canary archipelago, all islands exhibit atmosphere classes higher than τ_2, distributed as 37.8% with class τ_3, 45.9% with class τ_4, and 16.2% with class τ_5. This feature evidences the high humidity atmospheres occurring in the archipelago due to the action of the trade winds.

When the corrosivity categories were re-evaluated using the new ranges established by ISO 9223:2012, the distribution of categories exceeding the ranges of the category C5 as established in the first edition are listed in Table 8. Thus, for the entire archipelago, the introduction of a new CX category should account for all the cases found. However, this does not happen as much for zinc as for copper, where it is observed that 32.4% of cases exceed CX category for zinc and 11.0% for copper. It is observed for these two metals, Zn and Cu, that more than 44.4% and 38.4% of cases, respectively, have the highest category in the norm or higher, this effect being even more noticeable for the eastern islands. XRD analyses carried out for zinc samples reveal the absence of the protective layer of zinc hydroxosulphate ($Zn_4SO_4(OH)_6·H_2O$), being the majority compound a basic chloride $Zn_5Cl_2(OH)_8·H_2O$, typical of marine atmospheres and of a less protective nature. This fact justifies the high corrosion rate values found for zinc in the complete Canary archipelago. A special mention is deserved by corrosion exposure sites 15, 22, 59, 68, and 69, that were located in very windy areas with high salinity values. The corrosion rates exceed by far those determined in the remaining exposure sites, due to the combined effect of erosion that breaks the passive layer of corrosion products. On the other hand, the corrosion products found in copper were mostly a patina composed of cuprite (Cu_2O), hydroxyl-chloride dimorphs, atacamite and paratacamite ($CuCl_2·3Cu(OH)_2$), and malachite ($Cu_2CO_3(OH)_2$). The presence of atacamite in most exposure sites throughout the islands must be highlighted. Finally, in those exposure sites where the chlorides deposition rate exceeded 30 mg/(m^2 day) (namely, stations 3, 4, 5, 8, 13, 15, 22, 59, 65, 68, 69, and 71), atmospheric aggressiveness hindered the formation of a passive layer even after three years of exposure.

Table 8. Percentage distribution of atmosphere corrosivity categories in the Canary archipelago.

%	ISO 9223:1992 [15]		
	Carbon Steel	Zinc	Copper
C1	0	0	0
C2	57.5	2.7	1.4
C3	30.1	16.2	6.8
C4	5.5	21.6	16.4
C5	2.7	14.9	37.0
>C5	4.1	44.6	38.4
	ISO 9223:2012 [11]		
CX	4.1	14.9	27.4
>CX	-	32.4	11.0

4.2. ANOVA Analysis

The following observations were made with respect to the analysis of the ANOVA variance, for a significance level of 5%, and taking as reference the information indicated by the p-values in the last column of Tables 3–5:

- The corrosion rates, r_{corr}, showed a strong dependence with S_d for the three metals studied (carbon steel, zinc, and copper), since the p-values were below the level of significance of the study.
- The relative humidity (RH) only had an influence on the corrosion rate of zinc, because this was the only metal with p-value smaller than 5%.
- The deposition of sulfur dioxide (P_d) only showed a weak influence with the corrosion rate of steel, since its p-value slightly exceeded the level of significance 5.8%, being unable to associate such an influence in the cases of zinc and copper.
- Regarding the temperature (T) data, no influence was observed on the corrosion rates for any of the three metal systems.
- The wetting time (τ) exhibited a major effect on the corrosion rate for zinc, whereas its influence was small for steel, and almost negligible for copper.

4.3. Normative Corrosivity Estimation Based on Calculated First-Year Corrosion Losses

Regarding the estimation of corrosion rates using dose–response functions, Table 6 evidences that the results obtained using the functions given in ISO 9223:2012 [11], represented by Equation (1), delivered the worst fitting values to the experimental data. Conversely, when constants estimated especially for the experimental data were used, better fits were obtained, as indicated by the smaller values of the sum of the squared errors (SSE), which were listed in Table 6 under the label of Equation (1) *. Even better results were obtained in the case of the newly proposed Equation (3). In addition, the best behavior in terms of the plots of the residuals shown in Figure 5, corresponded again to the fits made using the new determined constants, for most of sampling exposure sites. These improvements were indicated in Table 6 under the labels for Equations (1) * and (3) for the estimation of the corrosion rates, and resulted mainly from the fact that the constants provided by ISO 9223:2012 have been determined in atmospheres with characteristics considerably different from those occurring in the Canary Islands. Altogether, it can be inferred that it would be desirable to establish a new set of models that adequately consider the parameters showing a greater influence on the corrosion rates, these being determined by means of analysis of influence, such as the variance analysis.

5. Conclusions

Due to the big microclimatic variability existing in the Canary Islands and the subtropical conditions determined by the trade winds, the corrosion rates for carbon steel, copper, and zinc from 74 corrosion exposure sites exceeded the ranges contained in the ISO 9223:2012 standard.

In the case of carbon steel, corrosivity categories were observed to range between C2 and CX. For this metal, the second edition of the ISO 9223 standard satisfactorily described all the situations found in the archipelago.

In the case of zinc and copper, the high salinity and TOW caused high corrosion rates in many locations. In the case of zinc, the CX corrosivity category was assigned to 14.9% of the exposure sites distributed throughout the archipelago, whereas 32.4% exhibited corrosion rates higher than those corresponding to the category CX. Copper exhibited a general behavior similar to that described for zinc. Thus, 27.4% of the total number of exposure sites exhibited a CX corrosivity category, and 11.0% were higher than CX. These results show that either it would be necessary to readjust the upper limit of the category CX for metals such as Zn and Cu, or even to introduce a new corrosivity category to describe the greater aggressivity of subtropical climatologies.

With respect to the analysis of the variance for the three metals, it was found that the most influential environmental parameter affecting the corrosion rates was the chloride deposition rate (S_d), with a p-value of 0.49% in the worst case (i.e., copper), well below the level of significance of the study, namely 5%. On the contrary, the environmental temperature (T) showed the smallest influence, with a p-value of 8.9% in the best case (copper), which was clearly above the level of significance of the study, 5%.

Regarding the dose–response functions associated with the corrosion rates, for the three metals, it was found that the proposed modifications of these functions, given in the form of new sets of constants, delivered better fits than those sets of constants given by the ISO 9223 standard. In the worst case, a decrease in the sum of squared errors, SEE, of approximately 25% was observed, with respect to the standard function for carbon steel, whereas in the best case, a decrease in the SEE of approximately 48% occurred with respect to the ISO 9223 standard function for copper.

Supplementary Materials: The following are available online at http://www.mdpi.com/2075-4701/9/10/1105/s1. Table S1 lists the location, characteristics, and corrosivity categories according to ISO 9223:2012 Norm [11] of the 74 corrosion exposure sites considered in this study.

Author Contributions: Conceptualization, J.J.S. and R.M.S.; Data curation, J.J.S., A.R., A.R.-G., V.M. and B.M.F.-P.; Formal analysis, J.J.S., A.R. and R.M.S.; Funding acquisition, J.J.S. and R.M.S.; Investigation, J.J.S. and R.M.S.; Methodology, J.J.S., H.C.V. and R.M.S.; Project administration, J.J.S., A.R. and R.M.S.; Resources, J.J.S. and R.M.S.; Software, J.J.S., A.R., A.R.-G., V.M. and B.M.F.-P.; Supervision, J.J.S. and .M.S.; Validation, J.J.S., H.C.V. and R.M.S.;

Visualization, J.J.S. and R.M.S.; Writing—original draft, J.J.S., A.R., A.R.-G. and H.C.V.; Writing—review & editing, J.J.S., A.R., A.R.-G. and R.M.S.

Funding: This research was funded by UNELCO-ENDESA (Las Palmas de Gran Canaria, Spain) and by the Canarian Agency for Research, Innovation and Information Society (Las Palmas de Gran Canaria, Spain) and the European Social Fund (Brussels, Belgium) under grant ProID2017010042. V.F.M. is grateful to Universidad de La Laguna and Obra Social "La Caixa" for a research contract.

Conflicts of Interest: The authors declare no conflict of interest.

References

1. Gramberg, U. Korrosionsschutz-Antworten auf eine Herausforderung. Zur Gründung der Gesellschaft für Korrosionsschutz GfKORR. *Mater. Corros.* **1996**, *47*, 139–145. [CrossRef]
2. Dean, S.W.; Hernández-Duque Delgadillo, G.; Bushman, J.B. *Marine Corrosion in Tropical Environments*, 1st ed.; ASTM Stock Number STP 1399; American Society for Testing Materials: West Conshohocken, PA, USA, 2000; pp. 1–305.
3. Leygraf, C.; Odnevall Wallinder, I.; Tidblad, J.; Graedel, T. *Atmospheric Corrosion*, 2nd ed.; The Electrochemical Society, John Wiley & Sons, Inc.: Pennington, NJ, USA, 2016; pp. 1–397.
4. Lopesino, P.; Alcántara, J.; de la Fuente, D.; Chico, B.; Jiménez, J.A.; Morcillo, M. Corrosion of copper in unpolluted chloride-rich atmospheres. *Metals* **2018**, *8*, 866. [CrossRef]
5. Díaz, I.; Cano, H.; Lopesino, P.; de la Fuente, D.; Chico, B.; Jiménez, J.A.; Medina, S.F.; Morcillo, M. Five-year atmospheric corrosion of Cu, Cr and Ni weathering steels in a wide range of environments. *Corros. Sci.* **2018**, *141*, 146–157. [CrossRef]
6. Díaz, I.; Cano, H.; Crespo, D.; Chico, B.; de la Fuente, D.; Morcillo, M. Atmospheric corrosion of ASTM A-242 and ASTM A-588 weathering steels in different types of atmosphere. *Corros. Eng. Sci. Technol.* **2018**, *53*, 449–459. [CrossRef]
7. LeBozec, N.; Thierry, D.; Persson, D.; Riener, C.K.; Luckeneder, G. Influence of microstructure of zinc-aluminium-magnesium alloy coated Steel on the corrosion behavior in outdoor marine atmosphere. *Surf. Coat. Technol.* **2019**, *374*, 897–909. [CrossRef]
8. Morcillo, M.; Díaz, I.; Cano, H.; Chico, B.; de la Fuente, D. Atmospheric corrosion of weathering steels. Overview for engineers. Part I: Basic concepts. *Constr. Build. Mater.* **2019**, *213*, 723–737. [CrossRef]
9. Morcillo, M.; Díaz, I.; Cano, H.; Chico, B.; de la Fuente, D. Atmospheric corrosion of weathering steels. Overview for engineers. Part II: Testing, inspection, maintenance. *Constr. Build. Mater.* **2019**, *222*, 750–765. [CrossRef]
10. Liu, H.; Cao, F.; Song, G.-L.; Zheng, D.; Shi, Z.; Dargusch, M.S.; Atrens, A. Review of the atmospheric corrosion of magnesium alloys. *J. Mater. Sci. Technol.* **2019**, *35*, 2003–2016. [CrossRef]
11. ISO 9223:2012. *Corrosion of Metals and Alloys—Corrosivity of Atmospheres—Clasification, Determination and Estimation*, 2nd ed.; International Organization for Standardization: Geneva, Switzerland, 2012.
12. ISO 9224:2012. *Corrosion of Metals and Alloys—Corrosivity of Atmospheres—Guiding Values for the Corrosivity Categories*, 2nd ed.; International Organization for Standardization: Geneva, Switzerland, 2012.
13. ISO 9225:2012. *Corrosion of Metals and Alloys—Corrosivity of Atmospheres—Measurement of Environmental Parameters Affecting Corrosivity of Atmospheres*, 2nd ed.; International Organization for Standardization: Geneva, Switzerland, 2012.
14. ISO 9226:2012. *Corrosion of Metals and Alloys—Corrosivity of Atmospheres—Determination of Corrosion Rate of Standard Specimens for the Evaluation of Corrosivity*, 2nd ed.; International Organization for Standardization: Geneva, Switzerland, 2012.
15. ISO 9223:1992(E). *Corrosion of Metals and Alloys—Corrosivity of Atmospheres—Classification*, 1st ed.; International Organization for Standardization: Geneva, Switzerland, 1992.
16. Ramanauskas, R.; Muleshkova, L.; Maldonado, L.; Dobrovolskis, P. Characterization of the corrosion behaviour of Zn and Zn alloy electrodeposits: Atmospheric and accelerated tests. *Corros. Sci.* **1998**, *40*, 401–410. [CrossRef]
17. Mendoza, A.R.; Corvo, F. Outdoor and indoor atmospheric corrosion of carbon steel. *Corros. Sci.* **1999**, *41*, 75–86. [CrossRef]

18. Santana, J.J.; Santana, J.; González, J.E.; de la Fuente, D.; Chico, B.; Morcillo, M. Atmospheric corrosivity map for steel in Canary Isles. *Br. Corros. J.* **2001**, *36*, 266–271. [CrossRef]
19. Veleva, L.; Kane, R. Atmospheric corrosion. In *Corrosion, Fundamentals, Testing and Applications, ASM Handbook Series*, 1st ed.; Cramer, S.D., Covino, B.S., Jr., Eds.; American Society for Testing Materials International: Columbus, OH, USA, 2003; Volume 13A, pp. 196–209.
20. Mikhailov, A.A.; Tidblad, J.; Lucera, V. The classification system of ISO 9223 Standard and the dose-response functions assessing the corrosivity of outdoor atmospheres. *Prot. Met.* **2004**, *40*, 541–550. [CrossRef]
21. Morales, J.; Martín-Krijer, S.; Díaz, F.; Hernández-Borges, J.; González, S. Atmospheric corrosion in subtropical areas: Influences of time of wetness and deficiency of the ISO 9223 norm. *Corros. Sci.* **2005**, *47*, 2005–2019. [CrossRef]
22. Veleva, L.; Acosta, M.; Meraz, E. Atmospheric corrosion of zinc induced by runoff. *Corros. Sci.* **2009**, *51*, 2055–2062. [CrossRef]
23. Sperling, N.F.; Washington, R.; Whittaker, R.J. Future climate change of the subtropical North Atlantic: Implications for the cloud forests of Tenerife. *Clim. Change* **2004**, *65*, 103–123. [CrossRef]
24. *Caracterización Climática de las Islas Canarias para la Aplicación del Código Técnico de la Edificaci*; CLIMCAN-010 y de su Aplicación Informática, prCTE-DR/CC.AA-008/10; Gobierno de Canarias: Las Palmas, Spain, 2010.
25. *Real Decreto 314/2006*; BOE No. 74; Ministry of Housing of Spain: Madrid, Spain, 2006; pp. 11816–11831.
26. Santana Rodríguez, J.J.; Santana Hernández, F.J.; González González, J.E. The effect of environmental and meteorological variables on atmospheric corrosion of carbon Steel, zinc and aluminium in a limited geographic zone with different types of environment. *Corros. Sci.* **2003**, *45*, 799–815. [CrossRef]
27. ASTM G1-90. *Standard Practice for Preparing, Cleaning, and Evaluating Corrosion Test Specimens*; American Society for Testing Materials: Philadelphia, PA, USA, 1990.
28. ISO/TC 156 N 250: Corrosion of Metals and Alloys. *Aggressivity of Atmospheres. Methods of Measurement of Pollution Data*; International Organization for Standardization: Geneva, Switzerland, 1986.
29. *ASTM D 2010-85: Standard Method for Evaluation of Total Sulfation Activity in the Atmosphere by the Lead Dioxide Candle*; American Society for Testing Materials: Philadelphia, PA, USA, 1985.
30. Nelder, J.A.; Mead, R. A Simplex method for function minimization. *Comput. J.* **1965**, *7*, 308–313. [CrossRef]
31. Cabanelas, I.; Collazo, A.; Izquierdo, M.; Nóvoa, X.R.; Pérez, C. Influence of galvanised surface state on the duplex systems behavior. *Corros. Sci.* **2007**, *49*, 1816–1832. [CrossRef]
32. Pillai, R.; Ackermann, H.; Lucka, K. Predicting the depletion of chromium in two high temperature Ni alloys. *Corros. Sci.* **2013**, *69*, 181–190. [CrossRef]

© 2019 by the authors. Licensee MDPI, Basel, Switzerland. This article is an open access article distributed under the terms and conditions of the Creative Commons Attribution (CC BY) license (http://creativecommons.org/licenses/by/4.0/).

Article

Influence of the Alkaline Reserve of Chloride-Contaminated Mortars on the 6-Year Corrosion Behavior of Corrugated UNS S32304 and S32001 Stainless Steels

Asunción Bautista *, Francisco Velasco and Manuel Torres-Carrasco

Department of Materials Science and Engineering—IAAB, Universidad Carlos III de Madrid, Avda. Universidad 30, 28911 Leganés, Madrid, Spain; fvelasco@ing.uc3m.es (F.V.); matorres@ing.uc3m.es (M.T.-C.)
* Correspondence: mbautist@ing.uc3m.es; Tel.: + 34-91-6249914

Received: 19 May 2019; Accepted: 13 June 2019; Published: 14 June 2019

Abstract: The durability of two lean corrugated duplex stainless steel (UNS S32304 and S32001) bars manufactured for concrete reinforcement was studied in four different corrosive conditions. These duplex stainless steels are more economical than the most traditional, well-known duplex grade steels (UNS S32205). The research was carried out in mortar samples for six years. In half of the samples, the alkaline reserve had been previously decreased, and their pH was slightly below 12, while in the other half, the pH close to the bars remained as-manufactured. Moreover, there were samples with modified and non-modified alkaline reserve where chlorides had been previously added to the mortar which were exposed to high relative humidity. In other samples—which were partially immersed in 3.5% NaCl—the chlorides entered through the mortar by natural diffusion. The electrochemical behavior of the reinforcements in these conditions was periodically monitored through corrosion potential (E_{corr}) and electrochemical impedance spectroscopy (EIS) measurements during the whole testing period. The samples were anodically polarized at the end of the exposure. The results prove that the decrease in the alkaline reserve of the mortars can affect the corrosion behavior of the studied lean duplex in environments with high chloride concentrations. The duplex microstructure of the reinforcements makes it so that the corrosion proceeds by selective attack of the phases.

Keywords: corrosion; mortar; reinforcement; lean duplex; stainless steel; chloride; alkalinity; microstructure; EIS; anodic polarization

1. Introduction

Stainless steels are increasingly used as reinforcement bars for concrete structures exposed to corrosive environments or with a long design life, as they have already shown high chloride concentration threshold [1] and high chemical durability when they are embedded in mortar [2,3]. Their high corrosion resistance is due to the formation of a passivating layer on their surface, which has a duplex structure [4], and whose protective properties increase with the pH of the media [5]. Moreover, stainless steels generally retain more high-temperature strength than carbon steel at temperatures above 550 °C, which is interesting for the structural design against fire [6].

As stainless-steel bars are more expensive than the usual carbon steel bars, the use of the former is limited to the most exposed regions of the structure and to high-risk elements. They are also used for the restoration of structures with artistic interest or that have already suffered severe corrosion damage a long time before the end of their expected service life [7].

Relatively good results related to the possible use of low-Cr ferritic stainless steels were published decades ago [8]. At present, the performance of low-Cr ferritic reinforcements is again under study, due to their reduced price in comparison to other grades of stainless steel [9–12]. Ferritic stainless steels can be an interesting option to assure the durability of structures in moderately corrosive environments, but other alternatives should be considered for highly aggressive conditions.

For very corrosive environments, austenitic stainless-steel grades such as UNS S30400 and S31600 were initially proposed and used as reinforcements [13]. Then, the duplex UNS S32205 appeared in the market, proving its excellent corrosion resistance in simulated concrete solutions [14,15] and in mortars [2,3].

However, several years ago, the high cost of Ni fostered the study of low-Ni grades as an alternative. Initially, low-Ni, high Mn-N austenitic stainless steels were considered [14,16,17], and more recently, the interest shifted towards the use of lean duplex grades—which also have less Ni than traditional austenitic grades. Though at this exact moment the Ni price is not as high as a decade ago and the huge economic advantages that lean duplex had against the traditional austenitic grades have been temporally reduced, the study of these grades is still interesting, because the instability in their price is always a disadvantage of high-Ni austenitic stainless steels.

Some results have suggested that UNS S32304 lean duplex steel has clearly better corrosion resistance than the austenitic S30400 steel in simulated pore solution tests [18] and in highly porous mortar [19]. Moreover, in studies carried out in solutions with high-strength bars, the results obtained for the S32304 were also better than those obtained for S30400 and S31600 [19], the lean duplex grade being cheaper than the S31600. The improvement of the durability that can be achieved using S31600 instead of S30400 [14] is foreseeably lower than that achieved using S32304 instead of S30400 [1,20]. On the other hand, the lean duplex grade UNS S32001 has proved to have a similar corrosion resistance to pitting onset than S30400 in tests in simulated concrete solutions [19], the current price of the former being slightly lower.

Lean duplex steel can also have other advantages over austenitic reinforcements. Tests carried out in solution indicate that the duplex microstructure in corrugated stainless bars favors a lower penetration of the corrosive attack and seems to be able to decrease their progress rate [20]. When formed through hot working, lean duplex steels exhibit ductility properties that give these materials advantages over carbon steel or cold-worked stainless steel reinforcements for structures located in seismic zones [21]. Moreover, duplex stainless-steel reinforcements have demonstrated suitable stress corrosion cracking behavior in tests carried out in simulated pore solutions with chlorides [15,22].

It is true that results have also been published where the corrosion behavior of lean duplex steel seemed worse than that of the traditional austenitic steels such as S30400 [23–25], especially when S32101 is considered. So, the interest in clarifying this controversy fosters the need for carrying out more research about the durability of these types of reinforcements. Bearing in mind that a great part of the previous research on the corrosion behavior of lean duplex steel has been carried out in solution, it is interesting to obtain more information through tests in concrete or mortar. There are factors that can meaningfully affect the corrosion behavior of reinforcements which cannot be simulated in solution tests, as has been discussed in previous publications [2].

The alkaline reserve is a key factor for the durability of the reinforcements, as the corrosion resistance of the carbon steel has proved to vary following the $[Cl^-]/[OH^-]^3$ ratio [26]. The pH of the pore solution can not only determine the nature of the protective layer formed on the stainless-steel reinforcements [5], but also control the acidification process associated with the pit development [27]. In practice, the alkaline reserve tends to decrease progressively due to the reaction of some hydration products of the mortar with the CO_2 of the air [28], and new, more ecologically-friendly cementitious materials (as many hybrid cements) have lower alkalinity in their pore solutions due to a lack of meaningful amounts of portlandite [29]. In environments with chlorides, duplex stainless steels embedded in concrete manufactured with pure CEM I have proved to show higher corrosion resistance than those embedded in concrete with 36% pozzolan in the cementitious material [30].

To date, previous studies have been carried out in fully carbonated media or in media whose pHs correspond to the one given by a saturated $Ca(OH)_2$ solution or higher, but it is also relevant to confirm the performance of the reinforcements at intermediate alkalinity conditions that can be normal in practice in aged structures and/or structures manufactured with some ecologically friendly cementitious materials.

2. Experimental

Two different lean duplex stainless steel grades were considered in the study: UNS S32304 (also known as SAF 2304 or EN 1.4362 grade) and UNS S32001 (also known as SAF 2001 or EN 1.4482 grade). The bars were manufactured by Roldan (Acerinox Group, Ponferrada, Spain) to be used as reinforcements in concrete structures. Both types of corrugated bars were manufactured through hot working, and had 16 mm diameter. Tables 1 and 2 show the mechanical properties of the stainless-steel bars and their chemical compositions, respectively. The S32304 differs from the traditional UNS S32205 grade mainly because of its very low Mo content, while the S32001 is an even less-alloyed grade that also has a lower Ni content.

Table 1. Mechanical properties of the duplex stainless-steel bars.

UNS Grade	Ultimate Tensile Strength (MPa)	Yield Strength (MPa)	Elongation (%)
S32304	769	568	38
S32001	824	553	44

Table 2. Chemical composition (w/w) of the duplex stainless-steel bars.

UNS Grade	Chemical Composition (%)									
	C	S	Si	Mn	Cr	Ni	Mo	N	Cu	Fe
S32304	0.017	0.002	0.57	1.68	23.7	4.32	0.24	0.153	0.186	Bal.
S32001	0.025	0.002	0.75	4.39	20.6	1.74	0.22	0.124	0.073	Bal.

The corrugated stainless-steel reinforcements were partly embedded in mortars that were manufactured using a cement/sand/water ratio of 1/3/0.6 (w/w). The cement used was CEM IV/B-(P-V) 32.5 V, and the sand was standardized CEN-NORMSAND (according to DIN EN 196-1 standard). Part of the samples were manufactured with 3% $CaCl_2$ additions (i.e., 1.9% Cl), weighed in relation to the cement amount. As a reference, it has become a common practice to limit the tolerable chloride content to around 0.4% of the weight of cement in European countries and in North America [31], so the added chloride content is clearly over the limit fixed for carbon steel reinforcements (4.75 times higher).

Cylindrical mortar samples were used (Figure 1), being 1.5 cm the thickness of the mortar cover. The corrugated surfaces of the bars were studied in as-received, industrially passivated condition, without any mill scale. The length of the bar exposed to the mortar was always 3 cm. The surface of bars exposed to mortars was delimited using an isolating tape. All cross-sections of the bars embedded in mortar were previously polished to 320# and passivated in the laboratory with 12% (w/w) HNO_3 for 2 min in order to reproduce the process carried out in the industry for improving the passivity of corrugated stainless steels.

After their manufacturing, the reinforced mortar samples were cured for 30 days at 20 ± 1 °C at 92–93% relative humidity, and then half of the cured reinforced mortar samples were submitted to a partial carbonation. The process was carried out in a chamber where 10% CO_2 enriched air was injected. The temperature in the chamber was 18 ± 1 °C and the relative humidity ranged between 75–80%. The exposure in the chamber lasted 15 days. The duration of the process had been previously explored using unreinforced control samples with the aim of achieving a slight decrease in the pH.

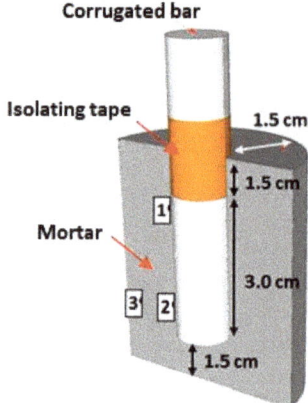

Figure 1. Scheme of the reinforcing mortar samples used in the study.

The reinforced samples were exposed at room temperature to four different aggressive conditions for 6 years:

- HH: Non-carbonated samples manufactured without chlorides and exposed to high relative humidity (between 90% and 95%).
- PI: Non-carbonated samples manufactured without chlorides and partially immersed in 3.5% (w/w) NaCl solution and at high relative humidity. In this case, the level of the solution was kept coinciding with the middle of the exposed length of the bars embedded in the mortar.
- HH-PC: Partially carbonated samples manufactured without chlorides and exposed to high relative humidity (90–95%).
- PI-PC: Partially carbonated samples manufactured without chlorides, partially immersed in 3.5% (w/w) NaCl solution and at high relative humidity. In this case, the level of the solution was also kept coinciding with the middle of the exposed length of the bars embedded in the mortar.

Three samples reinforced with S32304 bars and three with S32001 bars were exposed to each of the four considered conditions.

Corrosion potential (E_{corr}) and electrochemical impedance spectroscopy (EIS) measurements were used to monitor the corrosion behavior of all the samples during the 6-year exposure. A saturated calomel electrode (SCE) was used as reference to obtain the E_{corr}. For the EIS measurements, a three-electrode configuration was employed. The surface of corrugated duplex stainless steels exposed to the mortar acted as a working electrode, the reference electrode was a SCE, and the counter-electrode was a copper cylinder with a diameter slightly higher than that of the mortar samples. A wet pad was used to ensure a good contact between the mortar and the counter-electrode. Additional information about the arrangement of the electrodes can be found in [2]. The EIS spectra were acquired using a perturbation signal of 10 mV$_{rms}$ of amplitude, from 10^4 to 10^{-3} Hz. Five points per decade were measured. EIS measurements were performed using a potentiostat/galvanostat Solartron Modulab (Ametek Scientific Instruments, Oak Ridge, TN, USA) with Modulab ECS software.

After the 6-year exposure period, the reinforced mortar samples were submitted to anodic polarization tests. The tests started from the E_{corr} and potential was increased in steps of 20 mV, with duration of 10 min. When a potential of about 100 mV vs. SCE was reached, the length of the steps increased up to 1 h. The increasing length of the steps was needed due to higher difficulties in stabilizing the current signal after the pulses when the anodic overpotential increases. The polarization steps finished at 700 mV vs. SCE. The current densities plotted in the anodic polarization curves correspond to the stabilization values of the current after each potential step. This strategy was specifically designed to easily discriminate the interferences of the mortar resistivity in the measurements, and to obtain more reliable current values at high anodic overpotentials, where the transitories become more difficult to stabilize. More details about this type of test can be found in [2]. The anodic pulses were performed using the same potentiostat/galvanostat as for EIS measurements.

After the polarization tests, the samples were broken in half and the pH of the studied tested mortars was checked again after exposure using indicators (phenolphthalein and yellow alizarin). Phenolphthalein is commonly used to determine the advance of the carbonation front, but it is sensitive only to very extreme carbonation, and so its use has been questioned [32]. Phenolphthalein is colorless at pH < 8.3 and violet at pH > 10.0. Thus, the mortar becomes colorless with phenolphthalein only when a very large amount of CO_2 from the atmosphere has reacted, and the carbonate/bicarbonate buffer (that keeps pH around 10.2) has already been broken. Yellow alizarin is yellow at pH < 10.2 and red at pH > 12.0, so it changes its color at more alkaline pHs than phenolphthalein. In our study, part of the surface of each broken sample was sprayed with phenolphthalein and part with yellow alizarin to check the variation of the alkalinity of the mortar caused by the exposure in the carbonation chamber.

The total chloride content in different regions of the tested samples was measured by X-ray fluorescence spectrometry (XRF) [33]. The measurements were carried out after the 6-year exposure. Specimens coming from three different regions, labeled as 1, 2, and 3 in Figure 1, were evaluated. The equipment used was a SPECTRO XEPOS III X-ray Spectrometer (Spectro, Kleve, Germany) with XLabPro 4.5 Software. The given values for Cl^- concentration are the average of four measurements in each region of different mortar samples and they are expressed in relation to the mortar weight.

After cleaning the surfaces of the bars, the morphology and localization of the attack on them were studied. The possible preferential corrosion of the phases in duplex stainless-steel bars was analyzed by scanning electron microscopy (SEM, Philips, Eindhoven, Netherlands) using a Philips XL30, under voltages ranging between 15 and 17 kV.

3. Results and Discussion

In Figures 2 and 3, information about the pH and the amount of chlorides determined for the mortar samples at the end of the 6-year corrosion tests is shown, with the aim of facilitating a better understanding of the corrosion results that could be seen immediately afterwards.

Figure 2. Image of a broken reinforced sample previously exposed to CO_2, after the use of phenolphthalein (left) and yellow alizarin (right) indicators to check pH.

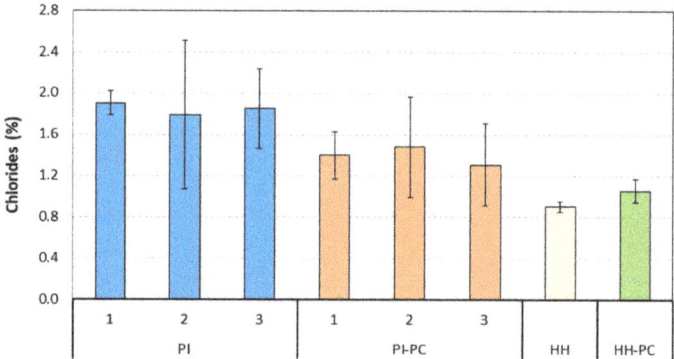

Figure 3. Chloride contents determined at the end of the 6-year exposure for reinforced mortar samples tested in different conditions. The numbers on the x-axis correspond to the regions marked in Figure 1.

Figure 2 shows an example of the results obtained after the use of phenolphthalein (on the left part of the broken mortar) and yellow alizarin (on the right part). The use of both indicators indicates that the mortar samples which were previously exposed inside the carbonation chamber had a pore solution pH similar to that given by the carbonate/bicarbonate buffer (around 10.2) or higher, as the surface wetted with phenolphthalein became violet (i.e., pH > 10) and the surface wetted with yellow alizarine was yellow and slightly reddish (i.e., pH < 12). So, the samples had a reduced alkaline reserve and their pHs can be initially considered as between 10 and 12.

The amount of chlorides determined by XRF can be seen Figure 3. In the samples where the chlorides were added during manufacturing (HH and HH-PC), the chloride distribution was uniform throughout the mortar, as expected. The obtained results also show that there were no meaningful differences among the chloride contents of studied regions (Figure 1) in the PI and PI-PC samples after the 6-year immersion period. Though it is feasible to assume that those differences would exist initially, as the chlorides penetrated from the surface of the lowest half of the mortar to the bulk of the mortar and subsequently diffused to the upper half, the concentrations of this depassivant ion seemed to be equalized at the end of the exposure on all the mortars covering the tested surface of the bars.

The data in Figure 3 also show that the long-term partial immersion exposure increased the chloride contents in the mortar to values higher than those of the HH and HH-PC samples. The PI-PC samples tended to have lower chloride concentration than the PI samples. This has also been observed in previous research, when the chloride concentrations in non-carbonated mortars [2] were compared to those in carbonated mortars [3] after similar long-term natural diffusion studies. The precipitation of some carbonates inside the pores of the mortar due to the reaction of the CO_2 of the chamber with the hydroxides of the pore solution should obviously represent a slight blockage for the diffusion of chloride ions [34].

The chloride concentrations determined for the PI and PI-PC samples (Figure 3) were about 3.5% and 3%, respectively when they were related to the weight of cement. Bearing in mind the previous results reported by Gastaldi et al. [24], they imply a clear corrosion risk at room temperature under potentiostatic polarizations at 200 mV for non-aged concrete samples. The chloride concentrations for the HH and HH-PC (about 2% in relation to the weight of cement) would not imply corrosion for lean duplex bars in the conditions of the cited study [24].

The results from the E_{corr} measurements carried out during the exposure of mortar samples are summarized in Figures 4 and 5. In these figures, the relationship between E_{corr} and corrosion probability proposed in the ASTM C876 standard is included. The standard proposes this criterion for non-coated carbon steel reinforcements, but previous results of our group [2,3,35,36] prove its utility also for monitoring stainless steel reinforcements.

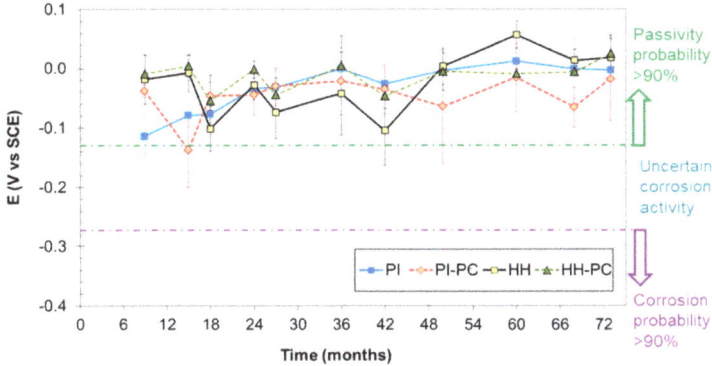

Figure 4. Corrosion potential (E_{corr}) values obtained for S32304 steel tested under different conditions. SCE: saturated calomel electrode.

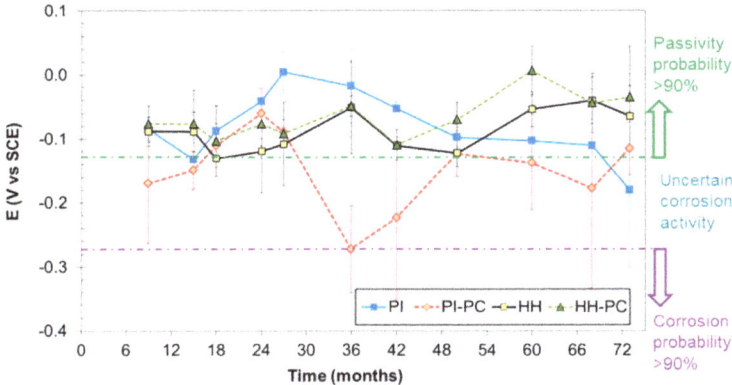

Figure 5. E_{corr} values obtained for S32001 steel tested under different conditions.

In Figure 4, the average E_{corr} values obtained for the S32304 bars in each exposure condition are plotted at different times. The vast majority of the measured E_{corr} were clearly in the region with passivity probability over 90%. Only one S32304 bar exposed to PI-PC exhibited twice the E_{corr} values in the region of uncertain corrosion activity (between −125 and −275 mV vs. SCE) during the monitoring period. This behavior was detected for the same bar S32304 at 15 and at 48 months of exposure, while the other two S32304 bars exposed in the same condition always kept their E_{corr} in the passivity region. The slight dispersion observed among the performances of different but theoretically identical samples is easy to understand, as corrosion is a phenomenon very sensitive to small, difficult-to-control variables related to metal surface and the surrounding medium. This experimental dispersion is especially foreseeable when real corrugated surfaces are considered in a heterogeneous medium such as mortar. The results plotted in Figure 4 suggest that S32304 could stand the four aggressive conditions tested, though there is a small possibility that its performance could be endangered if they were embedded in porous mortars with low alkaline reserve and exposed to high chloride concentrations and/or differential aeration cells (i.e., PI-PC condition).

The results from the E_{corr} monitoring of the least-alloyed steel (S32001) are plotted in Figure 5. The samples exposed to HH and HH-PC always showed E_{corr} values characteristic of passivity, with very few random values in the region of uncertain corrosion activity. The same occurred for the S32001 samples exposed to PI, though one of the three studied samples had its E_{corr} in the region with a corrosion probability >90% after 73 months of exposure. The PI-PC condition again seemed to be the most aggressive, often shifting the sample E_{corr} from the passivity to uncertainty region and vice versa. Moreover, one bar of S32001 in PI-PC had its E_{corr} in the corrosion probability >90% region after 9 months of exposure, other after 36 months and the third after 68 months, though these low E_{corr} values always increased again after a short time in the considered monitoring period. These E_{corr} oscillations suggest that if corrosion onset occurs, repassivation will probably take place later.

With the aim of checking the validity of the corrosion trend suggested by the E_{corr} and obtaining more information about the real corrosion rate of the samples and their passivation mechanism, a complete EIS study was carried out at the same time during the testing period. Spectra such as those shown in the examples in Figure 6a were obtained during the study. To achieve a proper simulation of this experimental behavior, it was checked that the equivalent circuit plotted in Figure 6b was necessary. This equivalent circuit comprises (in series with the mortar resistance (R_m)) three time constants. The lowest-frequency time constant has often been identified with the charge transfer process taking place on the surface of stainless steel rebars, R_t being the charge transfer resistance and CPE_{dl} the constant phase element used to simulate the non-ideal capacitive behavior of the double layer. The medium-frequencies time constant is identified with the electrical behavior of the passive layer formed on the reinforcements, R_{pl} and CPE_{pl} being the resistance and the constant phase element used to simulate the resistive and the non-ideal capacitive behaviors of this layer. The identification of the phenomenon taking place at low and medium frequencies is a controversial point in the previous literature, as some studies have identified the charge transfer process with that occurring at medium frequencies [25], while others have identified it with that occurring at high frequencies [12]. The authors consider that if R_{pl} would be on the order of magnitude of the resistance at low frequency, the ohmic drop at the surface of the passive materials would be huge, and this is not the case.

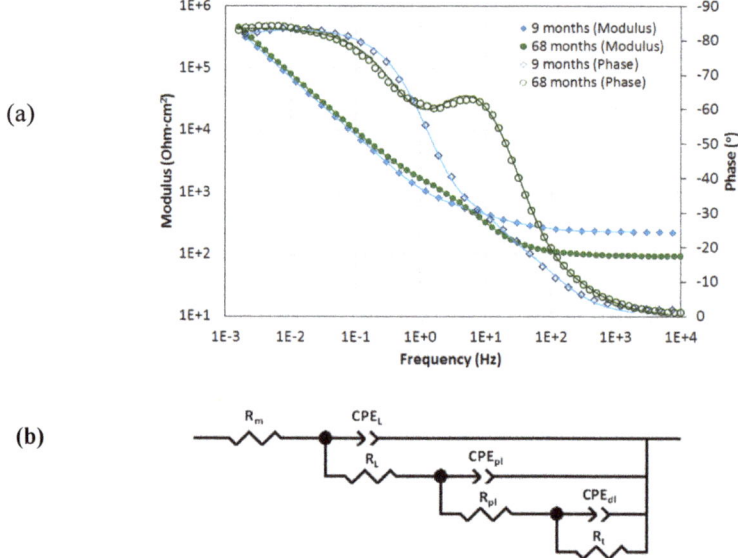

Figure 6. (a) Examples of electrochemical impedance spectroscopy (EIS) spectra, corresponding to S32001 steel tested under HH condition at two different times. Dots correspond to experimental data, and lines correspond to fitting. (b) Equivalent circuit used to simulate all the EIS spectra.

The high–medium frequencies time constant was identified with the resistance (R_L) and capacitive behavior (CPE_L) of the layer of Ca-rich hydration products formed on the surface of the bars, whose presence and characteristics have been previously studied [37–39]. A circuit similar to that in Figure 6b—but without the high–medium frequencies time constant—has been used in other studies to simulate the electrochemical performance of stainless-steel reinforcements in simulated pore solutions [5,12] and in mortar [2,20]. The length of the study and the high-quality search for the simulation have proved the interest in including this new time constant that has been identified with the layer of hydration products.

All the results obtained from the simulation of the EIS spectra for the different exposure conditions are summarized in Tables 3–6. The values of parameters related to the mortar (R_m, R_L, and CPE_L and its associated n_L) correspond to the mean values of measurements carried out in samples reinforced with S32001 and S32304. The values corresponding to parameters related to the electrochemical behavior of the stainless steels (R_{pl}, CPE_{pl}, n_{pl}, R_t, CPE_{dl}, and n_{dl}) are shown bearing in mind the nature of the reinforcement. The standard deviations corresponding to the experimental dispersion of the measurements are included in the tables.

Table 3. Results from simulations of the EIS spectra of reinforced samples tested under PI condition.

EIS Simulation Parameters		Months								
		9	15	24	36	42	50	60	68	73
R_m ($\Omega \cdot cm^2$)		268 ± 36	202 ± 70	247 ± 146	263 ± 112	202 ± 79	240 ± 86	256 ± 92	218 ± 108	205 ± 73
R_L ($\Omega \cdot cm^2$)		112 ± 42	65 ± 27	225 ± 119	527 ± 112	482 ± 85	802 ± 531	838 ± 497	1159 ± 355	1379 ± 286
CPE_L ($\mu F \cdot cm^{-2} \cdot s^{n-1}$)		30 ± 19	34 ± 17	20 ± 12	19 ± 3	22 ± 3	30 ± 11	34 ± 10	30 ± 5	29 ± 10
n_L		0.76 ± 0.14	0.78 ± 0.12	0.81 ± 0.11	0.72 ± 0.05	0.74 ± 0.03	0.75 ± 0.08	0.78 ± 0.08	0.69 ± 0.07	0.69 ± 0.03
R_{pl} ($k\Omega \cdot cm^2$)	S32304	0.4 ± 0.0	0.3 ± 0.1	0.5 ± 0.2	0.4 ± 0.0	0.6 ± 0.2	1.3 ± 0.6	1.5 ± 0.7	1.5 ± 0.5	1.6 ± 0.4
	S32001	0.4 ± 0.1	0.2 ± 0.1	0.3 ± 0.2	0.7 ± 0.1	0.6 ± 0.0	1.9 ± 1.6	1.8 ± 0.9	2.0 ± 1.3	3.2 ± 0.2
CPE_{pl} ($\mu F \cdot cm^{-2} \cdot s^{n-1}$)	S32304	130 ± 19	141 ± 5	142 ± 49	89 ± 20	101 ± 16	111 ± 23	92 ± 26	106 ± 24	107 ± 16
	S32001	99 ± 10	88 ± 22	108 ± 64	32 ± 10	64 ± 7	65 ± 53	79 ± 23	81 ± 24	86 ± 9
n_{pl}	S32304	0.97 ± 0.03	0.97 ± 0.02	0.96 ± 0.00	0.96 ± 0.02	0.95 ± 0.01	0.96 ± 0.01	0.95 ± 0.02	0.96 ± 0.02	0.96 ± 0.02
	S32001	0.99 ± 0.01	0.95 ± 0.06	0.96 ± 0.02	0.89 ± 0.01	0.95 ± 0.01	0.92 ± 0.12	0.97 ± 0.03	0.99 ± 0.01	0.99 ± 0.01
R_t ($M\Omega \cdot cm^2$)	S32304	14 ± 8	5.2 ± 3.1	13 ± 8	67 ± 24	16 ± 7.7	37 ± 5	28 ± 10	78 ± 20	43 ± 11
	S32001	5.2 ± 1	0.7 ± 0.3	5.3 ± 2.7	9.7 ± 2	8.2 ± 7	8.4 ± 2.6	16 ± 9.5	36 ± 27	19 ± 8.8
CPE_{dl} ($\mu F \cdot cm^{-2} \cdot s^{n-1}$)	S32304	130 ± 33	127 ± 30	163 ± 20	170 ± 53	187 ± 22	141 ± 17	165 ± 24	160 ± 19	159 ± 20
	S32001	98 ± 20	140 ± 45	133 ± 48	171 ± 20	132 ± 6	106 ± 40	87 ± 22	75 ± 18	64 ± 1
n_{dl}	S32304	0.94 ± 0.04	0.93 ± 0.02	0.94 ± 0.01	0.94 ± 0.01	0.94 ± 0.01	0.94 ± 0.01	0.94 ± 0.01	0.94 ± 0.02	0.94 ± 0.02
	S32001	0.98 ± 0.01	0.95 ± 0.01	0.95 ± 0.02	0.94 ± 0.01	0.94 ± 0.01	0.94 ± 0.03	0.92 ± 0.05	0.93 ± 0.05	0.97 ± 0.03

Table 4. Results from simulations of the EIS spectra of reinforced samples tested under PI-PC condition.

EIS Simulation Parameters		Months								
		9	15	24	36	42	50	60	68	73
R_m ($\Omega \cdot cm^2$)		268 ± 35	249 ± 62	247 ± 94	263 ± 59	202 ± 84	240 ± 89	256 ± 98	218 ± 87	205 ± 118
R_L ($\Omega \cdot cm^2$)		102 ± 16	101 ± 69	112 ± 71	182 ± 92	153 ± 67	313 ± 74	257 ± 111	201 ± 172	611 ± 290
CPE_L ($\mu F \cdot cm^{-2} \cdot s^{n-1}$)		72 ± 41	29 ± 15	65 ± 40	44 ± 29	32 ± 25	22 ± 8	21 ± 12	15 ± 11	13 ± 14
n_L		0.74 ± 0.09	0.86 ± 0.16	0.83 ± 0.11	0.84 ± 0.07	0.80 ± 0.09	0.67 ± 0.12	0.66 ± 0.11	0.75 ± 0.10	0.64 ± 0.06
R_{pl} ($k\Omega \cdot cm^2$)	S32304	0.8 ± 0.2	0.8 ± 0.5	0.8 ± 0.3	0.3 ± 0.0	0.4 ± 0.1	0.5 ± 0.1	0.4 ± 0.1	0.4 ± 0.2	0.5 ± 0.2
	S32001	0.6 ± 0.2	0.3 ± 0.1	0.9 ± 0.6	0.5 ± 0.0	0.4 ± 0.2	0.4 ± 0.3	0.8 ± 0.7	0.3 ± 0.0	0.9 ± 0.8
CPE_{pl} ($\mu F \cdot cm^{-2} \cdot s^{n-1}$)	S32304	114 ± 18	118 ± 85	131 ± 23	126 ± 22	133 ± 19	70 ± 21	154 ± 80	140 ± 66	145 ± 75
	S32001	89 ± 9	110 ± 27	157 ± 89	106 ± 13	131 ± 29	92 ± 31	109 ± 72	152 ± 29	111 ± 83
n_{pl}	S32304	0.97 ± 0.05	0.97 ± 0.04	0.95 ± 0.08	0.97 ± 0.01	0.96 ± 0.00	0.93 ± 0.06	0.91 ± 0.03	0.92 ± 0.04	0.91 ± 0.06
	S32001	0.99 ± 0.02	0.96 ± 0.04	0.93 ± 0.06	0.97 ± 0.04	0.93 ± 0.02	0.96 ± 0.04	0.97 ± 0.06	0.91 ± 0.13	0.94 ± 0.09
R_t ($M\Omega \cdot cm^2$)	S32304	7.3 ± 4.1	1.3 ± 1.0	3.8 ± 2.2	2.9 ± 0.3	15 ± 7	5 ± 2	10 ± 6.5	13 ± 9	4.7 ± 2.5
	S32001	1.1 ± 0.7	1.2 ± 0.8	17 ± 3	0.7 ± 0.1	0.1 ± 0.1	19 ± 10	16 ± 10	0.4 ± 0.1	0.2 ± 0.1
CPE_{dl} ($\mu F \cdot cm^{-2} \cdot s^{n-1}$)	S32304	55 ± 2	79 ± 34	92 ± 33	129 ± 0	154 ± 6	170 ± 55	148 ± 34	158 ± 7	170 ± 32
	S32001	80 ± 50	101 ± 27	29 ± 26	110 ± 33	133 ± 79	80 ± 55	110 ± 85	118 ± 40	103 ± 90
n_{dl}	S32304	0.98 ± 0.02	0.93 ± 0.07	0.95 ± 0.06	0.93 ± 0.00	0.93 ± 0.00	0.94 ± 0.02	0.94 ± 0.02	0.93 ± 0.01	0.92 ± 0.01
	S32001	0.92 ± 0.10	0.89 ± 0.06	1.00 ± 0.00	0.94 ± 0.04	0.94 ± 0.05	0.94 ± 0.06	0.95 ± 0.07	0.92 ± 0.12	0.95 ± 0.04

After several months of exposure, when the monitoring started, the mortar had already become a quite stable material, and so the R_m hardly changed during the testing period. Any decrease of R_m due to chloride diffusion was masked by the curing of the mortar. The HH-PC mortar (Table 6) shows R_m that tended to increase with time and that became clearly higher than those determined for other conditions (Tables 3–5) after 2 years of exposure.

The values related to the layer of hydration products formed on the surface show that, after some time, it was more resistive in PI mortars (Table 3). In PI-PC (Table 4), the partial carbonation seemed to reduce their electrochemical influence in the spectra, the determined R_L values being lower than those obtained for PI materials (Table 3). Moreover, when the first measurements were carried out, the n_L values obtained for partially immersed samples (PI and PI-PC) were already lower than those obtained for samples exposed at high relative humidity (HH and HH-PC) (Tables 5 and 6). It is also clear that while n_L kept its value relatively constant in HH and HH-PC samples during the exposure, this parameter decreased its value with time in PI and PI-PC samples. This suggests that when the pores are saturated with water and/or the chloride content is very high, the deposits tend to become less homogeneous with time.

Table 5. Results from simulations of the EIS spectra of reinforced samples tested under HH condition.

EIS Simulation Parameters		Months								
		9	15	24	36	42	50	60	68	73
R_m ($\Omega \cdot cm^2$)		282 ± 53	227 ± 47	276 ± 98	234 ± 41	197 ± 25	261 ± 64	295 ± 50	205 ± 58	208 ± 59
R_L ($\Omega \cdot cm^2$)		120 ± 31	72 ± 40	150 ± 25	55 ± 39	116 ± 96	90 ± 63	191 ± 95	84 ± 57	71 ± 21
CPE_L ($\mu F \cdot cm^{-2} \cdot s^{n-1}$)		33 ± 16	26 ± 23	39 ± 31	28 ± 19	34 ± 33	17 ± 12	39 ± 14	19 ± 22	14 ± 8
n_L		0.85 ± 0.07	0.86 ± 0.05	0.82 ± 0.10	0.87 ± 0.05	0.88 ± 0.12	0.80 ± 0.06	0.88 ± 0.02	0.83 ± 0.08	0.82 ± 0.07
R_{pl} ($k\Omega \cdot cm^2$)	S32304	1.1 ± 0.2	1.0 ± 0.1	2.5 ± 0.4	2.1 ± 1.2	1.8 ± 1.3	3.7 ± 1.4	7.8 ± 1.4	2.9 ± 1.8	5.5 ± 3.3
	S32001	0.8 ± 0.5	0.7 ± 0.4	2.6 ± 0.6	6.5 ± 5.5	0.9 ± 0.7	1.8 ± 1.5	15.0 ± 8.1	2.6 ± 2.3	5.6 ± 1.0
CPE_{pl} ($\mu F \cdot cm^{-2} \cdot s^{n-1}$)	S32304	54 ± 10	95 ± 18	68 ± 20	113 ± 5	121 ± 8	122 ± 9	78 ± 46	103 ± 24	77 ± 8
	S32001	61 ± 18	70 ± 18	75 ± 0	90 ± 35	62 ± 48	71 ± 13	95 ± 11	62 ± 26	42 ± 19
n_{pl}	S32304	0.89 ± 0.02	0.93 ± 0.01	0.91 ± 0.02	0.90 ± 0.00	0.94 ± 0.03	0.90 ± 0.00	0.90 ± 0.03	0.89 ± 0.02	0.88 ± 0.03
	S32001	0.96 ± 0.01	0.94 ± 0.07	0.91 ± 0.12	0.95 ± 0.02	0.94 ± 0.01	0.91 ± 0.05	0.94 ± 0.01	0.94 ± 0.01	0.96 ± 0.05
R_t ($M\Omega \cdot cm^2$)	S32304	30 ± 8	19 ± 13	68 ± 3	2 ± 0	17 ± 10	29 ± 15	60 ± 43	150 ± 77	270 ± 179
	S32001	10 ± 2	27 ± 20	12 ± 4	8 ± 5	14 ± 6	28 ± 7	330 ± 140	23 ± 6	77 ± 35
CPE_{dl} ($\mu F \cdot cm^{-2} \cdot s^{n-1}$)	S32304	145 ± 10	97 ± 24	114 ± 25	100 ± 17	79 ± 32	75 ± 17	87 ± 28	92 ± 48	94 ± 25
	S32001	87 ± 20	83 ± 8	67 ± 25	74 ± 64	106 ± 31	96 ± 16	38 ± 18	78 ± 29	98 ± 9
n_{dl}	S32304	0.89 ± 0.03	0.93 ± 0.01	0.91 ± 0.02	0.90 ± 0.00	0.94 ± 0.03	0.90 ± 0.00	0.90 ± 0.03	0.89 ± 0.02	0.88 ± 0.03
	S32001	0.96 ± 0.01	0.94 ± 0.07	0.95 ± 0.01	0.95 ± 0.02	0.94 ± 0.01	0.91 ± 0.05	0.94 ± 0.01	0.94 ± 0.01	0.93 ± 0.01

On the other hand, the values corresponding to R_{pl} tended to increase with time when the mortar was not carbonated (PI and HH) (Tables 3 and 5), and the trend was not clear for passive layers in less alkaline media (PI-PC and HH-PC) (Tables 4 and 6). The R_{pl} values obtained for the lean duplex in this study were somewhat lower than those reported for austenitic stainless steels in studies carried out in similar conditions [2,3]. The CPE_{dl} values obtained for PI and PI-PC (Tables 3 and 4) seemed higher than those obtained for HH and HH-PC (Tables 5 and 6), which could imply that the fuller the saturation of the pores, or perhaps the higher chloride content of their solution (Figure 3), the thinner—that is, more capacitive—the passive layers.

The most relevant parameter for the corrosion kinetics of the reinforcements is the low-frequency resistance. The R_t values were orders of magnitude higher than the other resistances identified in the spectra (Tables 3–6), so, the charge transfer process was the controlling rate step, and R_t can be directly related with the polarization resistance, R_p, in the Stern-Geary equation [40] and can be used to calculate the corrosion rate (i_{corr}). If a value for the B constant of 50 mV is assumed (as is usually done for passive steels in concrete and mortar [41]), the i_{corr} of each reinforcement for each time and condition can be easily calculated. The average i_{corr} values obtained for S32304 and S32001 lean duplex reinforcements in the different testing conditions considered can be seen in Figures 7 and 8, respectively.

Table 6. Results from simulations of the EIS spectra of reinforced samples tested under HH-PC condition.

EIS Simulation Parameters		Months								
		9	15	24	36	42	50	60	68	73
R_m ($\Omega \cdot cm^2$)		312 ± 57	309 ± 70	442 ± 35	463 ± 54	324 ± 37	463 ± 101	549 ± 57	422 ± 73	472 ± 53
R_L ($\Omega \cdot cm^2$)		127 ± 105	125 ± 51	324 ± 126	308 ± 139	186 ± 69	241 ± 70	346 ± 206	250 ± 223	271 ± 157
CPE_L ($\mu F \cdot cm^{-2} \cdot s^{n-1}$)		32 ± 30	28 ± 11	35 ± 15	28 ± 8	33 ± 15	17 ± 8	19 ± 8	17 ± 7	16 ± 9
n_L		0.87 ± 0.07	0.89 ± 0.03	0.88 ± 0.04	0.86 ± 0.05	0.87 ± 0.03	0.84 ± 0.02	0.85 ± 0.05	0.84 ± 0.06	0.83 ± 0.05
R_{pl} ($k\Omega \cdot cm^2$)	S32304	1.1 ± 1.0	0.5 ± 0.2	1.8 ± 0.6	1.8 ± 0.8	1.1 ± 0.5	1.3 ± 0.6	1.6 ± 0.4	1.4 ± 0.6	2.3 ± 1.4
	S32001	0.5 ± 0.4	0.5 ± 0.2	3.1 ± 1.5	1.6 ± 1.0	1.5 ± 1.4	1.1 ± 0.3	0.7 ± 0.2	0.6 ± 0.1	0.7 ± 0.1
CPE_{pl} ($\mu F \cdot cm^{-2} \cdot s^{n-1}$)	S32304	84 ± 20	80 ± 6	81 ± 19	81 ± 28	96 ± 16	80 ± 12	79 ± 4	86 ± 9	88 ± 7
	S32001	67 ± 13	67 ± 15	85 ± 9	68 ± 15	83 ± 15	48 ± 11	63 ± 15	69 ± 19	60 ± 21
n_{pl}	S32304	0.94 ± 0.00	0.94 ± 0.00	0.93 ± 0.01	0.92 ± 0.01	0.93 ± 0.01	0.91 ± 0.01	0.92 ± 0.01	0.93 ± 0.01	0.92 ± 0.02
	S32001	0.95 ± 0.02	0.95 ± 0.01	0.93 ± 0.01	0.89 ± 0.02	0.92 ± 0.02	0.90 ± 0.02	0.91 ± 0.03	0.89 ± 0.02	0.89 ± 0.02
R_t ($M\Omega \cdot cm^2$)	S32304	50 ± 27	32 ± 28	20 ± 22	130 ± 72	200 ± 130	80 ± 42	190 ± 120	82 ± 64	28 ± 10
	S32001	23 ± 14	23 ± 16	69 ± 48	140 ± 30	77 ± 97	24 ± 17	180 ± 97	40 ± 10	170 ± 105
CPE_{dl} ($\mu F \cdot cm^{-2} \cdot s^{n-1}$)	S32304	77 ± 22	101 ± 20	84 ± 21	96 ± 12	82 ± 11	97 ± 14	91 ± 17	86 ± 7	78 ± 16
	S32001	112 ± 14	105 ± 21	69 ± 15	75 ± 20	79 ± 28	75 ± 7	85 ± 36	81 ± 26	86 ± 30
n_{dl}	S32304	0.91 ± 0.00	0.91 ± 0.00	0.91 ± 0.01	0.91 ± 0.01	0.90 ± 0.00	0.90 ± 0.00	0.90 ± 0.01	0.91 ± 0.01	0.90 ± 0.00
	S32001	0.94 ± 0.00	0.92 ± 0.01	0.92 ± 0.01	0.87 ± 0.02	0.89 ± 0.02	0.88 ± 0.02	0.91 ± 0.03	0.90 ± 0.04	0.90 ± 0.03

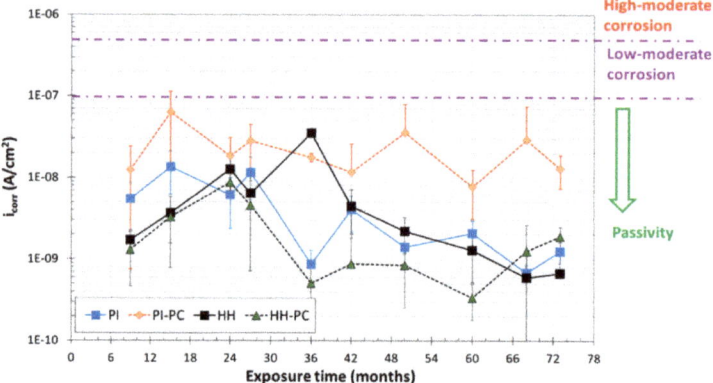

Figure 7. Corrosion rate (i_{corr}) values calculated from EIS experiments for S32304 steel under different testing conditions.

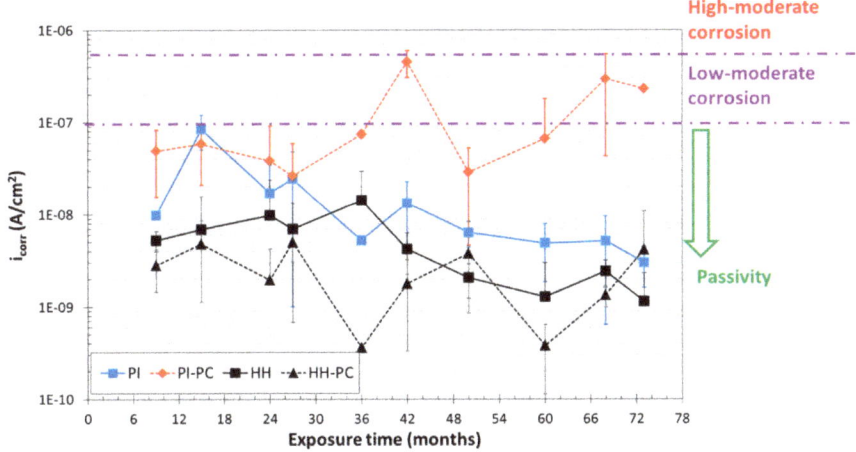

Figure 8. Corrosion rate (i_{corr}) values calculated from EIS experiments for S32001 steel under different testing conditions.

In Figure 7, it can be seen that S32304 stainless steel always kept its average i_{corr} inside the range of values typical of passive steel [42,43]. The PI-PC condition was confirmed to be the most aggressive one, and one of the three bars tested in this condition, at three different times during the testing period, had i_{corr} approaching or slightly surpassing the 0.1 μA/cm² limit. These three i_{corr} values correspond to E_{corr} in the range of uncertain corrosion activity (Figure 4). It is important to stress that, although some very low-intensity corrosion activity occasionally occurs in S32304 reinforcements when they have been exposed to very high chloride concentration and the mortar pores are water saturated, when this low-intensity attack took place during the testing period, it was always repassivated. The results obtained for S32304 were similar to those obtained for austenitic S30400 in mortars exposed in similar non-carbonated conditions [2], and they agreed with previous results about the corrosion resistance of this lean duplex grade obtained in solution tests [18].

The results regarding the i_{corr} measured for S32001 reinforcements can be seen in Figure 8. The PI-PC condition was confirmed to be especially dangerous for the cheapest grade under study. Meaningful i_{corr} values were occasionally measured for this low-alloyed duplex stainless steel in this condition (after 42 months and at the end of exposure for the three tested rebars, and also after 68 months for two of them). The S32001 steel in PI condition had a single i_{corr} value slightly higher than 0.1 μA/cm², and this point corresponds to an E_{corr} in the range of uncertain corrosion activity, although there were other E_{corr} values for S32001 in PI treatment in the range of uncertain corrosion activity whose i_{corr} clearly corresponded to passive reinforcements. Previous results in fully carbonated solutions have already suggested that S32001 grade steel could suffer chloride-induced corrosion in this media [18]. However, the values obtained for S32001 reinforcements in PI suggest that the higher chloride content in the mortar than in the HH-PC could more strongly affect the i_{corr} than the decrease in the alkaline reserve performed in this study before the exposures. Pitting corrosion has already been reported for S32001 when the bars were embedded in extremely porous chloride-contaminated mortar [19].

The anodic polarizations carried out at variable, very low rate in all the samples after the exposure allowed us to obtain the results shown in Figures 9 and 10. Although carried out until high anodic polarizations, these studies were always performed in the region of water stability. None of the samples of both studied lean duplex—when they were previously exposed to HH and HH-PC conditions—seemed to pit during the polarization tests. The increases in the current density at potential about 200 mV vs. SCE have been previously identified in tests carried out in alkaline solutions, with the current corresponding to the oxidation of the Cr_2O_3 present in the passive layer of the stainless steels into CrO_4^{2-} [15,21]. The fact that these peaks did not suffer a high shift for the samples that had been initially exposed to CO_2 [15] indicates that the decrease caused in the pH of the mortar was small. If these results are complemented with those in Figure 2, it can be concluded that the pH of the pore solution in mortar previously exposed to CO_2 was slightly below 12. Hence, the decrease in the alkaline reserve of the mortar is the main effect achieved by the exposure in the carbonation chamber that allows us to explain the differences observed between HH and HH-PC and among samples.

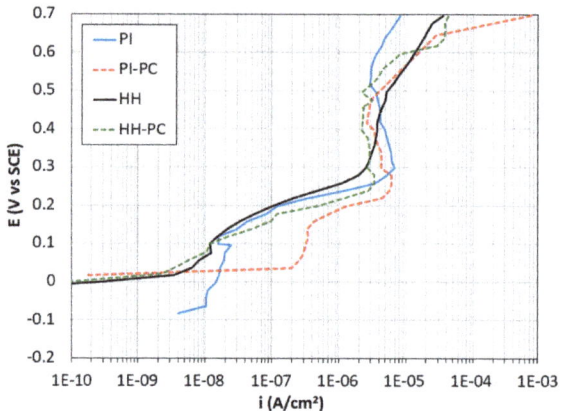

Figure 9. Anodic polarization curves obtained for S32304.

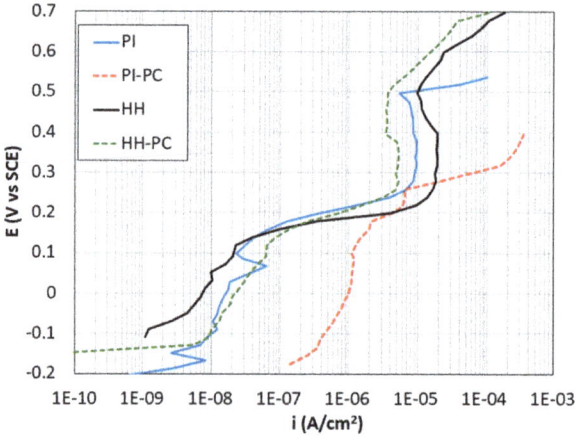

Figure 10. Anodic polarization curves obtained for S32001.

After six years of exposure in HH, HH-PC, and PI treatments, S32304 samples exhibited a very stable passivity (Figure 9). However, after exposures in PI-PC, the samples increased their current density when they were submitted to very small anodic overpotentials, reaching values about 0.3–0.5 µA/cm^2. This could be considered a worrisome situation if very-long-term durability is expected, as the values are close to corrosion rates that could be considered high or moderate [43]. Moreover, intense pitting activity was also detected at very high anodic overpotentials (+625 mV vs. SCE in the example in Figure 9), when a partial oxidation and dissolution of the passive layer had already taken place (at about 200 mV vs. SCE). This intense pitting attack appeared at such high anodic polarizations that its probability is very remote in practice.

The polarization studies of the S32001 (Figure 10) confirmed the lower stability of its passivity, as already suggested by the results in Figure 8. After exposure in PI-PC, the bars showed i_{corr} values characteristic of active corrosion at E_{corr}, and the current densities increased at a moderate rate under increasing anodic polarizations. An intense pitting activity began under moderately high anodic overpotentials (+285 mV vs. SCE in the curve plotted in the figure). If the S32001 bars were embedded in PI mortar, they only pitted at very high anodic overpotentials (between +460 and +500 mV vs. SCE). The curves corresponding to S32001 embedded in mortars with lower chloride concentrations (HH and HH-PC) suggest that no corrosion-related phenomena occurred during the anodic polarization of the bars. After the polarization tests, the mortar cover of the samples was broken and the bars were cleaned to allow a study of the location and morphology of the pits. Besides some mortar adherences and small surface damages caused by the mechanical cleaning process, only visible pits were found in the bars that showed high anodic currents when they were anodically polarized (S32304 after PI-PC and S32001 after PI-PC and PI). The pits (Figure 11) appeared mainly in the lower part of the immersed area of the bars, suggesting that there is no special danger associated to partial immersion conditions, in contrast to what occurs with carbon steel reinforced samples, where macrocell corrosion cannot be considered negligible [44]. This difference could be related to the different corrosion mechanisms taking place in stainless steels and in carbon steel. Moreover, the location of the pits was not related to ribs or more strained regions of the bars. The fact that the studied bars were manufactured by hot working could influence the pit location, as it has already been proved that the working method (i.e., hot or cold) determines the corrosion behavior of stainless-steel reinforcements [45]. Besides these macroscopic pits, the presence of other microscopic pits cannot be discarded.

Figure 11. Example of visually detectable pits observed on the surface of tested steels.

The microstructural study of the tested bars allowed us to check that the duplex microstructure of the stainless steels made it so that the attack proceeded by selective corrosion of the phases. In duplex stainless steels, austenite always appeared as a discontinuous phase, surrounded by a continuous phase (ferrite). This type of microstructure can be seen in Figure 12, where the etching previously carried out made it so that the ferrite appeared on a lower plane than the austenite. The images in Figure 12 correspond to damaged regions of the surface. The SEM study revealed non-corroded regions, but also small localized attacks, in addition to the large pits already observed (Figure 11).

As can be seen in Figure 12a, in pits formed in regions of the corrugated surface which were only moderately strained, austenite was dissolved selectively, while ferrite tended to remain uncorroded. The higher tendency of ferrite to dissolve Cr can explain this observation. Ferrite should be more Cr-rich than austenite, and therefore more corrosion resistant.

However, in very hard strained regions of the surface (e.g., corrugations or nerves), the ferrite—which is substantially less ductile than the austenite—accumulated much more strain during the working process. These very high stress concentrations in the ferrite decreased the corrosion resistance of this phase, while the decrease caused in the austenite by the strain was not as high. Hence, the relative corrosion resistance of austenite and ferrite reversed in the most strained regions of the surface of the bars. In the pits formed in very strained regions, the ferrite corroded selectively, while the austenite (partially transformed into strain-induced martensite [46]) remained uncorroded (Figure 12b). The selective dissolution of ferrite has already been reported for high-strength lean duplex bars (i.e., high-strained) when they were tested in simulated pore solutions [18].

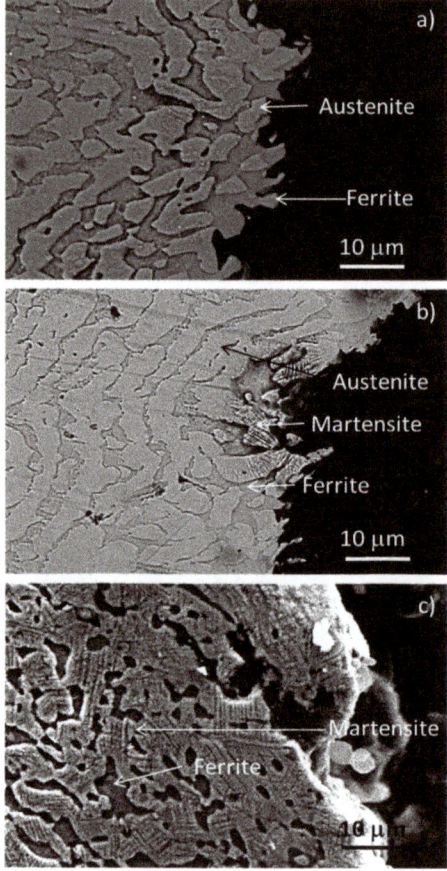

Figure 12. SEM images of S32001 steel after polarization tests carried out after: (**a**) PI-PC, (**b**) PI-PC, and (**c**) PI exposures. Images (a) and (b) were taken with backscattered electrons, while image (c) was taken with secondary electrons.

The change in the relative corrosion resistance of the phases in duplex stainless steels when submitted to strong deformations has already been observed and deeply discussed when they suffer general corrosion in acid medium [47]. In this study, this was checked for both lean duplex stainless steels tested in mortars with low alkaline reserve (Figure 12b), as well as in mortars that kept their initial alkaline reserve (Figure 12c).

The duplex microstructure seemed to be able to stop the deepening of the attack once it was initiated [21], as the most-corrosion-resistant phase can hinder the rate at which the corrosion progresses through the least-corrosion-resistant phase [47] without meaningful galvanic effects. So, the development of the attack through a selective corrosion mechanism can be considered as not especially dangerous for the durability of the structures.

4. Conclusions

In this work, interesting long-term results were obtained regarding the corrosion performance of corrugated lean duplex stainless-steel bars. Those grades are becoming the most relevant material as reinforcement for concrete structures exposed to marine environments, but the knowledge of their behavior is limited. The most innovative conclusions that can be drawn from the results obtained about the durability of S32304 and S32001 in chloride contaminated mortars are:

- S32304 steel was confirmed to be more corrosion resistant than S32001 (with lower Cr and Ni contents).
- The durability of S32001 in environments with very high chloride content can be limited, though other factors as the alkalinity of the cover also have a high impact.
- In mortars that keep their alkalinity, S32304 does not seem to have any corrosion risks for mortar chloride contents up to 1.8% (w/w). The decrease in the alkaline reserve of the mortars—with only a small decrease in their pH—could affect the corrosion behavior of both studied corrugated lean duplex stainless steels in chloride-rich environments.
- The duplex structure of the stainless steels makes the corrosion proceed by selective corrosion of the phases. Austenite corrodes preferentially except in the most-strained areas of the corrugated surface, where ferrite dissolves selectively.

Author Contributions: Conceptualization, A.B.; methodology, A.B.; validation, A.B., F.V. and M.T.-C.; formal analysis, A.B., F.V. and M.T.-C.; investigation, A.B., F.V. and M.T.-C.; resources, F.V.; data curation, A.B.; writing—original draft preparation, A.B.; writing—review and editing, A.B., F.V. and M.T.-C.; supervision, A.B.; funding acquisition, A.B. and F.V.

Funding: This work was founded by the Ministerio de Ciencia, Innovación y Universidades of Spain through the project RTI2018-096428-B-I00.

Conflicts of Interest: The authors declare no conflict of interest.

References

1. Pachón-Montaño, A.; Sánchez-Montero, J.; Andrade, C.; Fullea, J.; Moreno, E.; Matres, V. Threshold concentration of chlorides in concrete for stainless steel reinforcement: Classic austenitic and new duplex stainless steel. *Constr. Build. Mater.* **2018**, *186*, 495–502. [CrossRef]
2. Bautista, A.; Paredes, E.C.; Velasco, F.; Alvarez, S.M. Corrugated stainless steels embedded in mortar for 9 years: Corrosion results of non-carbonated, chloride-contaminated samples. *Constr. Build. Mater.* **2015**, *93*, 350–359. [CrossRef]
3. Bautista, A.; Alvarez, S.M.; Paredes, E.C.; Velasco, F.; Guzmán, S. Corrugated stainless steels embedded in carbonated mortars with and without chlorides: 9-Year corrosion results. *Constr. Build. Mater.* **2015**, *95*, 186–196. [CrossRef]
4. Fajardo, S.; Bastidas, D.M.; Ryan, M.P.; Criado, M.; McPhail, D.S.; Morris, R.J.H.; Bastidas, J.M. Low energy SIMS characterization of passive oxide films formed on a low-nickel stainless steel in alkaline media. *Appl. Surf. Sci.* **2014**, *288*, 423–429. [CrossRef]
5. Bautista, A.; Blanco, G.; Velasco, F.; Gutierrez, A.; Soriano, L.; Palomares, F.J.; Takenouti, H. Changes in the passive layer of corrugated austenitic stainless steel of low nickel content due to exposure to simulated pore solutions. *Corros. Sci.* **2009**, *51*, 785–792. [CrossRef]
6. Gardner, L.; Ng, K.T. Temperature development in structural stainless steel sections exposed to fire. *Fire Saf. J.* **2006**, *41*, 185–203. [CrossRef]
7. Pérez-Quiroz, J.T.; Terán, J.; Herrera, M.J.; Martínez, M.; Genescá, J. Assessment of stainless steel reinforcement for concrete structures rehabilitation. *J. Constr. Steel Res.* **2008**, *64*, 1317–1324. [CrossRef]
8. Callaghan, B.G. The performance of a 12% Cr chromium steel in concrete in severe marine environments. *Corros. Sci.* **1993**, *35*, 1535–1541. [CrossRef]
9. Sedar, M.; Meral, C.; Kunz, M.; Bjegovic, D.; Wenk, H.-R.; Monteiro, P.J.M. Spatial distribution of crystalline corrosion products formed during corrosion of stainless steel in concrete. *Cem. Concr. Res.* **2015**, *71*, 93–105. [CrossRef]

10. Kouril, M.; Novak, P.; Bojko, M. Threshold chloride concentration for stainless steel activation in concrete pore solutions. *Cem. Concr. Res.* **2010**, *40*, 431–436. [CrossRef]
11. Itty, P.-A.; Sedar, M.; Meral, C.; Parkinson, D.; MacDowell, A.A.; Bjegovic, D.; Monteiro, P.J.M. In situ 3D monitoring of corrosion on carbon steel and ferritic stainless steel embedded in cement paste. *Corros. Sci.* **2014**, *83*, 409–418. [CrossRef]
12. Luo, H.; Su, H.; Dong, C.; Xiao, K.; Li, X. Electrochemical and passivation behavior investigation of ferritic stainless steel in simulated concrete pore media. *Data Brief* **2015**, *5*, 171–178. [CrossRef] [PubMed]
13. Bertolini, L.; Pedeferri, P. Laboratory and field experience on the use of stainless steels to improve the durability of reinforced concrete. *Corros. Rev.* **2002**, *20*, 129–152. [CrossRef]
14. Bautista, A.; Blanco, G.; Velasco, F. Corrosion behaviour of low-nickel austenitic stainless steels reinforcements: A comparative study in simulated pore solutions. *Cem. Concr. Res.* **2006**, *36*, 1922–1930. [CrossRef]
15. Moser, R.; Singh, P.M.; Kahn, L.F.; Kurtis, K.E.; González-Niño, D.; McClelland, Z.B. Crevice corrosion and environmentally assisted cracking of high-strength duplex stainless steels in simulated concrete pore solutions. *Constr. Build. Mater.* **2019**, *203*, 366–376. [CrossRef]
16. Fajardo, S.; Bastidas, D.M.; Criado, M.; Bastidas, J.M. Electrochemical study of a new-low nickel stainless teel in carbonated solution in the presence of chlorides. *Electrochim. Acta* **2014**, *129*, 160–170. [CrossRef]
17. Freire, L.; Novoa, X.R.; Pena, G.; Vivier, V. On the corrosion mechanism of AISI 204Cu stainless steel in chlorinated alkaline media. *Corros. Sci.* **2008**, *50*, 3205–3212. [CrossRef]
18. Moser, R.D.; Singh, P.M.; Kahn, L.F.; Kurtis, K.E. Chloride-induced corrosion resistance of high-strength stainless steels in simulated alkaline and carbonated concrete pore solutions. *Corros. Sci.* **2012**, *57*, 241–253. [CrossRef]
19. Sedar, M.; Zulj, L.V.; Bjegovic, D. Long-term corrosion behaviour of stainless reinforcing steel in mortar exposed in chloride environment. *Corros. Sci.* **2013**, *69*, 149–157. [CrossRef]
20. Alvarez, S.M.; Bautista, A.; Velasco, F. Corrosion resistance of corrugated lean duplex stainless steel in simulated concrete pore solutions. *Corros. Sci.* **2011**, *53*, 1748–1755. [CrossRef]
21. Medina, E.; Medina, J.M.; Cobo, A.; Bastidas, D.M. Evaluation of mechanical and structural behavior of austenitic and duplex stainless steel reinforcements. *Constr. Build. Mater.* **2015**, *78*, 1–15. [CrossRef]
22. Briz, E.; Biezma, M.V.; Bastidas, D.M. Stress corrosion cracking of new 2001 lean–duplex stainless steel reinforcements in chloride contained concrete pore solution: An electrochemical study. *Constr. Build. Mater.* **2018**, *192*, 1–8. [CrossRef]
23. Bertolini, L.; Gastaldi, M. Corrosion resistance of low-nickel duplex stainless steel rebars. *Mater. Corros.* **2011**, *62*, 120–129. [CrossRef]
24. Gastaldi, M.; Bertolini, L. Effect of temperature on the corrosion behavior of new-low nickel stainless steel in concrete. *Cem. Concr. Res.* **2014**, *56*, 52–60. [CrossRef]
25. Duarte, R.G.; Castela, A.S.; Neves, R.; Freire, L.; Montemor, M.F. Corrosion behaviour of stainless steel rebars embedded in concrete: An electrochemical impedance spectroscopy study. *Electrochim. Acta* **2014**, *124*, 218–224. [CrossRef]
26. Mundra, S.; Criado, M.; Bernal, S.A.; Provis, J.L. Chloride-induced corrosion of steel rebars in simulated pore solutions of alkali-activated concretes. *Cem. Concr. Res.* **2017**, *100*, 385–397. [CrossRef]
27. Gonzalez, J.A.; Otero, E.; Feliu, S.; Bautista, A.; Ramírez, E.; Rodríguez, P.; López, W. Some considerations on the effect of chloride ions on the corrosion of steel reinforcements embedded in concrete structures. *Mag. Concr. Res.* **1998**, *50*, 189–199. [CrossRef]
28. Paul, S.C.; Panda, B.; Huang, Y.; Garg, A.; Peng, X. An empirical model design for evaluation and estimation of carbonation depth in concrete. *Measurement* **2018**, *124*, 205–210. [CrossRef]
29. Angulo-Ramírez, D.E.; Mejía de Gutiérrez, R.; Valencia-Saavedra, W.G.; de Medeiros, M.H.F.; Hoppe-Filho, J. Carbonation of hybrid concrete with high blast furnace slag content and its impact on structural steel corrosion. *Mater. Constr.* **2019**, *69*, e182. [CrossRef]
30. Alonso, M.C.; Luna, F.J.; Criado, M. Corrosion behavior of duplex stainless steelreinforcement in ternary binder concrete exposed to natural chloride penetration. *Constr. Build. Mater.* **2019**, *199*, 385–395. [CrossRef]
31. RILEM. Draft recommendation for repair strategies for concrete structures damaged by reinforcement corrosion. *Mater. Struct.* **1994**, *27*, 415–438. [CrossRef]
32. Czarnecki, L.; Woyciechowski, P. Concrete carbonation as a limited process and its relevance to concrete cover thickness. *ACI Mater. J.* **2012**, *109*, 275–282.

33. Dhir, R.K.; Jones, M.R.; Ahmed, H.E.H. Determination of total and soluble chlorides in concrete. *Cem. Concr. Res.* **1990**, *20*, 579–590. [CrossRef]
34. Song, H.-K.; Kwon, S.J. Permeability characteristics of carbonated concrete considering capillary pore structure. *Cem. Concr. Res.* **2007**, *37*, 909–915. [CrossRef]
35. Bautista, A.; Paredes, E.C.; Alvarez, S.M.; Velasco, F. Welded, sandblasted, stainless steel corrugated bars in non-carbonated and carbonated mortars: A 9-year corrosion study. *Corros. Sci.* **2016**, *103*, 363–372. [CrossRef]
36. Paredes, E.C.; Bautista, A.; Velasco, F.; Alvarez, S.M. Welded, pickled stainless steel reinforcements: Corrosion results after 9 years in mortar. *Mag. Concr. Res.* **2016**, *68*, 1099–1109. [CrossRef]
37. Page, C.L. Mechanism of corrosion protection in reinforced-concrete marine structures. *Nature* **1975**, *258*, 514–515. [CrossRef]
38. Glass, G.K.; Yang, R.; Dickhaus, T.; Buenfeld, N.R. Backscattered electron imaging of the steel-concrete interface. *Corros. Sci.* **2001**, *43*, 605–610. [CrossRef]
39. Chen, F.; Chun-Qing, L.; Baji, H.; Baogue, M. Quantification of steel-concrete interface in reinforced concrete using Backscattered Electron imaging technique. *Contr. Build. Mater.* **2018**, *179*, 420–429. [CrossRef]
40. Stern, M.; Geary, A. Electrochemical polarization I. A theoretical analysis of the shape of the polarization curves. *J. Electrochem. Soc.* **1957**, *104*, 56–58. [CrossRef]
41. Andrade, C.; Gonzalez, J.A. Quantitative measurements of corrosion rate of reinforcing steel embedded in concrete using polarization resistance measurements. *Mater. Corros.* **1978**, *29*, 515–519. [CrossRef]
42. Stefanoni, M.; Angst, U.; Elsener, B. Corrosion rate of carbon steel in carbonated concrete—A critical review. *Cem. Concr. Res.* **2018**, *103*, 35–48. [CrossRef]
43. Song, H.W.; Saraswathy, V. Corrosion monitoring of reinforced concrete structures—A review. *Int. J. Electrochem. Sci.* **2007**, *2*, 1–28.
44. Revert, A.B.; Hornbostel, K.; De Weerdt, K.; Geiker, M.R. Macrocell corrosion in carbonated Portland and Portland-fly ash concrete—Contribution and mechanism. *Cem. Concr. Res.* **2019**, *116*, 273–283. [CrossRef]
45. Paredes, E.C.; Bautista, A.; Alvarez, S.M.; Velasco, F. Influence of the forming process of corrugated stainless steels on their corrosion behaviour in simulated pore solutions. *Corros. Sci.* **2012**, *58*, 52–61. [CrossRef]
46. Monrrabal, G.; Bautista, A.; Guzman, S.; Gutierrez, C.; Velasco, F. Influence of the cold working induced martensite on the electrochemical behavior of AISI 304 stainless steel surfaces. *J. Mater. Res. Technol.* **2019**, *8*, 1135–1346. [CrossRef]
47. Bautista, A.; Alvarez, S.M.; Velasco, F. Selective corrosion of duplex stainless steel bars in acid. Part II: Effect of the surface strain and numerical analysis. *Mater. Corros.* **2015**, *66*, 357–365. [CrossRef]

© 2019 by the authors. Licensee MDPI, Basel, Switzerland. This article is an open access article distributed under the terms and conditions of the Creative Commons Attribution (CC BY) license (http://creativecommons.org/licenses/by/4.0/).

Article

Corrosion Behavior of Different Brass Alloys for Drinking Water Distribution Systems

Jamal Choucri [1,2], Federica Zanotto [1], Vincenzo Grassi [1], Andrea Balbo [1], Mohamed Ebn Touhami [2], Ilyass Mansouri [3] and Cecilia Monticelli [1,*]

[1] Centro di Studi sulla Corrosione e Metallurgia "A. Daccò", Dipartimento di Ingegneria, Università di Ferrara, 44122 Ferrara, Italy; j.choucri@libero.it (J.C.); zntfrc@unife.it (F.Z.); vincenzo.grassi@unife.it (V.G.); andrea.balbo@unife.it (A.B.)
[2] Laboratory of Materials Engineering and Environment: Modeling and Application, Faculty of Science, University Ibn Tofail, 14000 Kenitra, Morocco; m.ebntouhami@gmail.com
[3] International Institute for Water and Sanitation (IEA), National Office of Electricity and the Potable Water, 90000 Rabat, Morocco; xilyas72@gmail.com
* Correspondence: mtc@unife.it; Tel.: +39-0532-455136

Received: 31 March 2019; Accepted: 2 June 2019; Published: 4 June 2019

Abstract: Some α + β' brass components of drinking water distribution systems in Morocco underwent early failures and were investigated to assess the nature and extent of the corrosion attacks. They exhibited different corrosion forms, often accompanied by extensive β' dezincification. In order to offer viable alternatives to these traditional low cost materials, the corrosion behavior of two representative α + β' brass components was compared to that of brass alloys with nominal compositions CuZn36Pb2As and CuZn21Si3P, marketed as dezincification resistant. CuZn21Si3P is a recently developed eco-friendly brass produced without any arsenic or lead. Electrochemical tests in simulated drinking water showed that after 10 days of immersion CuZn21Si3P exhibited the highest polarization resistance (R_p) values but after longer immersion periods its R_p values became comparable or lower than those of the other alloys. After 150 days, scanning electron microscopy coupled to energy dispersive spectroscopy (SEM-EDS) analyses evidenced that the highest dezincification resistance was afforded by CuZn36Pb2As (longitudinal section of extruded bar), exhibiting dealloying and subsequent oxidation of β' only at a small depth. Limited surface dealloying was also found on CuZn21Si3P, which underwent selective silicon and zinc dissolution and negligible inner oxidation of both α and κ constituent phases, likely due to peculiar galvanic effects.

Keywords: brass; CuZn36Pb2As; CuZn21Si3P; corrosion; dezincification; simulated drinking water; EIS; SEM-EDS; long immersion

1. Introduction

Corrosion is an important technological economic and social problem which can be controlled after full knowledge of the alloy behavior under field conditions and proper choice of materials and design solutions [1–3]. In the case of brass corrosion, the large-scale use of these alloys in drinking water distribution systems for tube fittings, valves, and ancillaries may also determine human health concerns due to Pb and Cu release in drinking water [4–6].

Brasses are copper-zinc alloys that contain 5–40% Zn as the principal alloying element. Zinc concentrations up to 35% may dissolve in face-centered cubic (fcc) copper matrix to form a single solid solution (α brass). When an even higher zinc content is added, a zinc-rich β' second phase also forms, characterized by body-centred cubic (bcc) lattice. Duplex α + β' brasses are cheaper, easier to fabricate and exhibit higher mechanical strength than α brasses, but the zinc-rich β' phase exhibits a higher tendency to dezincification corrosion that leaves a porous layer of copper, structurally weakening the component and leading to brittle failures [7–9].

Beside zinc, many other elements are present as both alloying additions to improve specific properties, and impurities. Up to 3%, lead additions improve the alloy machinability, while beyond this percentage lead has deleterious effects on casting properties, namely on shrinkage and hot tearing [9]. However, in low cost commercial brasses its content is often greater than this value, due to the lack of strict control on alloying and impurity contents. Lead is insoluble in copper-zinc alloys and during solidification it precipitates forming globules both at grain boundaries and within the matrix. It produces little effects or some improvement of the corrosion resistance of the alloy, depending on the specific environment and alloy compositions [10–12].

The pressure to have lead removed from commercial brasses pushed industry and research studies to propose alternative alloys, either with reduced lead content or lead-free, and characterized by good dezincification resistance [13]. Recently, Bi and Si were used as alternatives to lead in free-cutting-brass [8]. Non toxic bismuth can play the role of lead in machinability without adverse health effects [14], and proved to be neutral towards dezincification resistance of duplex and nearly all α brasses [10]. Lead-free silicon brass containing 3% Si and 0.05% P was reported to offer good machinability [15] and significant corrosion resistance, due to the formation of a protective silicon-rich surface film (a sublayer of copper and zinc oxides topped by a protective zinc silicate film, according to thermodynamic considerations [16]). It also exhibited dezincification resistance thanks to the electrochemical phosphorus cycle, analogous to that afforded by arsenic. Actually, both arsenic and phosphorus are reputed capable to limit copper redeposition by reducing Cu^{2+} ions to Cu^+. As also reported for As^{3+}, P^{3+} ion leaching can also be avoided by its reaction with Cu metal which brings it back to its elemental state [17,18]. The positive influence of arsenic, phosphorus, and antimony against dezincification was well documented in various aggressive environments, mainly on α brasses [10,19–21]. Arsenic showed some positive effects on α + β′ alloys too [10,22]. However, the toxicity of arsenic and antimony should represent a limit to their wide use [23,24].

Dezincification of α [25,26] and α + β′ brasses was also prevented by tin addition, in the latter alloys especially in the presence of nickel [27]. The formation of a protective surface SnO_2 film was reputed as the origin of this beneficial effect [25,27]. Other elements inhibiting dezincification were aluminum which could improve brass corrosion behavior by forming a protective Al_2O_3 film [28] and niobium which operated by alloying with β′-phase to suppress the preferential dissolution of zinc and by forming a passivation layer that served as a barrier against chloride ions [29].

Most commercial brasses for drinking water distribution systems have an α + β′ structure due to their significant mechanical properties and easy forming capability, but they are often affected by an inadequate impurity control. Impurities of iron, manganese, nickel, and cobalt are usually detrimental to the alloy dezincification resistance [10]. All alloying elements change the stability range of the β′ phase; in particular, an increase in their content (except for nickel) shifts the range of existence of the β′ phase toward lower zinc content [28].

In the first part of this paper, the relevance of dezincification in some failed components of drinking water distribution systems in Morocco was evidenced. Then, the paper describes the results of tests performed to study the actual corrosion behavior and dealloying resistance of possible alternative alloys in comparison to two alloys selected among those of the failed service components. In the view of the importance of dezincification in the investigated components, the selected brass alloys were those characterized by the highest and the lowest dezincification resistance to the standard UNI EN ISO 6509 dezincification test. The alternative dezincification resistant alloys had nominal compositions CuZn36Pb2As and CuZn21Si3P of which the latter one is an interesting eco-friendly arsenic- and lead-free silicon brass, with still scarcely investigated corrosion behavior in drinking water [16,17]. The comparative tests used to assess the alloy corrosion resistance and to study the corrosion mechanism consisted of electrochemical tests (electrochemical impedance spectroscopy, electrochemical impedance spectroscopy (EIS), and polarization curve recording) carried out at different immersion times during

150 days of exposure to simulated drinking water (SDW). Moreover, surface investigations based on scanning electron microscope (SEM) observations, coupled to energy dispersive spectroscopy (EDS) analyses, were used to measure the dezincification depth and to further study the dealloying process.

2. Materials and Methods

Two types of brass samples were investigated, that is (i) forged brass components (F samples) including pipe fittings, connectors, and valves for drinking water distribution systems and (ii) bars (B samples). The first group included seven sample types (F1–7) extracted from drinking water distribution systems in Morocco after early failures (service lives from 8 months to 5 years). The quantitative chemical compositions of these samples obtained by optic emission spectrometry (OES) and their failure times are reported in Table 1. Samples F1–7 mainly have compositions in the range 54–57% Cu, 37–40% Zn, 2.3–4.6% Pb, and 0.5–1.4% Sn and exhibit significant Fe, Ni, and Si contents (up to 0.6%, 0.4%, and 0.08%, respectively).

Table 1. Service lives (SL; y = years; m = months) and chemical compositions (in weight %) of the examined forged brass components. Compositions were obtained by optic emission spectrometry (OES).

Code	SL	Cu	Zn	Pb	Sn	P	Mn	Fe	Ni	Si	As	Sb	Bi	Al	S
F1	3 y	57.42	38.89	2.63	0.450	0.0038	0.0080	0.388	0.208	0.0521	<0.0003	0.0310	0.00042	0.042	0.0011
F2	9 m	55.28	39.39	3.84	0.769	0.0058	0.0042	0.376	0.261	0.0095	<0.0003	0.0540	0.00278	0.072	0.0169
F3	8 m	54.34	40.49	3.57	0.509	0.0041	0.0160	0.379	0.225	0.0277	0.0217	0.0478	0.00227	0.476	0.0098
F4	5 y	57.42	37.63	2.39	1.127	0.0074	0.0380	0.605	0.325	0.0479	<0.0003	0.0537	0.00235	0.377	0.0019
F5	2 y	55.31	37.40	4.61	1.385	0.0066	0.0163	0.599	0.352	0.0436	<0.0003	0.1070	0.00490	0.271	0.0040
F6	2 y	56.59	40.33	2.28	0.370	0.0025	0.0073	0.259	0.129	0.0062	<0.0003	0.2310	0.00141	0.029	0.0287
F7	2 y	56.27	37.89	3.68	1.015	0.0057	0.0301	0.588	0.338	0.0348	<0.0003	0.0570	0.03543	0.270	0.0017

The second group of samples comprised two dezincification resistant alloys in bars, with nominal composition CuZn36Pb2As (code B1) and CuZn21Si3P (code B2). Table 2 reports their composition evaluated by OES: both of them contain small or negligible Fe, Mn, and Ni amounts. Si is absent in B1, while in B2 the Si content is 3.3%.

Table 2. Chemical compositions (in weight %) of dezincification resistant alloys, evaluated by OES. Nominal compositions of B1 and B2 are CuZn36Pb2As and CuZn21Si3P, respectively. (n.d. = not determined)

Code	Cu	Zn	Pb	Sn	P	Fe	Ni	Si	As	Al	Altro
B1	61.70	35.83	1.90	0.11	n.d.	0.120	0.030	<0.0010	0.09	0.02	<0.2
B2	77.10	19.52	0.00	0.01	0.05	0.02	0.00	3.3	0.00	0.00	-

2.1. Standard Dezincification Test

Since in most failed components dezincification was detected as the main corrosion form or was associated to other types of corrosion attack, the standard UNI EN ISO 6509 dezincification test was performed to assess the dezincification resistance of all brass types and to select the alloys of failed components to be further investigated by electrochemical tests. The exposed sample surfaces (area about 1 cm^2) were isolated by epoxy resin, prepared by grinding down to emery paper grade 500, washed by distilled water and degreased by alcohol. Then, vertically oriented surfaces were exposed to 1 wt. % CuCl$_2$·2H$_2$O solution for 24 h. B1 and B2 samples were prepared with both longitudinal (L) and transverse (T) orientation, because the bar microstructure was anisotropic.

2.2. Electrochemical Tests in Simulated Drinking Water

Electrochemical tests were performed in simulated drinking water (SDW), i.e., deionized water containing 400 ppm SO$_4^{2-}$ (4.2 mM), 400 ppm Cl$^-$ (11.2 mM), and 50 ppm NO$_3^-$ (0.8 mM), as sodium salts. This solution complies with drinking water composition requirements, according to

Moroccan standard NM 03.7.001. In fact, the sulfate and nitrate concentrations correspond to the maximum acceptance limits, while the chloride concentration is about one-half of the maximum accepted concentration (that is 750 ppm).

The electrodes were prepared by soldering a copper wire on the back of brass samples to ensure the electrical contact and then embedding them in epoxy resin (exposed surface area of 1 cm^2). For electrochemical tests, the final polishing was performed with 1 μm diamond paste.

Electrochemical tests were carried out in a conventional thermostated three-electrode cell by using a PARSTAT 2273 potentiostat/galvanostat (Ametek, Berwyn, PA, USA), piloted by PowerSuite software (v. 2.58, Advanced Measurement Technology, Inc., Oak Ridge, TN, USA). The reference and auxiliary electrodes were a saturated calomel electrode (SCE) and a Pt sheet, respectively. All the potential values quoted in the text are referred to the SCE. At least two experiments were carried out under each experimental condition.

Nondestructive electrochemical impedance spectroscopy (EIS) tests were performed at specific immersion times on the two alloys from failed components with the highest and the lowest susceptibility to dezincification according to the standard UNI EN ISO 6509 and on samples extracted from bars, with both transverse and longitudinal orientation (samples B1 (T and L) and B2 (T and L)). The tests were carried out at the open circuit potential, during 150 days of immersion in SDW. During the tests, a voltage perturbation amplitude of 10 mV (rms) was applied, in the frequency range 10^4–10^{-3} Hz, with five points per frequency decade. EIS spectra were fitted by SAI ZView v.3.5c software (Scribner Associates Inc., Southern Pines, NC, USA), according to the most suitable equivalent circuit, as described in the text.

After selected immersion times, cathodic and anodic polarization curves were recorded at a scan rate of 0.167 mV/s on separated electrodes always starting from the corrosion potential (E_{cor}) and corrosion rates were evaluated by the Tafel method from the cathodic polarization curves.

2.3. Surface Observations

The microstructure of the alloys was investigated by a Leica DMRM optical microscope (OM) (Leica Microsystems GmbH, Wetzlar, Germany) observations and, in some cases, also by scanning electron microscope (SEM) (Zeiss EVO MA15 (Zeiss, Oberkochen, Germany)) analysis, after etching with 10 wt. % FeCl$_3$ solution. OM observations and SEM-EDS (for EDS: Oxford Aztec energy dispersive X-ray spectroscopy system (Oxford, UK)) analyses (under an acceleration voltage of 20 kV) also permitted the characterization of the morphology and nature of the corrosion attacks, both on failed components and on samples exposed to SDW under free corrosion conditions.

3. Results

3.1. Microstructures of the Studied Materials

Samples F1–7 evidenced α + β′ microstructures containing dispersed Pb globules (as an example, the microstructures of F3 and F4 are shown in Figure 1). In F1–7, the α volume fractions ranged from 41% to 69% (calculated by image analysis software) and, due to the significant amounts of Fe, Mn, Ni, and Si impurities, also silicides of these metals were detected [29], better visible under SEM observation (not shown). α and β′ have a nominal atomic composition of Cu3Zn and CuZn, respectively [18].

Figure 2 shows the microstructures of B1 and B2 both in transverse (T) and longitudinal (L) sections. B1 showed an α fraction of 89% and in the figure the β′ phase is indicated by arrows. B2 exhibited comparable fractions of α (pale pink in Figure 2) and κ phase (grey colored Si-rich phase in Figure 2, with hexagonal lattice and nominal atomic composition Cu8Zn2Si [18]). This figure also evidences that in L sections of B1 and B2 samples, the β′ and κ phases are more continuous and elongated in the extrusion direction. In B2, SEM-EDS analysis also detected tiny particles of a third bright phase (Figure 3a), which was located at the grain boundaries and corresponded to the cubic γ phase with nominal composition Cu4ZnSi [18]. The compositions of α, κ, and γ phases (obtained by SEM-EDS as averages from at least six regions for each phase) are reported in Figure 3b. They result in good agreement with the expected values, except in the case of γ phase, where the low measured Si content was due to the small dimensions of this phase in comparison to the alloy volume analyzed under an acceleration voltage of 20 kV.

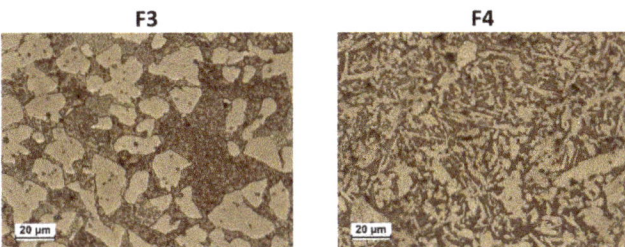

Figure 1. Optical Microscope (OM) microstructures of etched F3 and F4 brass samples. They have α + β′ microstructures (α is the pale brown phase and β′ is the dark brown one).

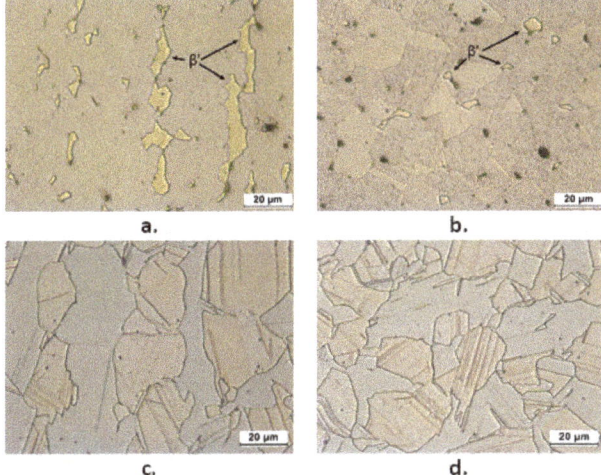

Figure 2. OM microstructures of etched longitudinal (L) (**a**,**c**) and transverse (T) (**b**,**d**) sections from B1 (CuZn36Pb2As: **a**,**b**) and B2 (CuZn21Si3P: **c**,**d**) samples. In (**a**) and (**b**), α is pale brown and β′ (indicated by arrows) is yellowish. In (**c**) and (**d**), α is pale pink and κ is grey.

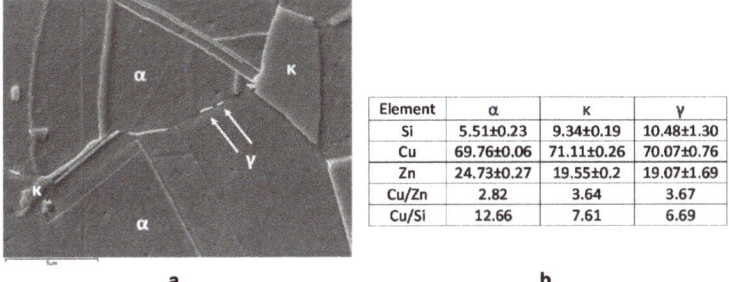

Element	α	κ	γ
Si	5.51±0.23	9.34±0.19	10.48±1.30
Cu	69.76±0.06	71.11±0.26	70.07±0.76
Zn	24.73±0.27	19.55±0.2	19.07±1.69
Cu/Zn	2.82	3.64	3.67
Cu/Si	12.66	7.61	6.69

Figure 3. Backscattered (BSD)-scanning electron microscope (SEM) micrograph of etched B2 (CuZn21Si3P) with indication of α, κ, and γ phases (**a**) and average SEM-energy dispersive spectroscopy (EDS) analysis (in atomic percentage) of the three phases (**b**).

3.2. Corrosion Forms Observed in Drinking Water Systems in Morocco

F1–7 components presented untimely corrosion attacks after relatively short permanence periods (from 8 months to 3 years) in contact with drinking water. Corrosion was essentially localized on the less corrosion-resistant Zn-rich β' phase and in most cases, it caused dezincification or dezincification-favored corrosion forms.

As an example, Figure 4a,b shows plug and layer dezincification of the β' phase, mainly detected in correspondence of crevices such as on threaded surfaces or under gaskets. Many components exhibited a selective attack with surface dissolution of the β' phase which left a rough porous surface (Figure 4c,d). In some components, this selective attack tended to penetrate inside the material where occluded cell conditions could develop, likely determining relatively high chloride concentrations and low pH values and stimulating both β' and α phase subsurface corrosion (Figure 4e,f).

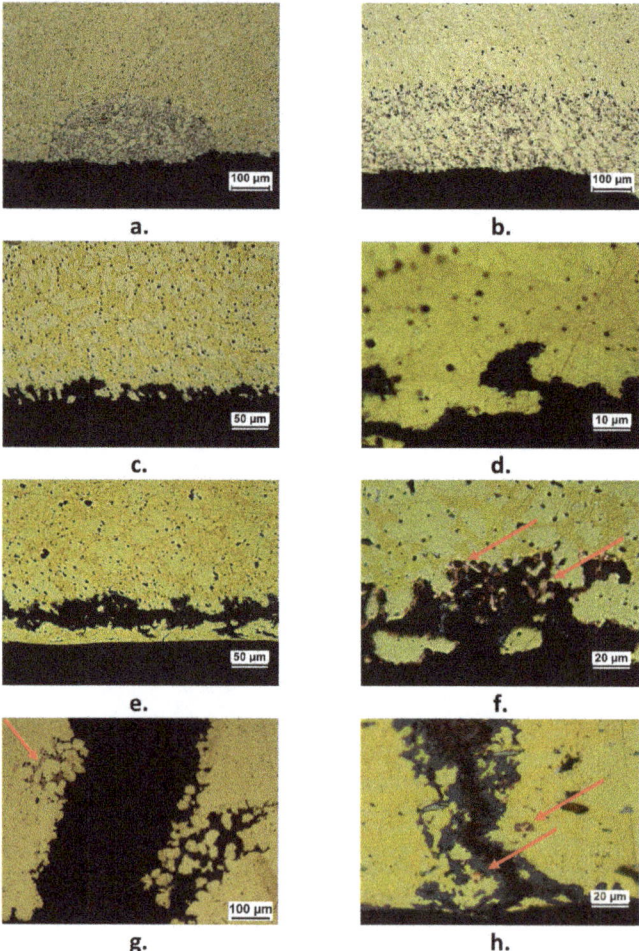

Figure 4. Forms of corrosion attack detected in failed components; (**a**) plug dezincification (F2 component); (**b**) layer dezincification (F6 component); (**c**) and (**d**): selective corrosion attack with dissolution of the β' phase (F3 component); (**e**) (F2 component) and (**f**) (F7 component): subsurface corrosion and dezincification; (**g**) and (**h**): cracks, accompanied by dezincification (F7 component). Arrows in (**f–h**) indicate porous copper produced by dezincification.

Three pipe fittings showed cracks accompanied by more or less extensive dezincification (Figure 4g,h). This could be due to the presence of Stress Corrosion Cracking (SCC), usually connected to residual stresses and presence of specific aggressive environments. Actually, SCC of various brass alloys was observed in sulfate solutions and drinking water installations [30–32].

3.3. Standard Dezincification Tests

In α + β′ samples, dezincification affected the less noble β′ phase. Therefore, given the relatively high β′ percentage in F1–7, these samples exhibited great average (d_{av}) and maximum (d_{max}) dezincification depths (Table 3). F3 and F4 showed, respectively, the lowest and the highest dezincification resistance and were selected for testing during the long time immersions in SDW.

As expected, B1 and B2 samples evidenced much better performances (Table 3). In the case of B1, mainly constituted by α phase, the addition of As in the range 0.06–0.09% efficiently protected the alloy by the so called "arsenic cycle" which prevents copper redeposition [19,20]. Similarly, an analogous "phosphorus cycle" is claimed to protect CuZn21Si3 alloy containing 0.05% P (B2) from corrosion [10,33]. The performances of B1 and B2 were better in the case of L samples because on these samples, the less noble β′ and κ phases were more discontinuous.

Table 3. Average dezincification depths (d_{av}) with standard deviations (σ) and maximum dezincification depths (d_{max}) after the standard UNI EN ISO 6509 dezincification test (24 h test).

Sample Code	$d_{av} \pm \sigma/\mu m$	$d_{max}/\mu m$	Sample Code	$d_{av} \pm \sigma/\mu m$	$d_{max}/\mu m$
F1	794 ± 110	923	F5	803 ± 23	826
F2	936 ± 27	967	F6	889 ± 94	988
F3	1110 ± 23	1140	F7	866 ± 121	948
F4	363 ± 23	382			
B1 (T)	32 ± 16	84.2	B2 (T)	25 ± 12	76.4
B1 (L)	20.1 ± 9.1	40.4	B2 (L)	14.6 ± 5.1	28.3

3.4. EIS Tests in SDW

The corrosion resistance of F3, F4, B1 (L and T), and B2 (L and T) in SDW was monitored by recording the EIS spectra at specific immersion times, during 150 days of immersion.

Figure 5a–h shows the EIS spectra of F3, F4, B1(T), and B2(T) in Nyquist and Bode (angle vs. frequency) forms, at specific immersion times. In F3 (Figure 5a,b) and F4 (Figure 5c,d), two time constants were detected at frequencies around 10 (medium frequency, mf, time constant) and 10^{-2} Hz (low frequency, LF, time constant). The latter time constant was just a hint in the Bode plots after 1 h immersion. On F4 a third time constant centered at about 10^4 Hz (high frequency, HF, time constant) occurred at long immersion times (82 and 150 days). In the case of B1(T) samples, again only the MF (10^2–10 Hz) and the LF (10^{-2}–10^{-1} Hz) time constants were detected after 1 h, 24 h, and 10 days of immersion, but just the beginning of the LF one was present after 1 h.

At longer immersion times, the hf time constant arose at 10^3 Hz, the MF time constant was scarcely evident and the LF time constant tended to shift to lower frequencies (10^{-3} Hz), in comparison to those after short exposure times (Figure 5d). B2(T) electrodes evidenced three times constants since the beginning of the immersion. The HF one tended to move from 10^2–10^3 Hz (after 1 and 24 h immersion) to 10^4 Hz (from 10 days on), while the mf and LF constants moved to lower frequencies at increasing immersion times (from 3–4 Hz to about 10^{-1} Hz for the mf one and from 10^{-1}–10^{-2} Hz to 10^{-2} Hz or lower frequencies in the case of the LF one).

The spectra were fitted with the equivalent circuits (ECs) reported in Figure 6a,b.

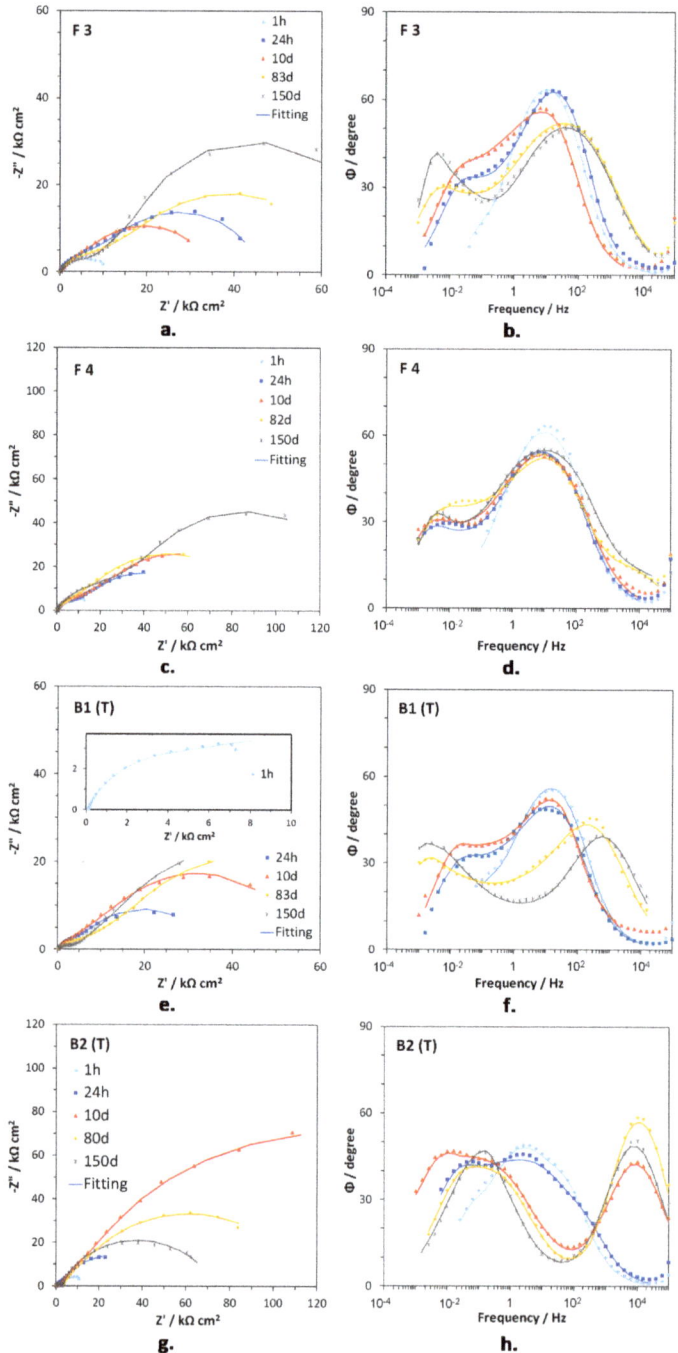

Figure 5. Nyquist (**a,c,e,g**) and Bode (phase angle vs. frequency, **b,d,f,h**) plots obtained on F3 (**a,b**), F4 (**c,d**), B1(T) (**e,f**) and B2(T) (**g,h**), during immersions in Simulated Drinking Water (SDW).

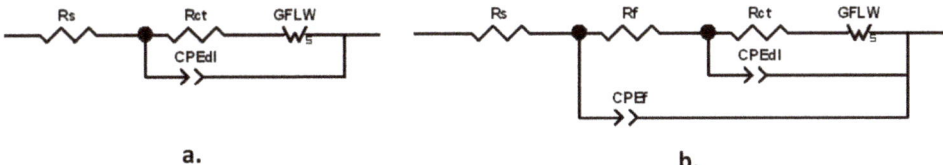

Figure 6. Equivalent circuits (EC) used to fit the Electrochemical Impedance Spectroscopy (EIS) spectra: (**a**) two-time-constant EC; (**b**) three-time-constant EC.

The former one, characterized by two times constants, can model a corrosion process with kinetics affected by both charge transfer and diffusion, while in the latter one, including three times constants, the corrosion process described in Figure 6a occurs at the bottom of the pores of a surface corrosion product film. This latter EC, also including the parameters of the previous EC, is composed by the electrolyte resistance, R_s, in series with two nested parallel R-CPE (Resistance-Constant Phase Element) couples. CPE is the Constant Phase Element, used instead of ideal capacitance to compensate for surface inhomogeneities, such as roughness, porosities and inclusions.

The analytical expression of the CPE impedance is:

$$Z_{CPE} = Y^{-1}(j\omega)^{-n}, \qquad (1)$$

where $\omega = 2\pi f$ is the angular frequency, $j = \sqrt{(-1)}$ is the imaginary unit, Y is a frequency independent value and n is a fit parameter with values in the range $0 \leq n \leq 1$, which measures the element deviation from the ideal capacitive behavior (exhibiting $n = 1$). For each R-CPE couple, the following general equation:

$$C = \left(R^{1-n} Y\right)^{\frac{1}{n}}, \qquad (2)$$

was used to convert the Y parameters of the CPE elements into the associated capacitances (with R being the corresponding couple resistance) [34–36]. The R_f–CPE_f couple described the dielectric properties of the surface corrosion product film (hf time constant), while R_{ct}–CPE_{dl} provided information about the charge transfer process at the metal/electrolyte interface (MF time constant). The previously quoted frequencies corresponding to the charge transfer process (MF time constant) and to the dielectric properties of the surface oxide film (HF time constant) are in agreement with those unequivocally attributed to these elements by erosion-corrosion tests carried out on brass in chloride solution [37].

In order to fit the LF time constant attributed to diffusion, a generalized finite length Warburg (GFLW) element was introduced in series to the charge transfer resistance, R_{ct}. Its mathematical impedance expression is:

$$Z_{GFLW} = R_W \frac{\tan h(j\omega T)^p}{(j\omega T)^p}, \qquad (3)$$

where T is a time constant, R_W is a resistance, and p is an exponent which can vary in the range $0 < p < 1$. For $p = 0.5$, $T = L^2/D$, where L is the thickness of the diffusion layer and D is the diffusion coefficient [38]. The ECs of Figure 6 or quite similar ones (with a Warburg element instead of a GFLW one) were already used to interpret EIS spectra of brass exposed to cooling water and chloride solutions [7,39–42].

The EIS fitting parameters for the spectra in Figure 5 are reported in Tables 4–7. The alloys indicated as dezincification resistant (B1(T) and B2(T)) resulted capable of forming a surface oxide film with a low but detectable protectiveness, particularly at long immersion times. In fact, increasing R_f values were achieved, while the generally low and decreasing C_f (with final values of the order of magnitude of 10^{-7}–10^{-8} F cm^{-2}) indicated a progressive increase in film thickness. No film was detected by EIS on F3 and only a small loop related to a surface film was found on F4 at long immersion times. Actually, on all alloys the corrosion process was largely dependent on the rates of charge transfer and particularly on mass transport, because even in the presence of measurable R_f values, R_{ct} and especially R_W values were higher than R_f by many orders of magnitude and their variations do not appear connected with the surface film protectiveness.

Table 4. Electrochemical Impedance Spectroscopy (EIS) fitting parameters for F3.

Immersion Time	1 h	24 h	10 d	83 d	150 d
R_s/Ω cm^2	103	98.9	94	77	82
$R_{ct}/k\Omega$ cm^2	7.69	5.30	1.16	8.27	7.80
$R_W/k\Omega$ cm^2	3.44	37.0	33.9	56.4	65.9
T/s	6.1	36	76	258	200
p	0.47	0.42	0.40	0.40	0.53
C_{dl}/μF cm^{-2}	27.8	16.4	32.8	19.1	14.9
n_{dl}	0.84	0.84	0.83	0.67	0.67

Table 5. EIS fitting parameters for F4.

Immersion Time	1 h	24 h	10 d	82 d	150 d
R_s/Ω cm^2	123	122	119	109	89
$R_f/k\Omega$ cm^2	-	-	-	0.097	0.098
$R_{ct}/k\Omega$ cm^2	8.00	8.55	11.5	6.13	28.0
$R_W/k\Omega$ cm^2	11.0	51.0	72.1	82.0	109
T/s	3.5	290	300	265	260
p	0.32	0.42	0.42	0.39	0.48
C_{dl}/μF cm^{-2}	16.2	34.1	29.9	16.5	13.4
n_{dl}	0.83	0.72	0.70	0.74	0.71
C_f/μF cm^{-2}	-	-	-	0.63	0.26
n_f	-	-	-	0.65	0.63

Table 6. EIS fitting parameters for B1(T).

Immersion Time	1 h	24 h	10 d	83 d	150 d
R_s/Ω cm^2	125	124	110	105	107
$R_f/k\Omega$ cm^2	-	-	0.079	2.61	2.90
$R_{ct}/k\Omega$ cm^2	5.80	3.79	2.40	17.60	7.06
$R_W/k\Omega$ cm^2	8.00	40.6	53.0	108	131
T/s	8.0	69	86	300	339
p	0.40	0.38	0.41	0.50	0.53
C_{dl}/μF cm^{-2}	20.2	14.3	10.0	19.6	273
n_{dl}	0.78	0.76	0.80	0.54	0.54
C_f/μF cm^{-2}	-	-	2.52	2.07	0.82
n_f	-	-	0.98	0.68	0.64

Table 7. EIS fitting parameters for B2(T).

Immersion time	1 h	24 h	10 d	80 d	150 d
R_s/Ω cm^2	138	135	100	101	98
$R_f/k\Omega$ cm^2	0.57	0.18	1.36	2.70	2.36
$R_{ct}/k\Omega$ cm^2	7.12	7.50	67.0	31.4	1.93
$R_W/k\Omega$ cm^2	7.03	42.0	197	82.5	70.3
T/s	9.4	30	63	19	2.2
p	0.48	0.44	0.51	0.42	0.64
C_{dl}/μF cm^{-2}	51.5	106	326	73.8	34.4
n_{dl}	0.69	0.58	0.60	0.65	0.66
C_f/μF cm^{-2}	11.6	4.10	0.084	0.036	0.056
n_f	0.839	0.84	0.75	0.85	0.80

R_W had a general tendency to increase with time on the studied brass alloys, except for B2(T) which showed a R_W maximum at 10 days of immersion (197 kΩ cm^2) and then a decrease to 70.3 kΩ

cm^2, at the end of the immersion time. Accordingly, the T parameter of the GFLW (the diffusion EC element) usually increased, suggesting that diffusion became slower with time, but on B2(T) it reached a maximum at 10 days (like R_W) and then decreased significantly.

On B2(T), R_{ct} had a trend similar to that of R_W with a maximum of 67.0 kΩ cm^2 at 10 days of immersion. Such a high maximum R_{ct} value was not achieved in concomitance with a maximum of R_f, so excluding that it was connected to a reduced film porosity. On the other alloys, rather oscillating R_{ct} values were detected, with the highest value of 28 kΩ cm^2 obtained on F4 at 150 days of immersion, again independently from the surface film protectiveness. A possible explanation for the R_{ct} oscillations could be the presence of a thin inner passive film, which underwent breakdown and subsequent heal events so justifying variations in charge transfer rate.

The analysis of C_{dl} values in Tables 5–8 evidenced that unexpectedly they often oscillated in phase with R_{ct}, that is, an increase in R_{ct} was usually accompanied by an increase in C_{dl}. This trend is difficult to explain. A possible explanation could be connected to the presence of breakdown and heal events of the inner passive layer. In fact, after a breakdown event the accumulation of metal cations at the metal-electrolyte interface, allowed by the slow diffusion rates (as suggested by the high detected R_W values), could subsequently hinder further anodic dissolution, so justifying the presence of high R_{ct} in correspondence of high C_{dl} values.

The polarization resistance values of the alloys, inversely proportional to the corrosion rates, were calculated as the sum: $R_p = R_f + R_{ct} + R_W$ and are collected in Figure 7, together with the corresponding E_{cor} values. In general, the alloys exhibited low initial R_p values tending to increase with time. Only for B2(T), R_p reached a maximum (266 kΩ cm^2) after 10 days of immersion, then decreased to rather low values (74.5 kΩ cm^2), suggesting an important loss of corrosion resistance. Interestingly, the time evolution of R_p for B1(L) and B2(L) were similar to those of the corresponding (T) electrodes. In fact, also for B2(L), R_p passed through a maximum at 10 days and then decreased significantly. B1(L) exhibited slightly higher corrosion resistance than B1(T). F3 was the least corrosion resistant alloy (lowest R_p values), but at long immersion time the behavior of B2 electrodes overlapped that of F3. Between 80 and 150 days of exposure, the overall R_p variations among traditional and dealloying-resistant alloys were quite limited, that is the highest and lowest values differed by a factor of 2–2.7.

The E_{cor} values of the alloys passed through a maximum (at 10 days of immersion for F3 and B2 electrodes; at about 80 days of immersion for the other alloys) and then became more active. In particular, on F3 and B2(L) samples, E_{cor} values as negative as −0.189 and −0.170 V_{SCE} were achieved, respectively. More negative E_{cor} are expected to favor zinc (and silicon in the case of B2) over copper dissolution, so stimulating dealloying.

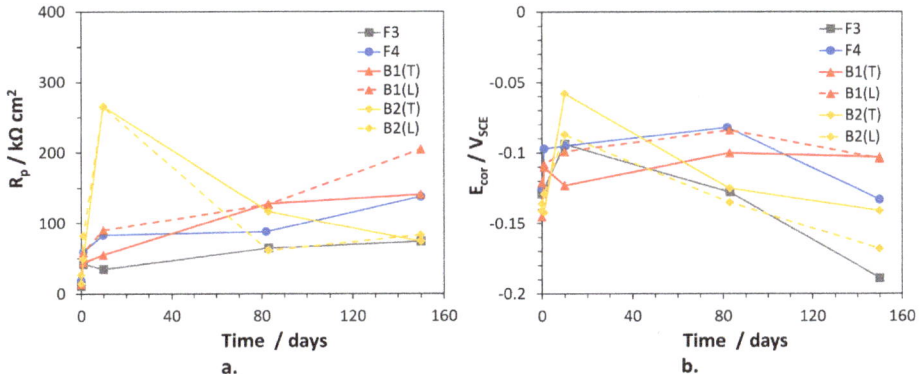

Figure 7. (a) Polarization resistance (R_p) and (b) corrosion potential (E_{cor}) values measured on selected brass alloys in SDW.

3.5. Polarization Curves

The polarization curves recorded on F3, F4, B2(T), and B3(T) after 1 h, 24 h, and 10 days of immersion in SDW are reported in Figure 8.

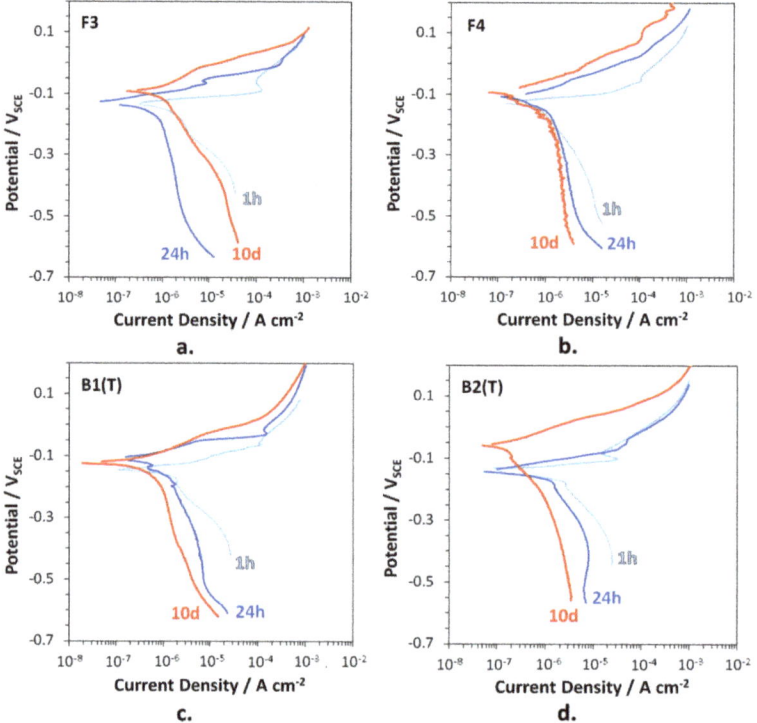

Figure 8. Polarization curves recorded on F3 (**a**), F4 (**b**), B2(T) (**c**), and B3(T) (**d**), after 1, 24 h, and 10 days of immersion in SDW.

At 1 h of immersion, the anodic curves of all alloys showed low anodic slopes associated to an active corrosion behavior, while the much higher cathodic slopes suggested that oxygen reduction was largely under diffusion control. Therefore, corrosion of the studied brasses was essentially under cathodic control after short immersion periods. At longer immersion times, mass transport also affected the anodic process as indicated by the increase in the anodic slopes. Therefore, the diffusion of oxidized species inside dealloyed regions towards the electrode surface and from the surface to the solution bulk became slower with time. Under these conditions, corrosion underwent a mixed control with both the anodic and mainly the cathodic reaction largely depending on diffusion.

Table 8 collects the E_{cor} and corrosion current densities (i_{cor}) obtained from the polarization curves by the Tafel method. The i_{cor} values are in agreement with the trend of the R_p values in this same time interval (Figure 7). In fact, with the exception of F3, until 10 days of immersion i_{cor} continuously decreased (and R_p increased), because of a progressive shift of both the anodic and the cathodic curves to lower current densities. In the case of F3, after 10 days the cathodic currents and i_{cor} accelerated in comparison to those after 24 h, again in agreement with a slight decrease in R_p values.

3.6. Characterization of Corrosion Attack after 150 Day Immersions in SDW

At the end of 150 days of immersion, all specimens exhibited dezincification attacks. In the case of B1, the maximum dezincification depths for (T) and (L) specimens were about 300 µm (Figure 9a) and 50 µm (Figure 9b), respectively. The good corrosion resistance of B1(L) samples agreed with their relatively high R_p values in comparison to those measured on B1(T) (Figure 7). Figure 9a and b clearly evidence that dezincification affected the β' phase. On B1(T), the β' phase was elongated perpendicularly to the exposed surface so favoring penetration of the corrosion attack, while on B1(L) the corrosion attack was more limited, although solution penetration along the grain boundaries induced dezincification [18] and even oxidation of β' grains at some distance from the exposed surface.

Table 8. Corrosion potentials (E_{cor}) and corrosion current densities (i_{cor}) from the polarization curves.

Time	F3		F4		B1(T)		B2(T)	
	E_{cor} V$_{SCE}$	i_{cor} µA cm^{-2}	E_{cor} V$_{SCE}$	i_{cor} µA cm^{-2}	E_{cor} V$_{SCE}$	i_{cor} µA cm^{-2}	E_{cor} V$_{SCE}$	i_{cor} µA cm^{-2}
1 h	−0.128	1.4	−0.129	1.0	−0.145	1.0	−0.137	1.5
24 h	−0.130	0.6	−0.098	0.8	−0.111	0.7	−0.144	1.1
10 days	−0.092	1.0	−0.086	0.7	−0.117	0.7	−0.058	0.3

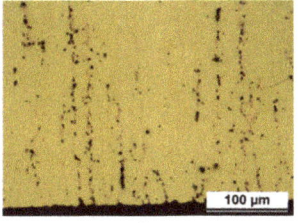

Figure 9. Sections of B1(T) (**a**) and B1(L) (**b**) samples after 150 days of exposure to SDW. Dezincification is only localized on the β' phase.

Figure 10 shows the SEM-EDS elemental mapping obtained on B1(T) cross sections. It confirms extensive oxidation of dezincified regions and reveals the penetration of chlorides well inside the dealloyed regions, with local concentrations up to 0.3 wt. %. Extensive plug dezincification of β' phase was detected on F3 and F4, with maximum depths of about 170 and 100 µm, respectively. Chloride penetration and inner oxidation were detected as well.

Figure 11 shows some cross sections of B2(T and L) obtained after 10 and 150 days. It shows that after 10 days corrosion initiates in correspondence of the κ phase (red arrows in Figure 11) and along grain boundaries (blue arrows in Figure 11). In fact, microelectrochemical tests showed that Si-rich κ and particularly γ phases are more active than α, so justifying the observed corrosion behavior [17].

At longer immersion times (Figure 11c,d), uniform dezincified layers were observed (maximum thickness about 60 µm), affecting both κ and α phases, with morphology and penetration depth independent of exposed surface orientation. The SEM-EDS elemental map on the cross section of a B2(T) sample after 150 days immersion in SDW is reported in Figure 12. In this sample, dealloyed regions contained only copper, negligible amounts of oxygen and silicon, and no chlorides, suggesting that on this alloy selective corrosion produced rather compact copper regions.

The SEM-EDS analyses of surface corrosion products on B1(T) specimens after 150 day immersions revealed the presence of significant surface arsenic concentrations (up to 0.69 wt. %), besides chlorine (up to 0.34 wt. %), oxygen, and sulfur (Figure 13a,b). This indicated the presence of surface chlorides, oxides, and perhaps sulfates and the incorporation of arsenate or arsenite among corrosion products [43]. On B2(T) (Figure 13c,d), SEM-EDS technique detected the same corrosion products (except arsenic species obviously), with addition of some silicates in agreement with literature information [16] and phosphates, as suggested by phosphorous concentrations up to 2.1 wt. %. Chloride concentrations reached 1.7 wt. %.

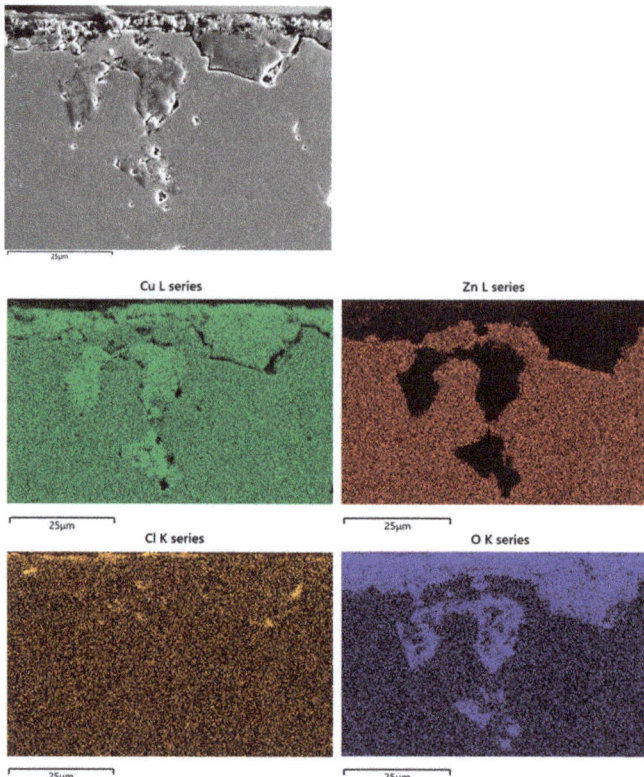

Figure 10. SEM-EDS elemental maps obtained on sections of B1(T) after 150 days of exposure to SDW.

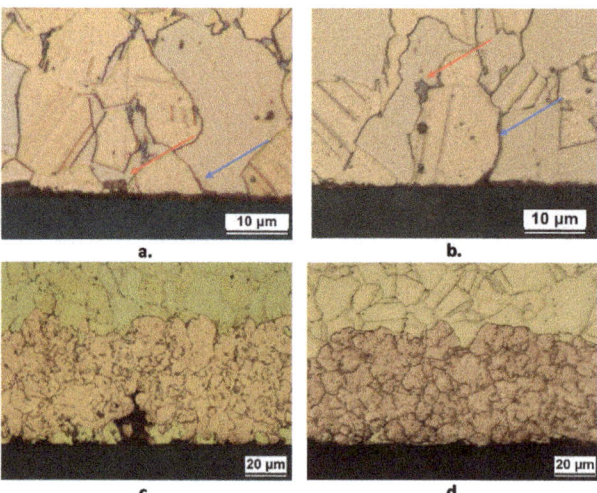

Figure 11. Sections of B2(T) (**a,c**) and B2(L) (**b,d**) samples after 10 days (**a,b**) and 150 days (**c,d**) of exposure to SDW. Surface etching by 10 wt. % FeCl$_3$ solution evidences the location of corrosion attack initiation with reference to the alloy microstructure: corrosion starts in correspondence with the κ phase (red arrows) and along grain boundaries (blue arrows).

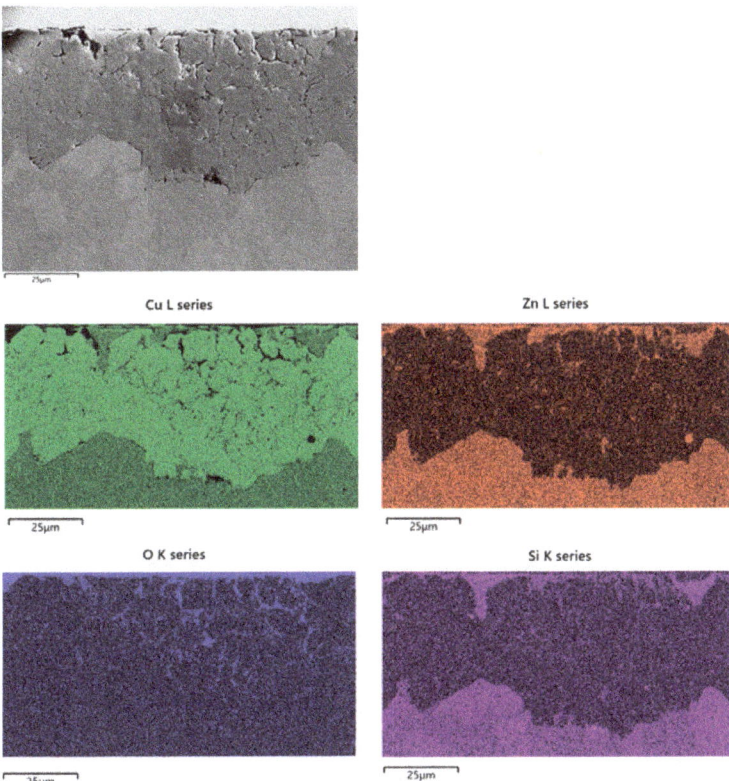

Figure 12. SEM-EDS elemental maps obtained on sections of B2(T) after 150 days of exposure to SDW.

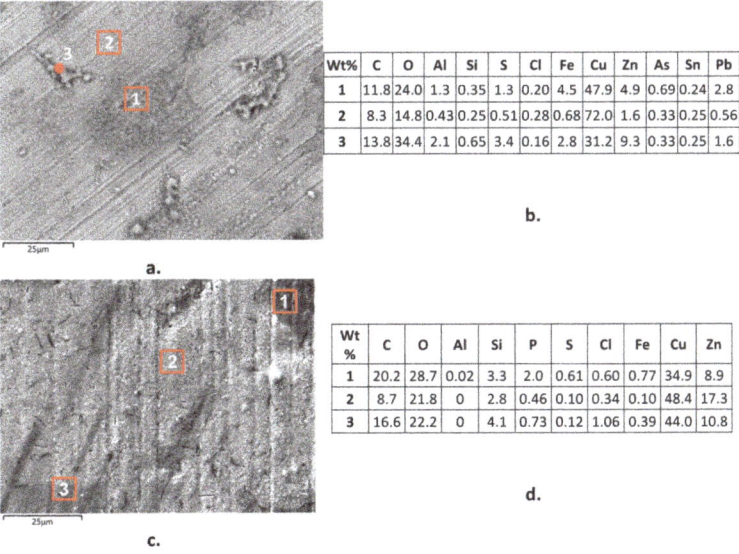

Figure 13. Surface corrosion products (**a**,**c**) and their local SEM-EDS analyses (**b**,**d**) on B1(T) (**a**,**b**) and B2(T) (**c**,**d**) samples after 150 days of exposure to SDW.

The presence of As/Cu and P/Cu ratios in the corrosion products higher or much higher than those in B1 and B2 alloys, respectively, denoted an alloy depletion in these inhibiting elements, so contributing to explain the significant dezincification attacks of these alloys.

4. Discussion

The early failures detected on the F1–7 components of drinking water distribution systems in Morocco are mostly connected to their high contents of β′ phase, which undergoes dezincification and selective dissolution attacks (Figure 4). On three failed brass fittings, dezincification was also detected in concomitance with cracks, suggesting that dealloying favors crack propagation (Figure 4). The standard dezincification test UNI EN ISO 6509 confirmed the high susceptibility of F1–7 alloys to this corrosion form, while both CuZn36Pb2As (B1(L and T)) and CuZn21Si3P (B2(L and T)) proved to be much more resistant.

During immersions in SDW, arsenic alloying in B1 could prevent α phase dezincification but did not completely avoid selective zinc dissolution in β′ phase, in agreement with many other authors' results [10,19–21]. The relatively high arsenic concentrations in surface corrosion products of B1(T) samples could have produced alloy depletion in this element and could contribute to decrease the alloy dezincification resistance. Instead, B1(L) confirmed an excellent overall dezincification resistance, because dezincification attack of β′ was detected only at a short distance from the exposed surface. On B1, the dezincified regions were composed of copper and oxidized copper with appreciable chlorine concentrations in the pores, suggesting easy solution penetration.

B2(T) and B2(L) showed quite similar corrosion behavior and proved to be more susceptible to dezincification than expected. In fact, although the maximum depth of dealloying attack was only slightly higher than that on B1(L), B2 samples exhibited complete dezincification of both α and κ. In this alloy too, the element capable to inhibit dealloying, that is phosphorous, was found to leach from the metal surfaces and resulted incorporated in the corrosion products at relatively high concentrations. In B2, selective dissolution initiated on the less noble κ phase and on grain boundaries (Figure 10) and then propagated to the α phase. This is in agreement with the low difference in practical nobility between these phases (50 mV) documented by other authors [17]. SEM-EDS analyses showed that the dealloyed regions produced from α and κ grains were constituted by pure copper with negligible zinc, oxygen, and silicon concentrations and no chlorine. Spreading of dealloying from κ to α grains could be due to the onset of a galvanic couple between the large copper areas produced by selective dissolution (acting as cathodic regions) and the α phase (acting as anodic region, due to its significant zinc and silicon contents which make it less noble than pure copper (Figure 3)). The absence of oxidized silicon species inside pure copper regions is surprising, because silicon is an active element which tends to oxidize to low solubility hydrated silicon oxides. The absence of this element inside dealloyed regions would suggest the occurrence of local solution alkalinization and silicate ion leaching, as a result of copper surface cathodic activity. Galvanic coupling would also justify the corrosion protection of copper from re-oxidation in dezincified regions.

The electrochemical tests showed that charge transfer and mainly mass transport controlled the corrosion rates of the studied brass. In particular, diffusion affected only the cathodic process at short immersion times, while with time it also slowed down the anodic reaction. In all cases, the surface films of corrosion products offered negligible if any protection from corrosion propagation, even in the case of B1 and B2 samples, where the surface film resistance reached the highest values (up to 2.9 kΩ cm^2). In addition, also in these cases, the film resistance remained much smaller than the diffusion resistance, R_W, by one or two orders of magnitude.

5. Conclusions

Early failures on α + β′ brass components of drinking water distribution systems in Morocco were analyzed. They consisted of severe corrosion attacks and cracks, often accompanied by significant dezincification. Dealloying or selective phase dissolution affected the β′ phase, present at high volume fractions (from 41% to 69%).

The corrosion behavior of two of these α + β′ alloys was compared to that of CuZn36Pb2As and CuZn21Si3P alloys marketed as dezincification resistant, during 150 day immersions in SDW. The electrochemical tests showed that at long immersion times, the general corrosion resistance of these alloys was similar, with polarization resistance values differing within a factor of 2–2.7.

Final dealloying depths on CuZn21Si3P (longitudinal and transverse sections of extruded bar) and CuZn36Pb2As (longitudinal section) were about 2–3 times lower than those on samples from failed components.

After 150 days of immersion in SDW, dezincification of CuZn36Pb2As and traditional alloys only affected the β′ phase, while on CuZn21Si3P dealloying produced zinc and silicon leaching from both α and κ phases. Dealloying started on the Si-rich κ phase and at grain boundaries, then it spread to α phase too, likely due to galvanic coupling with the nobler dealloyed copper regions.

Author Contributions: Conceptualization J.C., C.M., F.Z. and M.E.T.; investigation J.C., V.G., A.B.; resources I.M.; writing original draft J.C., C.M., F.Z. and M.E.T.

Funding: This research received no external funding.

Acknowledgments: The authors wish to thank the National Office of Electricity and the Potable Water, Rabat, Morocco for the cooperation and support during the starting of this research study.

Conflicts of Interest: The authors declare no conflict of interest.

References

1. Koch, G.H.; Brongers, M.P.H.; Thompson, N.G.; Virmani, Y.P.; Payer, J.H. Chapter 1—Cost of corrosion in the United States. In *Handbook of Environmental Degradation of Materials*, 1st ed.; Elsevier: Amsterdam, The Netherlands, 2005; pp. 3–24.
2. Brossia, S. Chapter 23—Corrosion of Pipes in Drinking Water Systems. In *Handbook of Environmental Degradation of Materials*, 3rd ed.; Elsevier: Amsterdam, The Netherlands, 2018; pp. 489–505.
3. Zanotto, F.; Grassi, V.; Balbo, A.; Monticelli, C.; Melandri, C.; Zucchi, F. Effect of brief thermal aging on stress corrosion cracking susceptibility of LDSS 2101 in the presence of chloride and thiosulphate ions. *Corros. Sci.* **2018**, *130*, 22–30. [CrossRef]
4. Latva, M.; Kaunisto, T.; Pelto-Huikko, A. Durability of the non-dezincification resistant CuZn40Pb2 brass in Scandinavian waters. *Eng. Fail. Anal.* **2017**, *74*, 133–141. [CrossRef]
5. Pantazopoulos, G.A.; Toulfatzis, A.I. Failure analysis of a machinable brass connector in a boiler unit installation. *Case Stud. Eng. Fail. Anal.* **2013**, *1*, 18–23. [CrossRef]
6. Mapelli, C.; Gruttadauria, A.; Bellogini, M. Analysis of the factors involved in failure of a brass sleeve mounted on an electro-valve. *Eng. Fail. Anal.* **2010**, *17*, 431–439. [CrossRef]
7. Galai, M.; Choucri, J.; Hassani, Y.; Benqlilou, H.; Mansouri, I.; Ouaki, B.; Ebn Touhami, M.; Monticelli, C.; Zucchi, F. Moisture content and chloride ion effect on the corrosion behavior of fitting brass (gate valves) used as a connection of PVC's conduits in aggressive sandy soil. *Chem. Data Collect.* **2019**, *19*, 100171. [CrossRef]
8. Doostmohammadi, H.; Moridshahi, H. Effects of Si on the microstructure, ordering transformation and properties of the Cu 60 Zn 40 alloy. *J. Alloys Compd.* **2015**, *640*, 401–407. [CrossRef]
9. Vilarinho, C.; Davim, J.P.; Soares, D.; Castro, F.; Barbosa, J. Influence of the chemical composition on the machinability of brasses. *J. Mater. Process. Technol.* **2005**, *170*, 441–447. [CrossRef]
10. Davies, D.D. *A Note on the Dezincification of Brass and the Inhibiting Effect of Elemental Additions*; Copper Development Association Inc.: New York, NY, USA, 1993.
11. Ismail, K.M.; Elsherif, R.M.; Badawy, W.A. Effect of Zn and Pb contents on the electrochemical behavior of brass alloys in chloride-free neutral sulfate solutions. *Electrochim. Acta* **2004**, *49*, 5151–5160. [CrossRef]
12. Kumar, S.; Narayanan, T.S.; Manimaran, A.; Kumar, M.S. Effect of lead content on the dezincification behaviour of leaded brass in neutral and acidified 3.5% NaCl solution. *Mat. Chem. Phys.* **2007**, *106*, 134–141. [CrossRef]

13. La Fontaine, A.; Keast, V.J. Compositional distributions in classical and lead-free brasses. *Mater. Charact.* **2006**, *57*, 424–429. [CrossRef]
14. You, S.-J.; Choi, Y.-S.; Kim, J.-G.; Oh, H.-J.; Chi, C.-S. Stress corrosion cracking properties of environmentally friendly unleaded brasses containing bismuth in Mattson's solution. *Mater. Sci. Eng. A* **2003**, *345*, 207–214. [CrossRef]
15. Nobel, C.; Hofmann, U.; Klocke, F.; Veselovac, D.; Puls, H. Application of a new, severe-condition friction test method to understand the machining characteristics of Cu–Zn alloys using coated cutting tools. *Wear* **2015**, *344–345*, 58–68. [CrossRef]
16. Nikolaychuk, P.A.; Tyurin, A.G. Thermodynamic evaluation of corrosion electrochemical behaviour of silicon brass CuZn17Si3. *Inorg. Mater.* **2013**, *49*, 457–467. [CrossRef]
17. Seuss, F.; Gaag, N.; Virtanen, S. Corrosion mechanism of CuZn21Si3P in aggressive tap water. *Mater. Corros.* **2017**, *68*, 42–49. [CrossRef]
18. Zhou, P.; Hutchison, M.J.; Erning, J.W.; Scully, J.R.; Ogle, K. An in situ kinetic study of brass dezincification and corrosion. *Electrochim. Acta* **2017**, *229*, 141–154. [CrossRef]
19. Lucey, V.F. The mechanism of dezincification and the effect of arsenic. I. *Br. Corros. J.* **1965**, *1*, 9–14. [CrossRef]
20. Lucey, V.F. The mechanism of dezincification and the effect of arsenic. II. *Br. Corros. J.* **1965**, *1*, 53–59. [CrossRef]
21. Zou, J.Y.; Wang, D.H.; Qiu, W.C. Solid-state diffusion during the selective dissolution of brass: Chronoamperometry and positron annihilation study. *Electrochim. Acta* **1997**, *42*, 1733–1737. [CrossRef]
22. Karpagavalli, R.; Balasubramaniam, R. Influence of arsenic, antimony and phosphorus on the microstructure and corrosion behavior of brasses. *J. Mater. Sci.* **2007**, *42*, 5954–5958. [CrossRef]
23. Gebel, T. Arsenic and antimony: Comparative approach on mechanistic toxicology. *Chem. Biol. Interact.* **1997**, *107*, 131–144. [CrossRef]
24. Copeland, R.C.; Lytle, D.A.; Dionysiou, D.D. Desorption of arsenic from drinking water distribution system solids. *Environ. Monit. Assess.* **2007**, *127*, 523–535. [CrossRef]
25. Kondrashin, V.Y. To the Theory of Anticorrosion Alloying of Brasses. *Prot. Met.* **2005**, *41*, 138–145. [CrossRef]
26. Tabrizi, U.; Parvizi, R.; Davoodi, A.; Moayed, M.H. Influence of heat treatment on microstructure and passivity of Cu–30Zn–1Sn alloy in buffer solution containing chloride ions. *Bull. Mater. Sci.* **2012**, *35*, 89–97. [CrossRef]
27. Sohn, S.; Kang, T. The effects of tin and nickel on the corrosion behavior of 60Cu–40Zn alloys. *J. Alloys Compd.* **2002**, *335*, 281–289. [CrossRef]
28. Pugacheva, N.B. Structure of commercial α + β brasses. *Met. Sci. Heat Treat.* **2007**, *49*, 67–74. [CrossRef]
29. Hung, C.-Y.; Lin, C.-M.; Hsieh, C.-C.; Li, C.-C.; Wu, P.T.-Y.; Chen, K.-T.; Wu, W. A novel approach to improving resistance to dezincification of diphasic brass. *J. Alloys Compd.* **2016**, *671*, 502–508. [CrossRef]
30. Kawashima, A.; Agrawal, A.K.; Staehle, R.W. Stress corrosion cracking of admiralty brass in nonammoniacal sulfate solutions. *J. Electrochem. Soc.* **1977**, *124*, 1822–1823. [CrossRef]
31. Torchio, S. The stress corrosion cracking of admiralty brass in sulphate solutions. *Corros. Sci.* **1986**, *26*, 133–151. [CrossRef]
32. Brandl, E.; Malke, R.; Beck, T.; Wanner, A.; Hack, T. Stress corrosion cracking and selective corrosion of copper-zinc alloys for the drinking water installation. *Mater. Corros.* **2009**, *60*, 251–258. [CrossRef]
33. Zucchi, F.; Trabanelli, G.; Fonsati, M.; Giusti, A. Influence of P, As and Sb on the susceptibility to SCC of α-β' brasses. *Mater. Corros.* **1998**, *49*, 864–869. [CrossRef]
34. Monticelli, C.; Balbo, A.; Esvan, J.; Chiavari, C.; Martini, C.; Zanotto, F.; Marvelli, L.; Robbiola, L. Evaluation of 2-(salicylideneimino) thiophenol and other Schiff bases as bronze corrosion inhibitors by electrochemical techniques and surface analysis. *Corros. Sci.* **2019**, *148*, 144–158. [CrossRef]
35. Bostan, R.; Varvara, S.; Găină, L.; Muresan, L.M. Evaluation of some phenothiazine derivatives as corrosion inhibitors for bronze in weakly acidic solution. *Corros. Sci.* **2012**, *63*, 275–286. [CrossRef]
36. Brug, G.J.; van den Eeden, A.L.G.; Sluyters-Rehbach, M.; Sluyters, J.H. The analysis of electrode impedances complicated by the presence of a constant phase element. *J. Electroanal. Chem.* **1984**, *176*, 275–295. [CrossRef]
37. Abedini, M.; Ghasemi, H.M. Corrosion behavior of Al-brass alloy during erosion—corrosion process: Effects of jet velocity and sand concentration. *Mater. Corros.* **2016**, *67*, 513–521. [CrossRef]
38. Trabanelli, G.; Monticelli, C.; Grassi, V.; Frignani, A. Electrochemical study on inhibitors of rebar corrosion in carbonated concrete. *Cem. Concr. Res.* **2005**, *35*, 1804–1813. [CrossRef]

39. Nagiub, A.; Mansfeld, F. Evaluation of corrosion inhibition of brass in chloride media using EIS and ENA. *Corros. Sci.* **2001**, *43*, 2147–2171. [CrossRef]
40. Rochdi, A.; Kassou, O.; Dkhireche, N.; Touir, R.; El Bakri, M.; Ebn Touhami, M.; Sfaira, M.; Mernari, B.; Hammouti, B. Inhibitive properties of 2,5-bis(n-methylphenyl)-1,3,4-oxadiazole and biocide on corrosion, biocorrosion and scaling controls of brass in simulated cooling water. *Corros. Sci.* **2014**, *80*, 442–452. [CrossRef]
41. Jie, H.; Xu, Q.; Wei, L.; Min, Y. Etching and heating treatment combined approach for superhydrophobic surface on brass substrates and the consequent corrosion resistance. *Corros. Sci.* **2016**, *102*, 251–258. [CrossRef]
42. Žerjav, G.; Milošev, I. Corrosion protection of brasses and zinc in simulated urban rain Part I: Individual inhibitors benzotriazole, 2-mercaptobenzimidazole and stearic acid. *Mater. Corros.* **2015**, *66*, 1402–1413. [CrossRef]
43. Pryor, M.J.; Giam, K. The Effect of Arsenic on the Dealloying of α-Brass. *J. Electrochem. Soc.* **1982**, *129*, 2157–2163. [CrossRef]

© 2019 by the authors. Licensee MDPI, Basel, Switzerland. This article is an open access article distributed under the terms and conditions of the Creative Commons Attribution (CC BY) license (http://creativecommons.org/licenses/by/4.0/).

Article

Atmospheric Corrosion Sensor Based on Strain Measurement with an Active Dummy Circuit Method in Experiment with Corrosion Products

Nining Purwasih [1,2], Naoya Kasai [1,*], Shinji Okazaki [3], Hiroshi Kihira [4] and Yukihisa Kuriyama [5]

1. Department of Risk Management and Environmental Sciences, Graduate School of Environmental and Information Sciences, Yokohama National University, Yokohama 240-8501, Japan; nining.purwasih@eng.unila.ac.id
2. Department of Electrical Engineering, Faculty of Engineering, Lampung University, Bandar Lampung 35141, Indonesia
3. Division of Material Science and Engineering, Graduate School of Engineering, Yokohama National University, Yokohama 240-8501, Japan; okazaki-shinji-yp@ynu.ac.jp
4. Nippon Steel Research Institute, Tokyo 100-0005, Japan; kihira.hiroshi@nsri.nssmc.com
5. RACE (Research into Artifacts, Center for Engineering), University of Tokyo, Japan; kuriyama@race.u-tokyo.ac.jp
* Correspondence: kasai-naoya-pf@ynu.ac.jp; Tel.: +81-45-339-3979

Received: 22 April 2019; Accepted: 16 May 2019; Published: 18 May 2019

Abstract: This study analyzed an atmospheric corrosion sensor using strain measurements (ACSSM) with an active dummy method for corrosion product experiments. An initial compensation thermal strain experiment was performed with elapsed time. Further analyses used dry-wet environments with salt water spray to investigate the thickness reduction performance of the corrosion product on low-carbon steel samples. The ACSSM with an active dummy method accurately measured signals induced by the specimen thickness reduction, despite the noise in the signal. Moreover, the effects of corrosion products on the signal were discussed.

Keywords: atmospheric corrosion; strain measurement; mild steel; corrosion product

1. Introduction

Because corrosion damages materials in structures located outside, studies on atmospheric corrosion processes have received special attention from researchers. Morcillo et al [1] studied the nature of corrosion products, the mechanisms, and kinetics of the corrosion process, the morphology of steel rust, and long-term atmospheric corrosion monitoring. Song and Saraswathy [2] reported on methods and durability problems for monitoring corrosion in reinforced concrete structures. Wen et al [3] studied carbon steel corrosion products and their classification, effects on subsequent corrosion processes, and the dependence of the initiation, growth, and transformation processes in H_2S environments on such products. Ahmad [4] investigated reinforcement corrosion in concrete structures including the mechanism of reinforcement corrosion, techniques for monitoring reinforcement corrosion, and methodologies to predict the remaining service lifetimes of structures.

In addition, many researchers have developed sensors to detect atmospheric corrosion using radio-frequency identification (RFID) sensors [5–7], passive wireless sensors [8], corrosion potential sensors [9], optic sensors, fibre Bragg gratings (FBGs) [10–15], atmospheric corrosion monitoring (ACM) methods [16,17], and atmospheric corrosion rate monitoring (ACRM) techniques [18].

Several techniques for atmospheric corrosion monitoring, such as weight and thickness loss measurement [19–22], electrochemical impedance spectroscopy (EIS) [23–25], microscopy and scanning electron microscopy (SEM) [26,27], and X-ray diffraction (XRD) [28,29] have been established as necessary.

A strain measurement method using strain gauges for in situ corrosion monitoring was proposed in our previous study [30,31]. However, precise monitoring was difficult because the strain measurement method was very sensitive to temperature drift. Thus, a highly accurate in situ method for monitoring atmospheric corrosion remains necessary. The strain-measurement circuit using the active dummy method for an atmospheric corrosion sensor based on strain measurement (ACSSM) with strain gauges accommodated such temperature drifts and showed good performance in the situ monitoring of specimen thinning under corrosion, as measured by galvanostatic electrolysis [32–34]. However, for actual applications, experiments with dry-wet method [35–38] to investigate the performance of the corrosion product/rust under strain behavior.

This study conducted experiments with the dry-wet method by applying a 5% NaCl solution to test pieces in order to investigate the performance of ACSSM via the active dummy method. Thickness changes, obtained from the weight loss of the specimens, were monitored simultaneously. Based on the experimental findings, the performance of the ACSSM and the effect of corrosion products on the signal were discussed.

2. Methods

2.1. Concept of the ACSSM

According to [30,31], the strain on the compressive surface of a low-carbon (mild) steel test piece under a bending moment can be expressed by:

$$\varepsilon = -h/2\rho \qquad (1)$$

where ε is the strain in the test piece ($\times 10^{-6}$ ε), ρ is the radius of curvature of the test piece (mm), h is the test piece thickness (mm), and $d\theta$ is the center angle of curvature of the test piece. Figure 1 shows the mechanical principle for the ACSSM sensor in normal and bent conditions. Figure 1a shows the condition of the test piece without thickness reduction. When the test piece thickness is decreased by corrosion, as shown in Figure 1b, being thickness change of the test piece Δh, using the assumption $\rho >> h$, $\left(\rho - \frac{h}{2}\right) = \rho$, the change in strain ($\varepsilon$) can be expressed as:

$$\Delta h = 2\rho \Delta \varepsilon \qquad (2)$$

Equation (2) is the fundamental equation to determine the thickness reduction of the test piece from the strain measurement in ACSSM.

2.2. Concept of the ACSSM with an Active Dummy Circuit Method

To observe the thickness reduction by corrosion of the test piece based on the strain measurement, accurate measurement is necessary, because the thickness reduction of the test piece in a short time period is generally $\leq 5 \times 10^{-6}$ m.

Environmental factors, such as temperature variations, during measurement affect not only the test piece and sensor, but also device elements such as the operation amplifier, causing measurement drift.

Therefore, to observe small thickness reductions caused by corrosion, the active dummy circuit method was proposed. In this study, the active dummy method consists of not only active and dummy sensors but also active and dummy circuits for strain measurement. Figure 2 shows the concept of ACSSM with an active dummy method. The active output (ε_A) of an active gauge of an active circuit included the drift of the sensor from environmental factors. Therefore, the dummy output (ε_D) of the dummy gauge of the active circuit was used to eliminate the effects of environmental factors during the measurement.

The output of the dummy circuit is the difference between the outputs of active and dummy strain gauges; to eliminate the drift from the circuit itself, the output of the dummy circuit was used. The strain (Δε) was finally obtained as the difference between the output of the active and dummy circuits with the difference circuit. In the study, two active and two dummy strain gauges were employed to enhance the resolution of the measured strain.

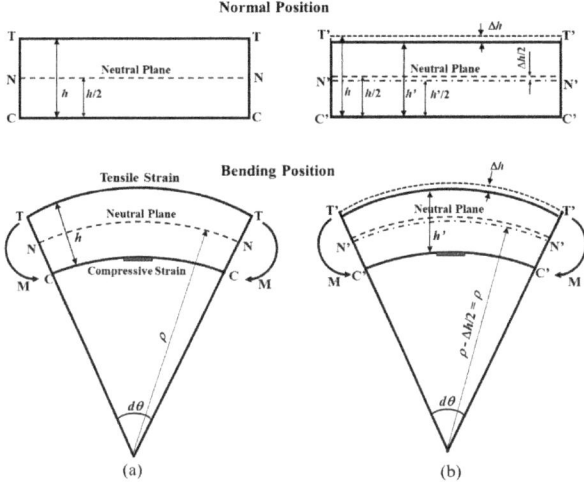

Figure 1. The illustration of test piece in normal position and bending position: (**a**) non-corroded test piece, (**b**) corroded test piece. T(T'), N(N'), and O(O') denote the tensile, neutral, and compressive planes. M(M') denotes the bending moment.

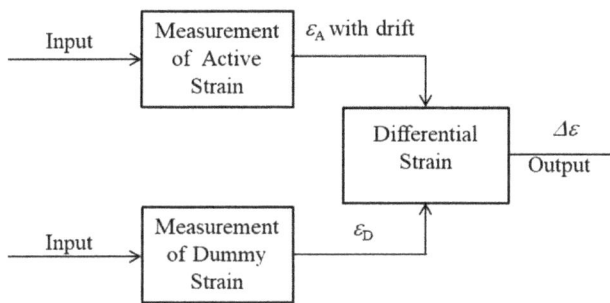

Figure 2. The concept of ACSSM with an active dummy method.

2.3. Design of the ACSSM with an Active Dummy Method

As mentioned above, two active and two dummy strain gauges were employed. The gain of the strain measurement circuit using the gain resistor (R_G) of 4 Ω with the active dummy method was 12,500. The relationship between the voltage and strain in the master curve was measured [31]. Using the R_G of 4 Ω, relationship between the strain and voltage was linear, following the equation $\Delta\varepsilon = -11 + 27\,\Delta V$, where ΔV is the output voltage of the strain measurement circuit before it is converted to $\Delta\varepsilon$. The slope of the equation was used to convert the output voltage of the circuit to strain, which was $27 \times 10^{-6}\,\varepsilon$ for 1 V.

2.4. Experimental Setup for Dry-Wet Measurement by ACSSM

The test piece of the sensor is 95 mm in length, 45 mm in width, and 0.5 mm in thickness, made of low-carbon steel, has a corroded area of 1350 mm^2, as shown in Figure 3. The ACSSM device comprises a base and cover with ρ = 430 mm, as in reference [30]. The test piece is placed in the apparatus and the edges of the test piece are fixed.

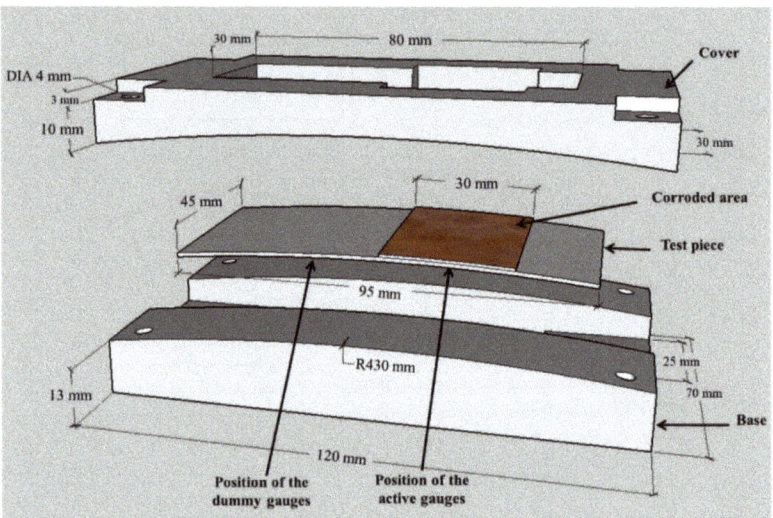

Figure 3. Test piece and apparatus for atmospheric corrosion sensor using strain measurements (ACSSM). The test piece is inserted in the apparatus, which consists of a base and cover with the corroded area measuring 45 mm in length and 30 mm in width.

Eight strain gauges were attached to the back side of the test piece in the configuration as shown in Figure 4. The two active gauges (R_{AA}) of the active circuit were attached beneath the corroded area and two dummy (R_{AD}) gauges of the active circuit are attached beneath the uncorroded area. In the dummy circuit, four strain gauges (R_{DA} and R_{DD}), equal to the number of strain gauges in the active circuit, were used and attached beneath the non-corroded area.

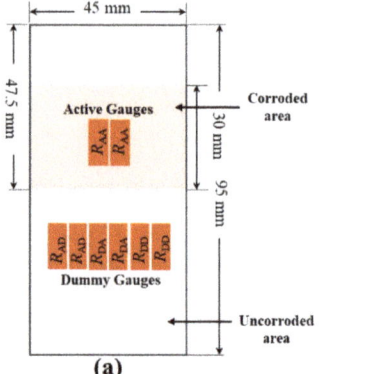 

Figure 4. (a) Configuration of strain gauges in test piece with 1350 mm^2 corroded area; the remaining area is not corroded. (b) Photograph of strain gauge attached to the test piece.

$\Delta\varepsilon$ between ε_A and ε_D is used to evaluate the thickness reduction from the corrosion. The strain measurement circuit having an active circuit, dummy circuit and differential circuit [31] were fabricated by authors. As strain gauge, FLA-5-11 (Tokyo Sokki, Tokyo, Japan), the ACSSM were used. The input voltage of bridge circuits of active and dummy circuits was 3 Volt.

Figure 5 shows the experimental set-up of the dry-wet condition using strain gauges. The data logger GL7000 (Graphtec, Yokohama, Japan) monitors the output voltage from the active, dummy, and differential circuits every 10 min and measures the temperature simultaneously with a thermocouple. Using the relationship between strain and voltage given in Section 2.3 and the relationship between strain and thickness using the mechanical calculation in [15], $\Delta h = 0.86\ \Delta\varepsilon \times 10^{-6}$ m, Δh is obtained with Equation (2).

In the experiment, a 5.0 wt % NaCl (salt) solution is periodically applied to the test piece in the sensor under dry-wet conditions. The experiment is performed in two stages: the initial measurement before spraying the salt solution to investigate the compensation of environmental factors, and the dry-wet condition with sprayed salt solution. To discuss thickness reduction based on the strain measurement, several specimens of test piece material as coupon were prepared and periodically sprayed with 5.0 wt % NaCl solution. The test piece and coupons were sprayed once a day using around 3.75 mL of 5.0 wt % NaCl solution.

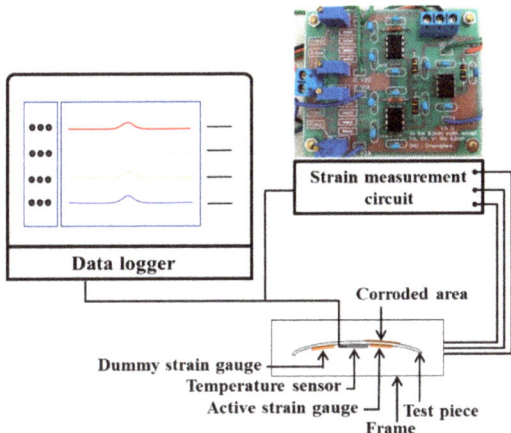

Figure 5. Experimental set-up of under dry-wet condition for ACSSM.

3. Results and Discussion

Figure 6 shows the experiment results under dry-wet condition using ACSSM for 15 days before salt solution is applied. The results indicate the effects and the compensation of environmental factors on the signal. The ε_A follows the temperature signal of the test piece (T_{TP}) and has the same tendency as the ε_D. T_{TP} varies over approximately 0 to 20°C for ε_A around $100 \times 10^{-6}\ \varepsilon$; the drift is $5 \times 10^{-6}\ \varepsilon/°C$. Despite using the active and dummy strain gauge method. Meanwhile, $\Delta\varepsilon$ is more constant around $25 \times 10^{-6}\ \varepsilon$ with a drift of is $1.25 \times 10^{-6}\ \varepsilon/°C$. This corresponds to a 75% decreased in drift from environmental factors. The ACSSM measurement system with the active dummy circuit method is therefore robust against environmental factors, such as temperature variations, during measurement.

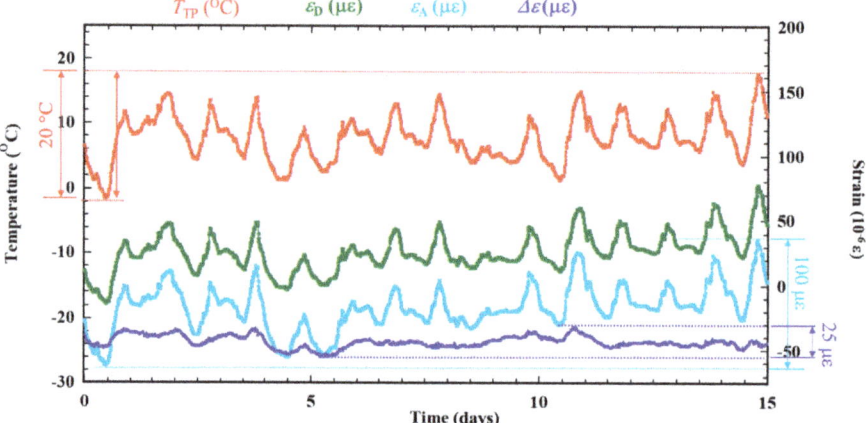

Figure 6. Compensation of thermal strain experiment signal (T_{TP}) before under dry-wet condition are applied.

Figure 7 shows the results of dry-wet exposure with ACSSM. Three stages are observed in the result. Stage I is the initial condition for 15 days before spraying with salt water. This stage is shown as Figure 6. Stage II is the condition after spraying salt water, in which corrosion products are generated. $\Delta\varepsilon$ shows a negative trend, indicating that the test piece thickness is increased by the corrosion products. Stage III is the condition of further corrosion progression, which causes thickness reduction by corrosion of the corrosion products. $\Delta\varepsilon$ shows a positive trend. According to Equation (2), this indicates that the test piece thickness is decreased by corrosion, including that of corrosion products. As the corrosion of the test piece continues after salt-solution spraying, the variations of $\Delta\varepsilon$ become large. It happened because of corrosion progress, two R_{AA} under the corroded area and the two R_{DD} under non-corroded area show different behavior, although these show the same behavior in Stage I. Compared to R_{DD}, R_{AA} under the corrosion product react slowly to temperature variations, because the corrosion product functions as a thermal insulator. Therefore, the difference in the balance of the measurement system bridge circuit is large, yielding a larger variation of $\Delta\varepsilon$.

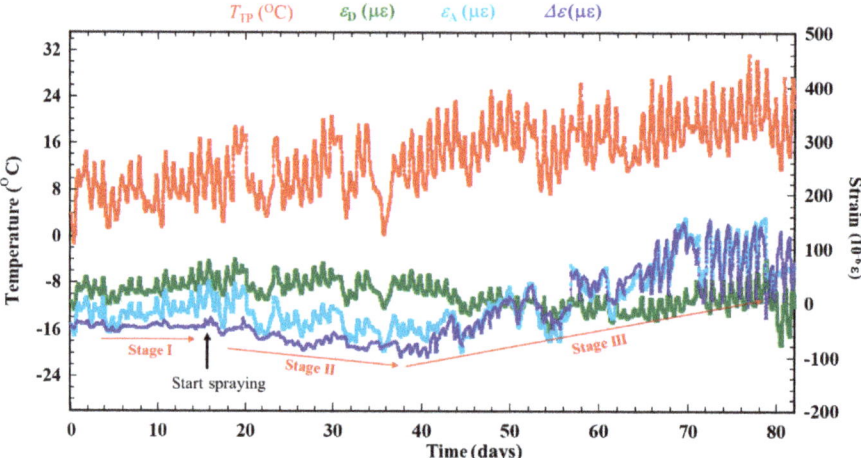

Figure 7. Result of a dry–wet cyclic exposure with ACSSM.

$\Delta \varepsilon$ is converted to Δh, and Δh is applied by a moving average analysis using intervals of 200 data sets to obtain the signal more clearly without deleting the trends of each stage. Moreover, Δh is moved to zero value become Δh_{offset} (10^{-6} m). The result is shown in Figure 8. Simultaneously, the test piece results are obtained, and the thickness reductions of the specimens calculated by the mass loss are indicated in Figure 8. The thickness reductions show good correlation that measured with ACSSM. However, the evaluation errors between the result measured with ACSSM and the thickness of specimen as shown in Table 1 were 14.5×10^{-6} m. This error is calculated using the average of the last four data points from each method.

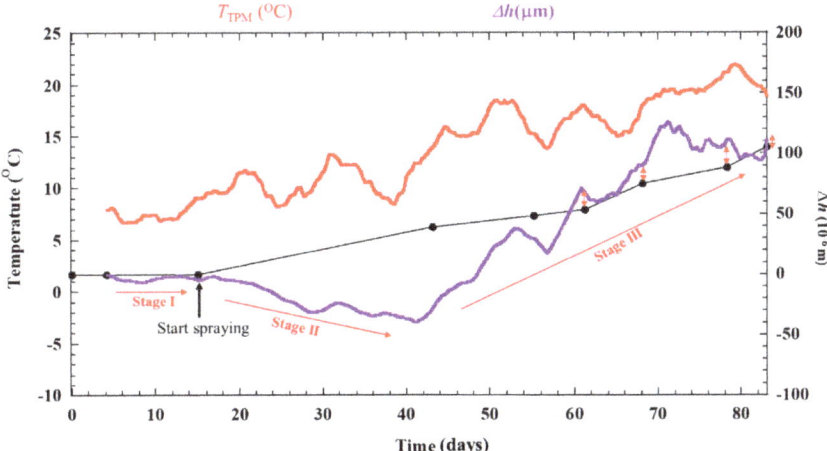

Figure 8. Result from dry-wet cyclic condition with ACSSM in 83 days after moving-average analysis and applying the offset value. The green line with double arrows indicated the differences of the last four dataset collected from the ACSSM and specimens.

Table 1. Error calculation between ACSSM and test specimens.

Δh_w (10^{-6} m)	Δh_{offset} (10^{-6} m)	Error (10^{-6} m)
54	68	14
75	90	15
88	111	23
105	111	6
Average		14.5

Figure 9 shows photograph of the evolution of the corrosion product on the test piece for 83 days. After spraying salt solution, the corrosion of the test piece progresses. These pictures are similar to the appearance of the specimen surfaces at the same times.

Although ACSSM can measure only the thickness reduction of the test piece, a schematic of the mechanism of corrosion behavior measured with ACSSM was illustrated. It is shown in Figure 10. Stage I is the initial stage, with no corrosion product on the surface of the test piece and constant $\Delta \varepsilon$.

In stage II, a tight corrosion product would be generated. $\Delta \varepsilon$ shows a slight decreasing tendency because of this tight corrosion product, indicating that the test piece thickness would be increased by the tight corrosion product. The increased thickness of the test piece measured with ACSSM is approximately 43×10^{-6} m for 25 days. In stage III, the corrosion of the test piece would yield a porous structure with continued corrosion, and the test piece thickness is decreased because the porous structure of corrosion product receives bending moment. The thickness reduction of the test piece measured with ACSSM is approximately 111×10^{-6} m in 43 days. As shown above, the monitoring with the ACSSM needs sufficient time after stage III to measure thickness reduction of the test piece.

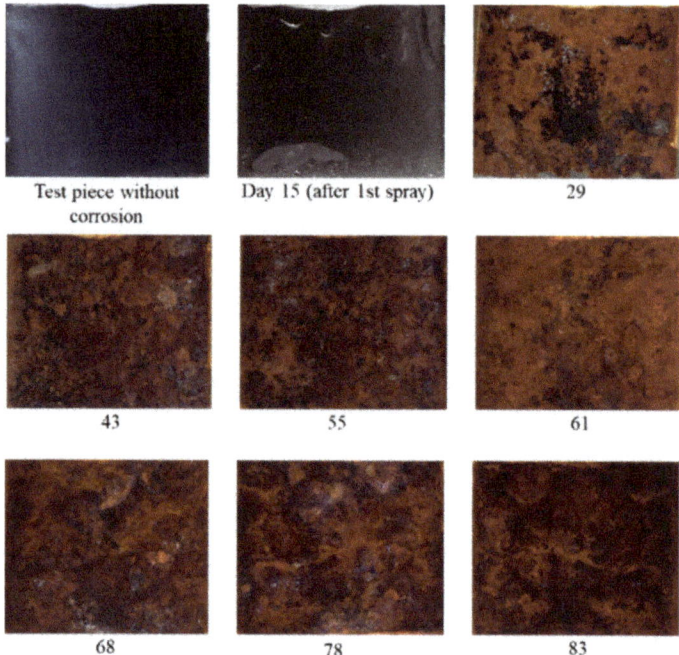

Figure 9. Picture of corrosion evolution for 83 days of measurement. Strain are from the sprayed salt water, shown in the picture at Day 15.

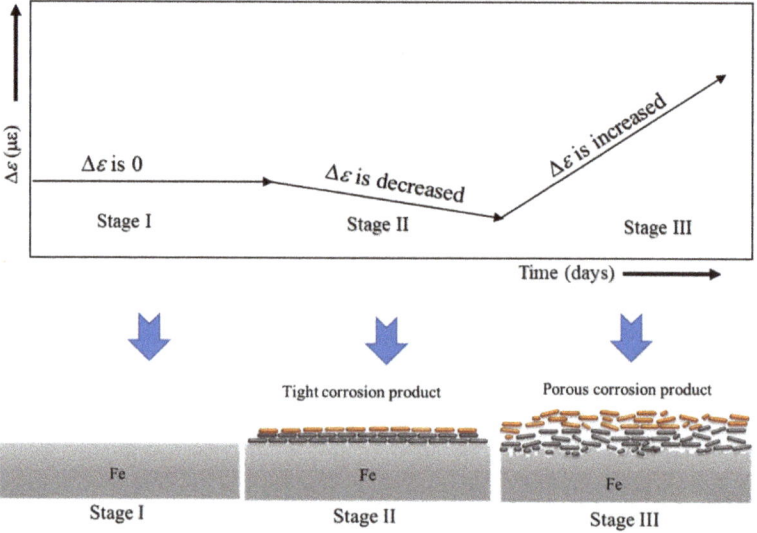

Figure 10. Mechanism of corrosion behavior based on strain measurements.

4. Conclusions

An ACSSM with an active dummy method, proposed by authors, was used for experimentation with the dry-wet method in long-term monitoring. The conclusions of this study are as follows:

- In stage I of the experiment, $\Delta\varepsilon$ had a relative constant signal, with drift decreasing from 5×10^{-6} $\varepsilon/°C$ to 1.25×10^{-6} $\varepsilon/°C$ under temperature variations.
- In stage II, $\Delta\varepsilon$ showed a negative trend, indicating the increased thickness of the test piece measured by ACSSM. This was attributed to the tight corrosion product formed on the test piece measured by ACSSM.
- In stage III, showed a positive trend because the produced corrosion product was porous. The accuracy of h was determined from the thickness reduction of the coupons. Thus, ACSSM can be used for atmospheric corrosion monitoring in the field.
- The sensor applies for atmospheric corrosion, for general corrosion. The sensor can be measured the strain that has relation with the thickness of test piece and finally we can calculate the corrosion rate. The local corrosion condition in the test piece of the sensor affects the accuracy of the corrosion rate estimation. This problem might be considered in a future study.

Author Contributions: Conceptualization, H.K.; methodology, S.O. and N.P.; software, N.K.; validation, H.K., N.K. and Y.K.; formal analysis, N.K. and N.P.; investigation, N.P.; resources, N.K.; data curation, Y.K. and S.O.; writing—original draft preparation, N.P.; writing—review and editing, N.K., Y.K. and N.P.; visualization, S.O.; supervision, N.K.; project administration, N.K.; funding acquisition, N.K.

Funding: This work was supported by a Grant-in-Aid for Scientific Research (B) JSPS KAKENHI Grant No. 16H03132 and Yokohama National University.

Acknowledgments: The authors would like to thank to the Indonesian Directorate General of Higher Education for their financial support of the author's study.

Conflicts of Interest: The authors declare there is no conflict of interest. The supporters did not have a role in the design of the study; in the collection, analyses, and interpretation of data; in the writing of the manuscript, or in the decision to publish the results.

Nomenclature

ρ	Radius of curvature of the test piece (m)
h	Test piece in thickness (m)
ε	Strain in the test piece (-)
$d\theta$	Centre angle of curvature of the test piece (°)
Δh	Change of thickness of the test piece (m)
h'	Test piece in thickness due to the corrosion (m)
ε_A	Strain of active circuit ($\times 10^{-6}$ ε)
ε_D	Strain of dummy circuit ($\times 10^{-6}$ ε)
$\Delta\varepsilon$	Difference in strain in the test piece due to the corrosion ($\times 10^{-6}$ ε) Difference in strain between ε_A and ε_D ($\times 10^{-6}$ ε)
E	Young's modulus of the test piece (Pa)
σ_y	Yield stress of the test piece (Pa)
V_{IN}	Input voltage for bridge circuit (V)
R_{AA}	Resistance of active gauge for active circuit (Ω)
R_{DA}	Resistance of dummy gauge for active circuit (Ω)
R_{AD}	Resistance of active gauge for dummy circuit (Ω)
R_{DD}	Resistance of dummy gauge for dummy circuit (Ω)
ΔV	Different output voltage of active and dummy circuit (V)
T_{TP}	Temperature of test piece (°C)
T_{TPM}	Temperature of test piece after applied moving average analysis (°C)
Δh_w	Actual thickness from weight analysis of coupons ($\times 10^{-6}$ m)

References

1. Morcillo, M.; De Fuente, D.; Diaz, I.; Cano, H. Atmospheric corrosion of mild steel. *Rev. Metal.* **2011**, *47*, 426–444.
2. Song, H.W.; Saraswathy, V. Corrosion monitoring of reinforced concrete structures—A review. *Int. J. Electrochem. Sci.* **2007**, 1–28.
3. Wen, X.; Bai, P.; Luo, B.; Zheng, S.; Chen, C. Review of recent progress in the study of corrosion products of steels in a hydrogen sulphide environment. *Corros. Sci.* **2018**, *139*, 124–140.
4. Ahmad, S. Reinforcement corrosion in concrete structures, its monitoring and service life prediction—A review. *Cem. Concr. Compos.* **2003**, *25*, 459–471. [CrossRef]
5. Alamin, M.; Tian, G.Y.; Andrews, A.; Jackson, P. Corrosion detection using low-frequency RFID technology. *Insight-Non-Destr. Test. Cond. Monit.* **2012**, *54*, 72–75.
6. Zhang, H.; Yang, R.; He, Y.; Tian, G.Y.; Xu, L.; Wu, R. Identification and characterization of steel corrosion using passive high frequency RFID sensors. *Measurement* **2016**, *92*, 421–427.
7. Zhang, J.; Tian, G.Y.; Marindra, A.M.J.; Sunny, A.I.; Zhao, A.B. A review of passive RFID tag antenna-based sensors and systems for structural health monitoring applications. *Sensors* **2017**, *17*, 265. [CrossRef]
8. Yasri, M.; Gallee, F.; Lescop, B.; Diler, E.; Thierry, D.; Rioual, S. Passive wireless sensor for atmospheric corrosion monitoring. In Proceedings of the 8th European Conference on Antennas and Propagation (EuCAP), The Hague, The Netherlands, 6–11 April 2014; pp. 2945–2949.
9. Perveen, K.; Bridges, G.E.; Bhadra, S.; Thomson, D.J. Corrosion potential sensor for remote monitoring of civil structure based on printed circuit board sensor. *IEEE Trans. Instrum. Meas.* **2014**, *63*, 2422–2431. [CrossRef]
10. Almubaied, O.; Chai, H.K.; Islam, M.R.; Lim, K.; Tan, C.G. Monitoring corrosion process of reinforced concrete structure using FBG strain sensor. *IEEE Trans. Instrum. Meas.* **2017**, *66*, 2148–2155. [CrossRef]
11. Tan, C.H.; Adikan, F.R.M.; Shee, Y.G.; Yap, B.K. Non-destructive fiber Bragg grating based sensing system: Early corrosion detection for structural health monitoring. *Sens. Actuators A Phys.* **2017**, *268*, 61–67. [CrossRef]
12. Hassan, M.R.A.; Bakar, M.H.A.; Dambul, K.; Adikan, F.R.M. Optical-based sensors for monitoring corrosion of reinforcement rebar via an etched cladding Bragg grating. *Sensors (Basel)* **2012**, *12*, 15820–15826. [CrossRef] [PubMed]
13. Hu, W.; Ding, L.; Zhu, C.; Guo, D.; Yuan, Y.; Ma, N.; Chen, W. Optical fiber polarizer with Fe-C film for corrosion monitoring. *IEEE Sens. J.* **2017**, *17*, 6904–6910.
14. Chen, W.; Dong, X. Modification of the wavelength-strain coefficient of FBG for the prediction of steel bar corrosion embedded in concrete. *Opt. Fiber Technol.* **2012**, *18*, 47–50. [CrossRef]
15. Al Handawi, K.; Vahdati, N.; Rostron, P.; Lawand, L.; Shiryayev, O. Strain based FBG sensor for real-time corrosion rate monitoring in pre-stressed structures. *Sens. Actuator B* **2016**, *236*, 276–285. [CrossRef]
16. Dara, T.; Shinohara, T.; Umezawa, O. The Behavior of corrosion of low carbon steel affected by corrosion product and Na2SO4 concentration under artificial rainfall test. *Zairyo Kankyo* **2016**, *C-114*, 298–302.
17. Monsada, A.M.; Margarito, T.; Milo, L.C.; Casa, E.P.; Zabala, J.V.; Maglines, A.S.; Basilia, B.A.; Harada, S.; Shinohara, T. Atmospheric corrosion exposure study of the Philippine historical all-steel Basilica. *Zair. Kankyo* **2016**, *C-107*, 267–270.
18. Mansfeld, F.; Jeanjaquet, S.L.; Kendig, M.W.; Roe, D.K. A new atmospheric corrosion rate monitor development and evaluation. *Atmos. Environ.* **1986**, *20*, 1179–1192.
19. Ridha, M.; Fonna, S.; Huzni, S.; Supardi, J.; Ariffin, A.K. Atmospheric corrosion of structural steels exposed in the 2004 tsunami-affected areas of Aceh. *IJAME* **2013**, *7*, 1014–1022. [CrossRef]
20. Parson, N.; Khamsuk, P.; Sorachot, S.; Khonraeng, W.; Wongpinkaew, K.; Kaewkumsai, S.; Pongsaksawad, W.; Viyanit, E.; Chianpairot, A. Atmospheric corrosion of structural steels in Thailand Tropical Climate. *Zairyo Kankyo* **2016**, *C-108*, 271–274.
21. Lien, L.T.H.; Hong, H.L.; San, P.T.; Hieu, N.T.; Nga, N.T.T. Atmospheric corrosion of carbon steel and weathering steel—Relation of corrosion and environmental factors. *Zairyo Kankyo* **2016**, *C-110*, 280–284.
22. Odara, T.; Tahara, A.; Dara, T. Atmospheric corrosion behaviors of steels in Japan. *Zairyo Kankyo* **2016**, *C-111*, 285–288.

23. Shitanda, I.; Okumura, A.; Itagaki, M.; Watanabe, K.; Asano, Y. Screen printed atmospheric corrosion monitoring sensor based on electrochemical impedance spectroscopy. *Sens. Actuators B Chem.* **2009**, *139*, 292–297.
24. Li, C.; Ma, Y.; Li, Y.; Wang, F. EIS monitoring study of atmospheric corrosion under variable relative. *Corros. Sci.* **2010**, *52*, 3677–3686. [CrossRef]
25. Thee, C.; Dong, J.; Ke, W. Corrosion monitoring of weathering steel in a simulated coastal-industrial environment. *Int. J. Environ. Chem. Ecol. Eng.* **2015**, *9*, 587–593.
26. Luo, J. Corrosion of the galvanized steel bolts of overhead catenary system in the tunnel areas. *Proc. JSCE Mater Environ.* **2016**, 293–297.
27. Hu, W.; Cai, H.; Yang, M.; Tong, X.; Zhou, C.; Chen, W. Fe-C-coated fiber Bragg grating sensor for steel corrosion monitoring. *Corros. Sci.* **2011**, *53*, 1933–1938.
28. Zang, N.; Chen, W.; Zheng, X.; Hu, W.; Gao, M. Optical sensor for steel corrosion monitoring based on etched Fiber Bragg Grating sputtered with iron film. *IEEE Sens. J.* **2015**, *15*, 3511–3556.
29. Ke, W.; Dong, J.H.; Chen, W. Corrosion evolution of steel simulated of SO2 polluted coastal atmospheres. *Zair. Kankyo.* **2016**, *C-112*, 289–292.
30. Kasai, N.; Hiroki, M.; Yamada, T.; Kihira, H.; Matsuoka, K.; Kuriyama, Y.; Okazaki, S. Atmospheric corrosion sensor based on strain measurement. *Meas. Sci. Technol.* **2016**, *28*, 015106.
31. Purwasih, N.; Kasai, N.; Okazaki, S.; Kihira, H. Development of amplifier circuit by active dummy method for atmospheric corrosion monitoring on steel. *Metals* **2018**, *8*, 1–12.
32. Abbas, Y.; Nutma, J.S.; Olthuis, W.; Van Den Berg, A. Corrosion monitoring of reinforcement steel using galvanostatically induced potential transients. *IEEE Sens. J.* **2016**, *16*, 693–698.
33. Strosnijder, M.F.; Brugnoni, C.; Laguzzi, G.; Luvidi, L.; De Cristofaro, N. Atmospheric corrosion evaluation of galvanised steel by thin layer activation. *Corros. Sci.* **2004**, *46*, 2355–2359.
34. Portella, M.O.G.; Portella, K.F.; Pereira, P.A.M.; Inone, P.C.; Brambilla, K.J.C.; Cabussú, M.S.; Cerqueira, D.P.; Salles, R.N. Atmospheric corrosion rates of copper, galvanized steel, carbon steel and aluminum in the metropolitan region of Salvador, BA, Northeast Brazil. *Proc. Eng.* **2012**, *42*, 171–185. [CrossRef]
35. Yadav, A.P.; Suzuki, F.; Nishikata, A.; Tsuru, T. Investigation of atmospheric corrosion of Zn using ac impedance and differential pressure meter. *Electrochim. Acta* **2004**, *49*, 2725–2729. [CrossRef]
36. El-Mahdy, G.A. Atmospheric corrosion of copper under wet / dry cyclic conditions. *Corro. Sci.* **2005**, *47*, 1370–1383.
37. Kiosidou, E.D.; Karantonis, A.; Sakalis, G.N.; Pantelis, D.I. Electrochemical impedance spectroscopy of scribed coated steel after salt spray testing. *Corros. Sci.* **2018**, *137*, 127–150.
38. Dillmann, P.; Mazaudier, F.; Hoerle, S. Advances in understanding atmospheric corrosion of iron Rust characterisation of ancient ferrous artefacts exposed to indoor atmospheric corrosion. *Corros. Sci.* **2004**, *46*, 1401–1429.

© 2019 by the authors. Licensee MDPI, Basel, Switzerland. This article is an open access article distributed under the terms and conditions of the Creative Commons Attribution (CC BY) license (http://creativecommons.org/licenses/by/4.0/).

Article

Electrochemical Corrosion Behavior of Fe₃Al/TiC and Fe₃Al-Cr/TiC Coatings Prepared by HVOF in NaCl Solution

Najmeh Ahledel [1], Robert Schulz [2], Mario Gariepy [3], Hendra Hermawan [1] and Houshang Alamdari [1,*]

[1] Department of Mining, Metallurgical and Materials Engineering, Université Laval, Québec, QC G1V 0A6, Canada; najmeh.ahledel.1@ulaval.ca (N.A.); hendra.hermawan@gmn.ulaval.ca (H.H.)
[2] Hydro-Quebec Research Institute, 1800 Boulevard Lionel Boulet, Varennes, QC J3X 1S1, Canada; Schulz.Robert@ireq.ca
[3] Wärtsilä Canada, Inc., American Hydro, 8600 St-Patrick, LaSalle (Montréal), QC H8N 1V1, Canada; mario.gariepy@ahydro.com
* Correspondence: Houshang.Alamdari@gmn.ulaval.ca; Tel.: +1-418-656-7666

Received: 20 December 2018; Accepted: 9 April 2019; Published: 13 April 2019

Abstract: Adding TiC particles into iron aluminide coatings has been found to improve its wear resistance, but its corrosion behavior is less known. In this study, the corrosion behavior of Fe₃Al/TiC and Fe₃Al-Cr/TiC composite coatings, prepared by high velocity oxy fuel (HVOF) spraying, was studied in 3.5 wt. % NaCl solution by means of electrochemical techniques and surface analysis. Results revealed that adding TiC particles into Fe₃Al matrix to improve the wear resistance does not deteriorate the corrosion behavior of Fe₃Al coating. It was also showed that addition of chromium to Fe₃Al/TiC composite provides a more protective layer.

Keywords: corrosion; electrochemical impedance spectroscopy; high velocity oxy fuel coatings; iron aluminide; titanium carbide

1. Introduction

Iron aluminides have received great attention as materials with high potential for a number of industrial applications. This is basically due to their low density, excellent oxidation resistance, hot corrosion resistance, high specific strength and low ductile to brittle transition temperature [1,2]. Their application is, however, limited by their low ductility, poor creep resistance [3] and low wear resistance [4]. It has been shown that the incorporation of ceramic particles into iron aluminide matrix improves its tribological properties [5,6]. For example, the composite coating of iron aluminide with tungsten carbide particles exhibited higher erosive wear-resistance than did the iron aluminide coating alone. Among all ceramic materials, titanium carbide (TiC), owing to its excellent mechanical, chemical and thermal properties, has been identified as a good reinforcing phase to improve the mechanical properties of the aluminide matrix [7,8]. Chen et al. [9] noted that Fe-Al intermetallics with TiC reinforcement have excellent dry sliding wear resistance. In addition to laser cladding, several other techniques, such as plasma spraying [10–13], wire arc spraying [14] and high-velocity oxy-fuel (HVOF) projection [15] have also been used to deposit Fe-Al alloys on carbon steels or stainless steel substrates. HVOF is a convenient process to deposit thick coatings on variety of substrates with superior properties at low cost [16]. The process utilizes a combination of oxygen with various fuel gases including hydrogen, propane, propylene and kerosene as the feed of the combustion chamber while the spray powder (mixed metal or oxides) comes through from other side. The combustion of the gases provides the temperature and pressure needed to flow the gases through the nozzle. With the flame temperature ranging between 2500 and 3200 °C, powder particles partially or completely melt during the flight through the nozzle and due to the presence of oxygen and high temperature, oxidation also occurs.

Although iron aluminides are mainly developed for applications at high temperatures, they also exhibit a high potential for low-temperature applications. These intermetallics are good candidates to replace stainless steel in several applications, i.e., pipes and tubes for heating elements and main components for distillation and desalination plants [17]. A number of studies were thus concentrated on the aqueous corrosion behavior of these materials to test their durability in various corrosive environments. The corrosion behavior of iron aluminides has been studied in several acidic and basic solutions [18–24]. Chiang et al. [18] evaluated the passivation behavior of Fe-Al alloys containing 3.4, 10.4, 18.7, 19.4, 29.5, and 41.7 at. % Al. They showed that Fe-Al alloys, with an Al content exceeding 19 at. %, have wide passivation regions and low passivation current in 0.1 N H_2SO_4. In addition, when the Al content of Fe-Al alloys exceeds this limit, further increment of Al content has only a slight influence on passivation. Sharma et al. [19] compared the stability of the passive film and pitting behavior of Fe-28Al (at. %) and Fe-28Al-3Cr (at. %) with AISI SS 304 under different pH conditions to evaluate their performance in acidic, basic, and neutral solutions. They reported that the presence of 3 at. % Cr in iron aluminides improves their aqueous corrosion resistance and makes them comparable to AISI SS 304. Porcayo-Calderon et al. [20] studied corrosion behavior of FeAl and Fe_3Al coatings prepared by two thermal spray techniques, flame spraying and HVOF using three different particle sizes and compared their corrosion behavior with that of the base alloys in 1.0 M NaOH solution at room temperature. The coatings prepared using medium particle size and flame spray and those using fine particle size and HVOF spray were shown to be more stable, uniform, denser, and with lower porosity, thus exhibited a greater corrosion resistance. Grosdidier et al. [21] prepared nanocrystalline Fe-40Al coatings by HVOF with varied amount of hard un-melted particles of feedstock powder and found that feedstock powder size has a strong effect on the coating hardness. The electrochemical response of the coatings revealed that these particles were the reason of poor corrosion resistance compared to bulk material. Analysis of corrosion damage showed a prevalent localized attack at intersplat boundaries or around un-melted powder particles, probably enhanced by galvanic phenomena. Amiriyan et al. [4,25] reported that the Vickers hardness and the dry sliding wear resistance of Fe_3Al/TiC composite coatings, at sliding speeds ranging from 0.04 to 0.8 m·s^{-1} and under a constant load of 5 N, increased as the amount of TiC particles in Fe_3Al matrix increased. They also compared the phase composition, microstructure, micro hardness and elastic modulus of Fe_3Al and Fe_3Al/TiC composite coatings, however the authors did not report the corrosion behavior of these coatings. Related to corrosion behavior, Rao [23] proposed a mechanistic model for re-passivation of iron aluminide in comparison to pure Al and Fe, based on his investigation using a rapid scratched electrode technique in 0.25 M H_2SO_4.

Both the mechanical or electrochemical properties of Fe-Al intermetallics can be improved by alloying elements in the form of solid solutions [26–30]. The addition of 6 at. % Cr improved the ductility of Fe_3Al up to 10% [31]. A few mechanisms were proposed to explain the effect of Cr on mechanical and corrosion properties of Fe_3Al. First, it has been postulated that Cr in solid solution facilitates the dislocation cross-slipping and solid solution softening [32,33]. Epelboin et al. [34] also showed that Cr affects the surface properties through the contribution of chromium oxides to the formation of passive layers and the decrement of reaction kinetics. The decrease of water reduction reaction rate may also lead to the reduction of hydrogen evolution and thus mitigating the hydrogen embrittlement [35].

In this work, we studied the corrosion behavior of Fe_3Al/TiC composite coatings, prepared by HVOF technique. These coatings had shown improved wear resistance [4,25] however, their corrosion behavior had not been reported. The effect of Cr on the passivity of Fe_3Al/TiC composite in 3.5 wt. % NaCl solution was revealed.

2. Materials and Methods

2.1. Coating Preparation

The starting materials in powder form were prepared by mechanically alloying commercial iron aluminide (Fe_3Al, 96% purity), titanium (Ti, 99.4% purity) and chromium (Cr, 99% purity) (Alfa Aesar, Haverhill, MA, USA), and graphite (C, 96% purity, Asbury Graphite Mills, Asbury, NJ, USA). The nominal composition of the feedstock powder before milling is given in Table 1.

Table 1. Nominal composition of the feedstock powder before milling.

Sample	at. % of Elements				
	Fe	Al	Ti	C	Cr
Fe_3Al	75	25	0	0	0
Fe_3Al/TiC	50	16.66	16.66	16.66	0
Fe_3Al-Cr/TiC	37.5	12.5	12.5	12.5	25

A 300 g batch of each powder mixture was placed in a high-energy ball mill (Zoz GmbH, Wenden, Germany, Simoloyer CM08) and milled under argon atmosphere for 6 h. Hardened steel balls (52100 heat-treated steel with a hardness of 60–64 Rockwell C) and jar were used and a ball to powder weight ratio of 10:1 was chosen for milling. The XRD patterns of the milled powder, were provided in other work by Amiriyan et al. [4,25]. The milled powders were then heat-treated at 1000 °C for 2 h under 10^{-6} mbar of vacuum. This post-treatment was performed in order to ensure the maximum reaction between the additive elements. In fact, the ball milling provides a mixture with a good homogeneity and the reaction between additive elements starts during milling. However, to complete these reactions, very long milling times are required. The post-treatment at 1000 °C helps the completion of these reactions. Based on the X-Ray diffraction patterns, the peaks related to the additive elements are disappeared after this post treatment. Titanium carbide starts forming by mechano-chemical reaction between titanium and graphite during milling and this reaction progresses during the subsequent heat treatment and the final product is a composite of Fe_3Al matrix and TiC particles [4]. The composite powders were deposited on mild steel plates (AISI 1020) using a JP-8000 HVOF spray system (Praxair Surface Technologies, Indianapolis, IN, USA) with spray parameters listed in Table 2. The substrates were sand-blasted and then washed with acetone and ethanol prior to HVOF deposition. Argon and kerosene were used as the carrier gas and the fuel during the HVOF deposition, respectively. A coating with 150–200 micron thickness was deposited on the substrate. Coated samples were rinsed with ethanol before corrosion test and no polishing was performed.

Table 2. High-velocity oxy-fuel (HVOF) spraying parameters.

Oxygen flow rate (m^3/s)	1.5×10^{-2}
Kerosene flow rate (m^3/s)	5.57×10^{-6}
Carrier gas	Argon
Spraying distance (m)	0.38
Number of deposition passes	5

2.2. Electrochemical Tests

The open circuit potential (OCP) and potentiodynamic polarization tests were carried out using a VersaSTAT3 Potentiostat (Ametek Princeton Applied Research, Oak Ridge, TN, USA) in a conventional three-electrode cell. A Princeton model K0235 flat cell, accommodating wide range of electrode shapes and sizes, was used in order to test different areas of a similar sample without cutting it. An Ag/AgCl was used as reference electrode and a platinum mesh as a counter electrode. All tests were performed in an electrolyte containing 3.5 wt. % NaCl with magnetic stirring at 60 rpm at room temperature

(25 ± 2 °C). The OCP was measured over 20 h of immersion. Potentiodynamic polarization was conducted from −0.7 to +2 V with respect to the OCP, using a scanning rate of 1 mV/s. Before starting the experiments, the electrodes were left for one hour in the solution to stabilize at their free corrosion potential. The surface of samples were washed with ethanol before testing. All experiments were repeated three times.

Electrochemical impedance spectroscopy (EIS) was carried out using a Reference 3000 potentiostat/galvanostat (Gamry Instruments, Warminster, PA, USA). Electrodes were mounted in epoxy resin and assembled in a three-electrode teflon holder. A platinum foil with a surface of 1 cm^2 was used as cathode and an Ag/AgCl (KCl sat.) as the reference electrode. Prior to EIS experiments, one-hour stabilization time was considered then the frequency scanning was performed from 10^{-2} to 10^5 Hz with a root-mean square potential amplitude of 5 mV. The interpretation of EIS results was made with the help of ZSimpWin software (Ametek Princeton Applied Research, Oak Ridge, TN, USA).

2.3. Surface Analysis

The cross-sectional morphology of the coatings was studied after performing the potentiodynamic polarization tests by using a scanning electron microscope (SEM, JSM-840A, JEOL Inc., Peabody, MA, USA) equipped with an energy dispersive X-ray spectroscopy (EDS, SwiftED 3000, Oxford Instruments, Concord, MA, USA). Further surface chemistry analysis was performed by using an X-ray photoelectron spectroscopy (XPS, Axis-Ultra, Kratos Analytical, Manchester, UK), using an incident X-ray radiation of Al Kα under vacuum (10^{-12} bar). The surface film was analyzed at a take-off angle of 30°. Narrow multiple scans were recorded with 160 and 1 eV step sizes. High resolution spectra were recorded at pass energies of 40 and 20 eV and step sizes of 0.1 and 0.05 eV. Apparent relative concentrations were calculated using the CasaXPS software (Casa Software, Teignmouth, UK) with the appropriate sensitivity factors.

3. Results and Discussion

3.1. Electrochemical Behavior

The variation of OCP of metallic materials as a function of time gives valuable information about film formation and passivation. A rise of potential in the positive direction indicates the formation of a protective film, a steady potential indicates the presence of an intact and protective film, and a drop of potential in the negative direction indicates breakage or dissolution of the film, or no film formation at all [36]. The OCP vs Ag/AgCl reference electrode plots of all tested coatings (Figure 1a) show that at the early moments of immersion the potential rapidly shifts towards negative, indicating the dissolution of the oxide layer existing before immersion. The potential then experienced a steady state and no change was observed until the end of the experiment over 20 h, indicating an equilibrium established between the corrosion and the formation of an oxide film on the surface. The Fe$_3$Al coating exhibited more negative potential value than did of Fe$_3$Al/TiC and Fe$_3$Al-Cr/TiC coatings. Adding TiC shifted the absolute potential of Fe$_3$Al to less negative value. This indicates that for composite coatings, corrosion via dissolving the electrode surface decreased.

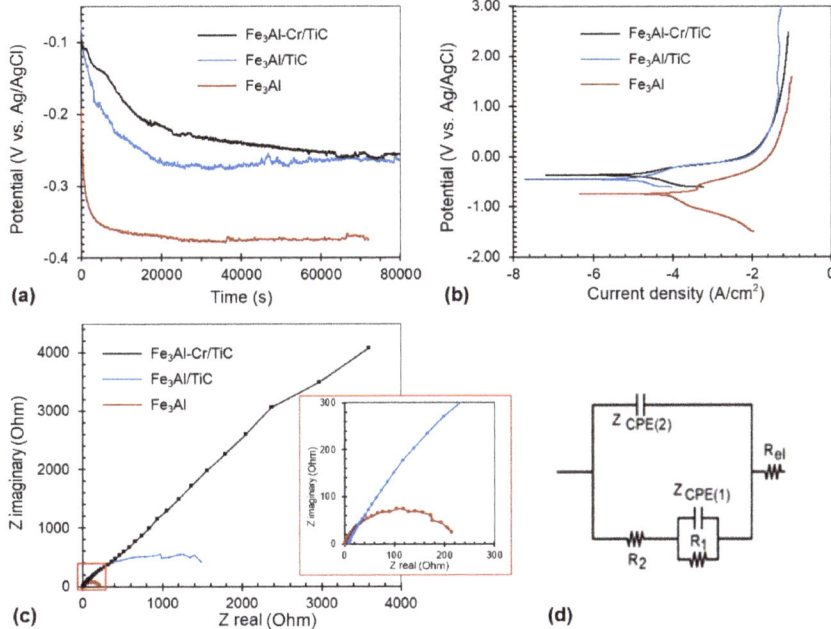

Figure 1. Electrochemical test results of Fe$_3$Al, Fe$_3$Al/TiC and Fe$_3$Al-Cr/TiC in freely aerated 3.5 wt. % NaCl solution: (**a**) open circuit potential measured for up to 20 h, (**b**) potentiodynamic polarization curves, (**c**) Nyquist plots, and (**d**) proposed equivalent circuit model.

The three coatings show a similar shape of polarization curves (Figure 1b). A difference is observed as the composite coatings (Fe$_3$Al/TiC and Fe$_3$Al-Cr/TiC) show a nobler corrosion potential (E_{corr}) and a lower corrosion current density (j_{corr}) than that of the Fe$_3$Al. This observation is in accordance with the OCP results. Although the current density remains constant beyond a certain potential range, the curves do not demonstrate an active-passive behavior, and no breakdown potential can be observed even at relatively high potentials. The values of corrosion parameters obtained from the polarization curve, such as cathodic Tafel (β_c) and anodic Tafel (β_a) slopes, and polarization resistance (R_p) are presented in Table 3. The results indicate that adding TiC particles to Fe$_3$Al matrix decreases the anodic and cathodic current density and the value of E_{corr} shifts to the less negative direction. In composite coatings, smaller values of j_{corr} and greater amounts of R_p compared to Fe$_3$Al, revealed that the corrosion performance of composite coatings in 3.5% NaCl solution has been improved by adding TiC. A decrease of calculated corrosion rate (CR) is also shown by the composite coating which further decreasing by the addition of Cr into the Fe$_3$Al/TiC.

Table 3. Corrosion parameters obtained from the polarization curves.

Sample	β_c (mV·dec^{-1})	E_{corr} (mV)	β_a (mV·dec^{-1})	j_{corr} (μA·cm^{-2})	R_p (Ω·cm^2)	CR (mm·y^{-1})
Fe$_3$Al	350	−780	250	83.8	15.3	7.7 × 10^{-9}
Fe$_3$Al/TiC	380	−480	380	13.9	300	1.4 × 10^{-9}
Fe$_3$Al-Cr/TiC	170	−350	250	9.9	205	0.4 × 10^{-9}

The electrochemical impedance spectroscopy (EIS) test results, represented in the form of Nyquist plots (Figure 1c), show two types of diagrams. The first type, as also shown in magnified inset of (Figure 1c) is related to Fe$_3$Al and Fe$_3$Al/TiC samples and consists of two overlapped capacitive loops. The time constant at high frequencies is attributed to the corrosion product layer while the second at intermediate and low frequencies is ascribed to the faradaic charge transfer process [37]. Compared to Fe_3Al, the composite coatings exhibit higher impedance values, indicating that the corrosion products layer formed on the surface is more protective than that formed on Fe_3Al [38]. This result is in good agreement with the results obtained by potentiodynamic tests. The second type (the black curve in the main diagram of Figure 1c) is related to Fe_3Al-Cr/TiC showing a much larger diameter of the arc, indicating larger impedance due to the addition of Cr into Fe_3Al/TiC. The diagram consists of a small semicircle, tended to be a straight line with a slope of around 45° (after Z_{real} = 500). Since a slope higher than 45° was reported to be the characteristic of a diffusion process, corresponding to a concentration gradient localized in a porous layer and in the solution [38], this response for Fe_3Al-Cr/TiC could therefore be the characteristic of a diffusion process while the semicircle curves of Fe_3Al and Fe_3Al/TiC are attributed to a charge transfer process. In order to provide quantitative support to the experimental EIS results, an equivalent circuit was proposed (Figure 1d) and the related impedance parameters were fitted and calculated (Table 4).

Table 4. Impedance fitting parameters.

Sample	R_{el} ($\Omega\cdot cm^2$)	R_2 ($\Omega\cdot cm^2$)	$(CPE_2)\ Y_{02}$ ($S\cdot cm^{-2}\cdot s^n$)	n_2	R_1 ($\Omega\cdot cm^2$)	$(CPE_1)\ Y_{01}$ ($S\cdot cm^{-2}\cdot s^n$)	n_1
Fe_3Al	1.9	187.3	7.9×10^{-3}	0.8	27.7	70×10^{-3}	1
Fe_3Al-TiC	1.8	402.1	4.2×10^{-3}	0.68	160	4.7×10^{-3}	0.6
Fe_3Al-Cr/TiC	7.2	1979	8.7×10^{-6}	0.9	16.8×10^3	0.8×10^{-3}	0.8

In the proposed equivalent circuit model, R_{el} corresponds to the resistance of the electrolyte while R_2 and CPE_2 represent the resistance and the capacitance of the corrosion product layer, respectively. R_1 represents the charge transfer resistance, and CPE_1 represents the double-layer capacitance. A constant phase element (CPE) was used to consider a deviation from an ideal capacitor. The origins of the CPE were summarized by Jorcin et al. [39], which includes distributed surface roughness and heterogeneity, slow adsorption reaction, non-uniform potential and current distribution. CPE_1 and CPE_2 are two constant phase elements of the equivalent circuit and n_1 and n_2 are corresponding exponents ($CPE = Y_0(j\omega)^n$). CPE can represent pure resistance ($n = 0$), pure capacitance ($n = 1$), Warburg impedance ($n = 0.5$) or inductance ($n = -1$) [28]. The calculated parameters indicate that the values of fractional exponent, n_2, for all specimens are close to 1, being near to that of a pure capacitance. The CPE_2 possesses physical meaning of the capacitance of the corrosion products layer ($C = \varepsilon_r \varepsilon_0 \frac{A}{d}$) that is inversely proportional to the thickness of the corrosion layer (d).

The CPE_2 values for composite coatings (Fe_3Al/TiC and Fe_3Al-Cr/TiC) are less than that of Fe_3Al. This result could be due to an increase of the thickness of the corrosion product layer or its composition change. R_2 value is also higher for the composite samples, compared to the pure Fe_3Al. This value significantly increases by adding Cr into the composite. The increase of the resistance of the corrosion products layer, R_2, for composite coatings indicates that the layer is more resistant to electron transfer. This could be an indication of the compactness of this layer, i.e., more compact corrosion product layer can block the dissolution reaction, providing an effective barrier against corrosion. Therefore, among the composite samples, the corrosion product resistance of Fe_3Al-Cr/TiC is much greater than that of Fe_3Al/TiC. The charge transfer resistance, R_1 of Fe_3Al/TiC is also higher than that of the two other samples, which is likely due to the blocking effect of the more compact corrosion product layer formed at the surface. R_1 for composite coatings (Fe_3Al/TiC and Fe_3Al-Cr/TiC) are considerably higher than that of Fe_3Al.

3.2. Surface Analysis

After potentiodynamic polarization test, three distinct layers are observed on the cross-section of the tested samples, i.e., substrate, coating and corrosion products (Figure 2). A thick layer of corrosion product uniformly covered the entire surface of the coatings. The coating consists of lamellar microstructure. This structure is typical in the HVOF process while partially molten particles hit the surface of the substrate and spread on, resulting in a lamellar microstructure called splats, which appears in the form of elongated grains at the cross-section view.

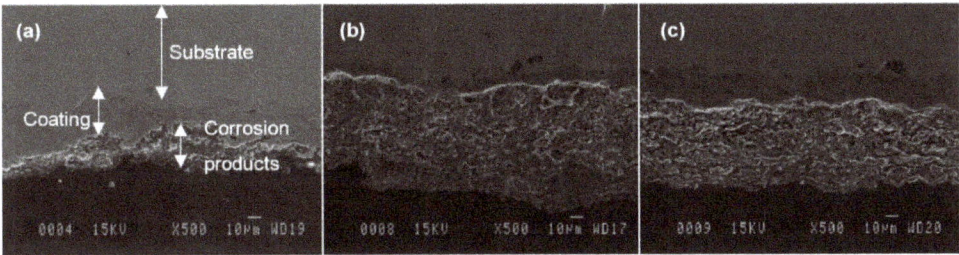

Figure 2. Cross-sectional SEM micrograph of the corrosion product layer formed on: (**a**) Fe$_3$Al, (**b**) Fe$_3$Al/TiC, and (**c**) Fe$_3$Al-Cr/TiC after subjected to the potentiodynamic polarization test.

The XPS analysis results confirm the presence of a mixture of FeO and Fe$_2$O$_3$ as well as Al$_2$O$_3$ on the top layer of all three coatings (Figures S1 and S2a, Supplementary File). Other detected corrosion products are Al$_2$O$_3$, Al(OH)$_3$ and AlCl$_3$ (Figure S2, Supplementary File). The oxides could be formed during HVOF process, similar to that reported by Frangini et al. [13]. They reported that the outer part of the oxide layer predominantly consists of mixed Al-Fe oxy-hydroxide whereas the inner part is of mostly an Al-rich oxide phase. The hydroxides and chlorides are formed during polarization tests due to the presence of sodium chloride in the solution. On the XPS spectra of the two composite coatings appear also the peaks associated with TiO, TiO$_2$ and Cr(OH)$_3$ (Figure S3, Supplementary File). These results as well as the EDS analysis of the corrosion products indicate that a corrosion product layer was formed on the composite coatings (Fe$_3$Al/TiC and Fe$_3$Al-Cr/TiC) and consists of a mixture of aluminum, iron and titanium oxide, aluminum hydroxide, titanium carbide and chromium hydroxide in case of Fe$_3$Al-Cr/TiC. The presence of hydroxide could be the result of hydration of product film that can happen by raising potential, as reported by Rao [23]. The elemental analysis, extracted from XPS spectra of the three coatings before and after polarization tests, reveals the composition of the top oxide layer (Table 5). In addition to the elements presented in Table 5, an appreciable amount of carbon and oxygen was also detected by XPS analysis. Oxygen in coatings before the test could be due to the oxidation of the feedstock during coating by HVOF technique or to the natural oxidation in contact with air. The presence of carbon is due to TiC. However, carbon is also present in the sample without TiC. One source of carbon and oxygen, detected by XPS, is usually the chemisorption of CO and CO$_2$ on the sample. The amount of the chemisorbed species could be significant on the porous materials with high surface area. As it is difficult to quantify the portion of analyzed carbon and oxygen coming from the chemisorption, the XPS data in Table 5 were presented only for the elements other than oxygen and carbon. Therefore, the at. % in this table is the relative amount of each element normalized on the sum of the analyzed elements (i.e., Fe, Al, Ti, Cr, Cl).

Table 5. XPS surface analysis of the samples (excluding carbon and oxygen) before and after polarization tests. The at. % is the normalized value for each element on the sum of all elements presented in the table.

at. %	Fe$_3$Al		Fe$_3$Al/TiC		Fe$_3$Al-Cr/TiC	
	Before Test	After Corrosion	Before Test	After Corrosion	Before Test	After Corrosion
Fe	62	22	54	31	36	29
Al	37	67	34	14	30	6.8
Ti	0.0	0.0	10	37	11	32
Cr	1.1	1	1.7	8.5	23	20
Cl	0.0	9.8	0.0	8.7	0.0	11

A notable increase of Al content is detected on Fe$_3$Al coating after corrosion, but not on the two composite coatings. Rao [23] stated that Al gets enriched on the surface of Fe$_3$Al coating due to the formation of oxide and chloride, whereas Fe gradually dissolves into the corrosive solution of H$_2$SO$_4$. A significant increase of Ti content is detected on both composite coatings, which could be related to the formation of TiO$_2$ and accumulation of TiC in the corrosion products. In addition, sample Fe$_3$Al-Cr/TiC exhibits a large amount of Cr in the corrosion products. As discussed by Zamanzade et al. [40], Cr^{3+} may substitute the Al^{3+} sites during the formation of the passive film by diffusion into the solution/oxide interface.

4. Conclusions

Adding TiC into Fe$_3$Al matrix, prepared by high-velocity oxy-fuel spraying, increases the corrosion resistance of Fe$_3$Al/TiC composite coating in 3.5 wt. % NaCl solution. The corrosion rates of Fe$_3$Al/TiC coating are about five times lower than that of Fe$_3$Al. The addition of Cr into Fe$_3$Al/TiC further decreases the corrosion rate of Fe$_3$Al-Cr/TiC coating to three times lower than that of Fe$_3$Al/TiC. A corrosion product layer consisting of a mixture of aluminum, iron and titanium oxide and aluminum hydroxide is formed on the samples. Chromium hydroxide as also formed in case of Fe$_3$Al-Cr/TiC. A more compact corrosion layer formed on the Fe$_3$Al-Cr/TiC leads to a corrosion mechanism via diffusion process, while the Fe$_3$Al/TiC and Fe$_3$Al corrode via a charge transfer mechanism. Finally, this work concludes that the benefit of adding TiC particles into Fe$_3$Al matrix to improve the wear resistance does not deteriorate its corrosion resistance, and further addition of Cr slightly improves its corrosion resistance even more.

Supplementary Materials: The following are available online at http://www.mdpi.com/2075-4701/9/4/437/s1, Figure S1: XPS spectra for Fe2p for: (**a**) Fe$_3$Al, (**b**) Fe$_3$Al/TiC, and (**c**) Fe$_3$Al-Cr/TiC, Figure S2: XPS spectra of Al2p for Fe$_3$Al/TiC: (**a**) before and (**b**) after polarization test, Figure S3: XPS spectra of Ti2p for: (**a**) Fe$_3$Al/TiC (**b**) Fe$_3$Al-Cr/TiC, and (**c**) Cr2p of Fe$_3$Al-Cr/TiC.

Author Contributions: N.A. conducted the research under the supervision of H.A. and H.H. R.S. and M.G. contributed in design of the experiments and result analysis during the research activities.

Funding: Natural Sciences and Engineering Research Council of Canada (NSERC) via the Discovery Grant.

Acknowledgments: The authors would like to acknowledge the financial support of Natural Sciences and Engineering Research Council of Canada, Fonds de Recherche du Québec-Nature et Technologies, and the Aluminum Research Centre–REGAL. The assistance of Silvio Savoie (Hydro-Quebec Research Institute, Montreal, QC, Canada) and Guillaume Gauvin (Université Laval) for conducting HVOF depositions and the chemical analyses is also gratefully acknowledged.

Conflicts of Interest: The authors declare no conflict of interest.

References

1. Rao, V.S. A review of the electrochemical corrosion behavior of iron aluminides. *Electrochim. Acta* **2004**, *49*, 4533–4542. [CrossRef]
2. Zamanzade, M.; Barnoush, A.; Motz, C. A review on the properties of iron aluminide intermetallics. *Crystals* **2016**, *6*, 10. [CrossRef]
3. Cinca, N.; Lima, C.R.C.; Guilemany, J.M. An overview of intermetallics research and application: Status of thermal spray coatings. *J. Mater. Res. Technol.* **2013**, *2*, 75–86. [CrossRef]
4. Amiriyan, M.; Alamdari, H.; Blais, C.; Savoie, S.; Schulz, R.; Gariépy, M. Dry sliding wear behavior of Fe$_3$Al and Fe$_3$Al/TiC coatings prepared by HVOF. *Wear* **2015**, *342*, 154–162. [CrossRef]
5. Xu, B.; Zhu, Z.; Ma, S.; Zhang, W.; Liu, W. Sliding wear behavior of Fe–Al and Fe–Al/WC coatings prepared by high velocity arc spraying. *Wear* **2004**, *257*, 1089–1095. [CrossRef]
6. Zhu, Z.-X.; Liu, Y.; Xu, B.-S.; Ma, S.-N.; Zhang, W. Influence of heat treatment on microstructure and wear behavior of Fe-Al/WC composite coatings. *J. Mater. Eng.* **2004**, *7*, 3–5.
7. Durlu, N. Titanium carbide based composites for high temperature applications. *J. Eur. Ceram. Soc.* **1999**, *19*, 2415–2419. [CrossRef]
8. Hussainova, I. Some aspects of solid particle erosion of cermets. *Tribol. Int.* **2001**, *34*, 89–93. [CrossRef]

9. Chen, Y.; Wang, H. Microstructure and wear resistance of laser clad TiC reinforced FeAl intermetallic matrix composite coatings. *Surf. Coat. Technol.* **2003**, *168*, 30–36. [CrossRef]
10. Kumar, S.; Selvarajan, V.; Padmanabhan, P.V.A.; Sreekumar, K.P. Characterization and comparison between ball milled and plasma processed iron-aluminium thermal spray coatings. *Surf. Coat. Technol.* **2006**, *201*, 1267–1275. [CrossRef]
11. Song, B.; Dong, S.; Coddet, P.; Hansz, B.; Grosdidier, T.; Liao, H.; Coddet, C. Oxidation control of atmospheric plasma sprayed Fe-Al intermetallic coatings using dry-ice blasting. *J. Therm. Spray Technol.* **2013**, *22*, 345–351. [CrossRef]
12. Luer, K.R.; DuPont, J.N.; Marder, A.R. High-Temperature sulfidation of Fe$_3$Al thermal spray coatings at 600 °C. *Corrosion* **2000**, *56*, 189–198. [CrossRef]
13. Frangini, S.; Masci, A. Intermetallic FeAl based coatings deposited by the electrospark technique: Corrosion behavior in molten (Li+K) carbonate. *Surf. Coat. Technol.* **2004**, *184*, 31–39. [CrossRef]
14. Pokhmurska, H.; Dovhunyk, V.; Student, M.; Bielanska, E.; Beltowska, E. Tribological properties of arc sprayed coatings obtained from FeCrB and FeCr-based powder wires. *Surf. Coat. Technol.* **2002**, *151–152*, 490–494. [CrossRef]
15. Szczucka-Lasota, B.; Formanek, B.; Hernas, A.; Szymanski, K. Oxidation models of the growth of corrosion products on the intermetallic coatings strengthened by a fine dispersive Al$_2$O$_3$. *J. Mater. Proc. Technol.* **2005**, *164–165*, 935–939. [CrossRef]
16. Sidhu, T.; Prakash, S.; Agrawal, R. Studies on the properties of high-velocity oxy-fuel thermal spray coatings for higher temperature applications. *Mater. Sci.* **2005**, *41*, 805–823. [CrossRef]
17. Rosalbino, F.; Carlini, R.; Parodi, R.; Zanicchi, G. Effect of silicon and germanium alloying additions on the passivation characteristics of Fe$_3$Al intermetallic in sulphuric acid solution. *Electrochim. Acta* **2012**, *62*, 305–312. [CrossRef]
18. Chiang, W.-C.; Luu, W.-C.; Wu, J.-K. Effect of aluminum content on the passivation behavior of Fe–Al alloys in sulfuric acid solution. *J. Mater. Sci.* **2006**, *41*, 3041–3044. [CrossRef]
19. Sharma, G.; Singh, P.R.; Sharma, R.K.; Gaonkar, K.B.; Ramanujan, R.V. Aqueous corrosion behavior of iron aluminide intermetallics. *J. Mater. Eng. Perform.* **2007**, *16*, 779–783. [CrossRef]
20. Porcayo-Calderon, J.; Luna, A.; Arrieta-Gonzalez, C.D.; Salinas-Bravo, V.M. Corrosion performance of Fe-Al intermetallic coatings in 1.0 M NaOH solution. *Int. J. Electrochem. Sci.* **2013**, *8*, 12205–12218.
21. Ji, G.; Elkedim, O.; Grosdidier, T. Deposition and corrosion resistance of HVOF sprayed nanocrystalline iron aluminide coatings. *Surf. Coat. Technol.* **2005**, *190*, 406–416. [CrossRef]
22. Liu, T.; Lau, K.-T.; Chen, S.; Cheng, S.; He, T.; Yin, Y. The electrochemistry corrosion behavior of Fe$_3$Al-type intermetallic with super-hydrophobic surfaces. *Mater. Manuf. Process.* **2010**, *25*, 298–301. [CrossRef]
23. Rao, V.S. Repassivation behaviour and surface analysis of Fe$_3$Al based iron aluminide in 0.25 M H$_2$SO$_4$. *Corros. Sci.* **2005**, *47*, 183–194.
24. Huape-Padilla, E.; Sanchez-Carrillo, M.; Flores-De los Ríos, J.P.; Espinosa-Medina, M.J.; Raúl Germán, B.M.; Ferrer-Sánchez, M.I.; Carbajal-de-la-Torre, G.; Bejar-Gómez, L.; Chacon-Nava, J.; Martinez-Villafane, A. Corrosion study of Fe-Al intermetallic alloys in simulated acid rain. *Int. J. Electrochem. Sci.* **2015**, *10*, 2141–2154.
25. Amiriyan, M.; Blais, C.; Savoie, S.; Schulz, R.; Gariepy, M.; Alamdari, H. Mechanical behavior and sliding wear studies on iron aluminide coatings reinforced with titanium carbide. *Metals* **2017**, *7*, 177. [CrossRef]
26. Rosalbino, F.; Carlini, R.; Parodi, R.; Zanicchi, G.; Scavino, G. Investigation of passivity and its breakdown on Fe$_3$Al–Si and Fe$_3$Al–Ge intermetallics in chloride-containing solution. *Corros. Sci.* **2014**, *85*, 394–400. [CrossRef]
27. Negache, M.; Taibi, K.; Souami, N.; Bouchemel, H.; Belkada, R. Effect of Cr, Nb and Zr additions on the aqueous corrosion behavior of iron-aluminide. *Intermetallics* **2013**, *36*, 73–80. [CrossRef]
28. Rosalbino, F.; Carlini, R.; Zanicchi, G.; Scavino, G. Effect of copper alloying addition on the electrochemical corrosion behavior of Fe$_3$Al intermetallic in sulphuric acid solution. *Mater. Corros.* **2016**, *67*, 1042–1048. [CrossRef]
29. Zamanzade, M.; Vehoff, H.; Barnoush, A. Cr effect on hydrogen embrittlement of Fe$_3$Al-based iron aluminide intermetallics: Surface or bulk effect. *Acta Mater.* **2014**, *69*, 210–223. [CrossRef]
30. Rao, V.S.; Baligidad, R.; Raja, V. Effect of carbon on corrosion behavior of Fe$_3$Al intermetallics in 0.5 N sulphuric acid. *Corros. Sci.* **2002**, *44*, 521–533. [CrossRef]

31. McKamey, C.; Horton, J.; Liu, C. Effect of chromium on properties of Fe$_3$Al. *J. Mater. Res.* **1989**, *4*, 1156–1163. [CrossRef]
32. Keddam, M.; Mattos, O.R.; Takenouti, H. Mechanism of anodic dissolution of iron-chromium alloys investigated by electrode impedances—I. Experimental results and reaction model. *Electrochim. Acta* **1986**, *31*, 1147–1158. [CrossRef]
33. Keddam, M.; Mattos, O.; Takenouti, H. Mechanism of anodic dissolution of iron-chromium alloys investigated by electrode impedances—II. Elaboration of the reaction model. *Electrochim. Acta* **1986**, *31*, 1159–1165. [CrossRef]
34. Epelboin, I.; Keddam, M.; Mattos, O.R.; Takenouti, H. The dissolution and passivation of Fe and FeCr alloys in acidified sulphate medium: Influences of pH and Cr content. *Corros. Sci.* **1979**, *19*, 1105–1112. [CrossRef]
35. Zamanzade, M.; Barnoush, A. An overview of the hydrogen embrittlement of iron aluminides. *Procedia Mater. Sci.* **2014**, *3*, 2016–2023. [CrossRef]
36. Arrieta-Gonzalez, C.; Porcayo-Calderon, J.; Salinas-Bravo, V.; Gonzalez-Rodriguez, J.; Chacon-Nava, J. Electrochemical behavior of Fe$_3$Al modified with Ni in Hank's solution. *Int. J. Electrochem. Sci.* **2011**, *6*, 4016–4031.
37. Osório, W.R.; Freitas, E.S.; Garcia, A. EIS and potentiodynamic polarization studies on immiscible monotectic Al–In alloys. *Electrochim. Acta* **2013**, *102*, 436–445. [CrossRef]
38. Devos, O.; Gabrielli, C.; Tribollet, B. Simultaneous EIS and in situ microscope observation on a partially blocked electrode application to scale electrodeposition. *Electrochim. Acta* **2006**, *51*, 1413–1422. [CrossRef]
39. Jorcin, J.-B.; Mark, M.O.; Nadine, P.; Bernard, T. CPE analysis by local electrochemical impedance spectroscopy. *Electrochim. Acta* **2006**, *51*, 1473–1479. [CrossRef]
40. Zamanzade, M.; Barnoush, A. Effect of chromium on the electrochemical properties of iron aluminide intermetallics. *Corros. Sci.* **2014**, *78*, 223–232. [CrossRef]

© 2019 by the authors. Licensee MDPI, Basel, Switzerland. This article is an open access article distributed under the terms and conditions of the Creative Commons Attribution (CC BY) license (http://creativecommons.org/licenses/by/4.0/).

Article

Simulation Approach for Cathodic Protection Prediction of Aluminum Fin-Tube Heat Exchanger Using Boundary Element Method

Yong-Sang Kim [1], In-Jun Park [2] and Jung-Gu Kim [2],*

[1] Department of Mechanical and Materials Engineering, Korea Institute of Nuclear Safety, 62, Gwahak-ro, Yuseong-gu, Daejeon 34142, Korea; kimys@kins.re.kr
[2] School of Advanced Materials Engineering, Sungkyunkwan University, 2066, Seobu-Ro, Jangan-Gu, Suwon-Si 16419, Gyonggi-Do, Korea; injunpark@gmail.com
* Correspondence: kimjg@skku.edu; Tel.: +82-312-907-360

Received: 12 February 2019; Accepted: 20 March 2019; Published: 23 March 2019

Abstract: The multi-galvanic effect of an Al fin-tube heat exchanger was evaluated using polarization tests, numerical simulation, and the seawater acetic acid test (SWAAT). Determination of the polarization state using polarization curves was well correlated with numerical simulations using a high-conductivity electrolyte. However, the polarization results did not match those of the low-conductivity electrolyte due to the lower galvanic effect. Although the polarization state is changed by electrolyte conductivity, the total net current of the tube is decreased in the case of the anodic joint. From SWAAT results, the leakage time of Al fin-tube heat exchanger assembled by anodic joint was longer than the case with cathodic joint.

Keywords: aluminum; heat exchanger; galvanic corrosion; simulation; polarization

1. Introduction

Aluminum (Al) is a light metal that has interesting properties for heat exchanger applications (e.g., its low density, high thermal conductivity [1], good corrosion resistance [2], and good mechanical properties [3]) in the HVACR (heating, ventilation, air conditioning, and refrigeration) industry for use in cooling systems and air-ventilated units [4–6]. In the mid-1990s, the mechanical assembly of automotive heat exchangers started using brazed Al alloys, and this trend is currently applied in the heat exchangers of air conditioners [7]. In air conditioners, AA 1xxx and AA 3xxx series alloys are used because automotive heat exchangers do not require high mechanical properties and these series have higher thermal transfer efficiency and are economic advantages [8].

To increase the wettability of the melted filler metal at the surface, fluxing is used to remove the natural oxide layer covering the Al surface [9]. Due to the high temperature and cladding with the flux material [10,11], the geometry and microstructure of the joint region, tube, and fin are modified [12–15]. The microstructural change can influence the corrosion behavior. Thus, the microstructural effects on the corrosion properties after brazing should also be considered. In addition, the Al series used in the tube, fin, and joint (filler metal) are different according to their applications. In this study, AA 1100, AA 3003, and AA 4343 or modified AA 4343 are used in the tube, fin, and filler metal, respectively. For assembly of the fin and tube, each part underwent brazing and the fin-tube was jointed as shown in Figure 1. Although Al is used in all parts of the fin-tube heat exchanger, the different alloying elements in the Al series influence the corrosion properties, resulting in galvanic corrosion [16,17].

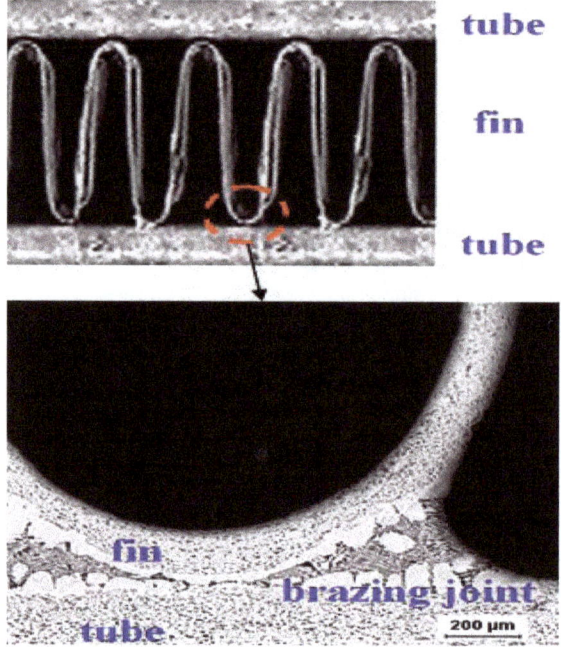

Figure 1. Cross-sectional view of the Al fin-tube heat exchanger assembled by the brazing process.

An important consideration related to the corrosion protection of Al fin-tube heat exchangers is the galvanic coupling between the fin and tube materials. Generally, the material used in the fin has a more negative potential than the tube materials in order to provide cathodic protection for the tube. However, in the case of the Al fin-tube heat exchangers that are assembled by the brazing process, the brazing joint influences the galvanic coupling; this is called multi-material galvanic corrosion. The basic principles of galvanic corrosion are well established and commonly accepted, but galvanic corrosion in realistic situations, such as with multi-material coupling and complex geometries, is hard to predict. Fortunately, the advance of computational simulations has made it possible to model many complex corrosion situations; thus, computational simulation can be directly used to solve engineering design problems [18]. Computational simulation approaches are one of the most effective methods for corrosion design of a product.

Generally, the outside part of the Al fin-tube heat exchanger is the main corroded region because the acid rain and outside pollutants such as sulphur oxides (SO_x) and nitrogen oxides (NO_x) cause a severe corrosion problem. Thus, a test for outside corrosion is mainly considered for Al fin-tube heat exchanger. One of the corrosion reliability tests of the Al fin-tube heat exchanger is the sea water acetic acid test (SWAAT), which is a cyclic spray test (as opposed to a full immersion test). Therefore, in SWAAT, wet and high humidity environments are produced in turn. This indicates that the corrosion condition of the Al fin-tube heat exchanger is alternatively exposed to the thin-layer electrolyte and full immersion. The conductivities of full immersion electrolytes and thin-layer electrolytes are significantly different, which affects the potential and current distribution of the galvanic coupled metals [19]. Thus, the solution conductivity needs to be taken into account when designing a system that protects against galvanic corrosion. In this study, we evaluate the corrosion of Al fin-tube heat exchangers with different brazing joints in high- and low-conductivity electrolyte conditions based on computational simulations. Also, to compare the result of computational results, the actual SWAAT was conducted for Al fin-tube heat exchanger applied different joint materials. Although it would be not a perfect validation of simulation results, it can be supporting the simulation results for application of actual corrosion reliability test.

2. Boundary Element Method (BEM)

A model of an Al fin-tube heat exchanger is shown in Figure 2; this has electrolyte domains (Ω_{bulk} and $\Omega_{thin\text{-}layer}$) and is surrounded by the surface of the electrolyte (Γ_n), the surface of the tube part (Γ_a), the surface of the fin part (Γ_b), and the surface of the joint part (Γ_c). The electrolyte conductivity (σ) is uniform in the whole domain and there is no current loss. The potential field in the electrolyte domain (Ω) can be modeled by Laplace's equation [20]:

$$\nabla^2 \Phi = 0 \tag{1}$$

Here, Φ is the electrical potential, which is the potential relative to a reference electrode, such as a saturated calomel electrode (SCE).

Laplace's equation is calculated using the following boundary conditions:

$$i = i_0, \text{ on } \Gamma_n \tag{2}$$

$$i_a = f_a(\Phi_a), \text{ on } \Gamma_a \tag{3}$$

$$i_b = f_b(\Phi_b), \text{ on } \Gamma_b \tag{4}$$

$$i_c = f_c(\Phi_c), \text{ on } \Gamma_c \tag{5}$$

Here, Γ is the entire surface of the electrolyte domain, which includes Γ_n (electrolyte surface), Γ_a (tube surface), Γ_b (fin surface), and Γ_c (joint surface). $f_a(\Phi_a)$, $f_b(\Phi_b)$, and $f_c(\Phi_c)$ are the non-linear functions on the surfaces of the tube, fin, and joint areas, respectively, which represent the experimentally achieved polarization curves. Thus, the boundary element method (BEM) can be used to calculate Laplace's Equation (1) when the tube, fin, and joint areas are prescribed on the Al surface and their polarization curves are known [21]. Based on this, Φ and i on the whole surface can be determined.

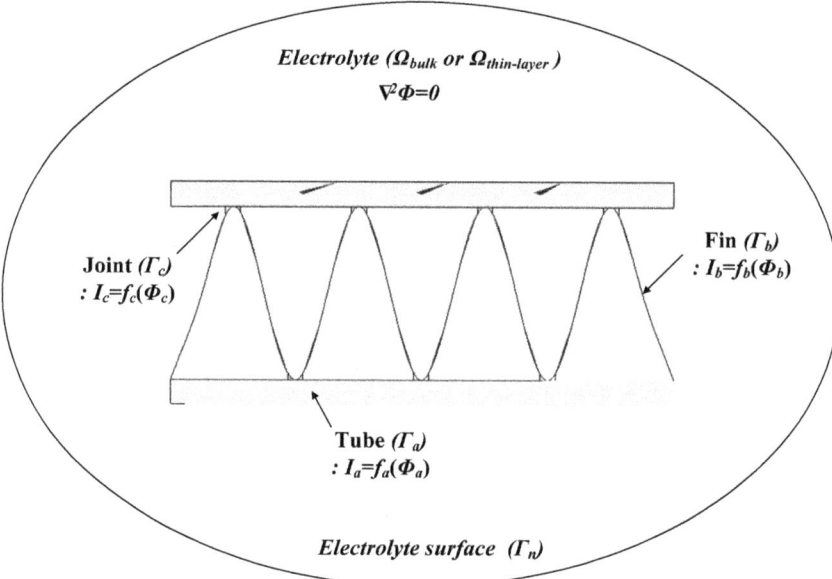

Figure 2. Boundary conditions for corrosion simulation of the Al fin-tube heat exchanger.

An Al fin-tube heat exchanger model was created using the Rhinoceros 4.0 (Robert McNeel & Associates, Seattle, USA)3D drawing software based on the real shapes and dimensions. The inner parts of the heat exchanger were not modeled to avoid computational errors because outside corrosion is the main consideration in this study. The 3D model of the Al fin-tube heat exchanger, with detailed dimensions, is shown in Figure 3. The joint part is modeled by rectangular shape between the tube and fin to simplify the heat exchanger model. Areas of the tube, fin, and joint are 870, 1115, and 40 mm^2, respectively. After modeling, the heat exchanger, the 3D model was imported into the program BEASY version 10.0r14, which is BEM-based software(BEASY, Southampton, England). Setting the boundary conditions is an essential step for corrosion simulation. Different electrolyte) conductivity values were applied for comparison of the bulk (0.4 S/m) and thin-layer (0.00004 S/m) electrolytes; this was done because the electrolyte conductivity is generally decreased in thin-layer electrolyte conditions. Although the oxygen concentration and ion transfer parameter can also influence the corrosion rate, we consider the electrolyte conductivity difference because the galvanic effect between the tube, fin, and joint is the main focus in this study. Also, the localized corrosion caused by galvanic corrosion is the important factor for corrosion reliability. However, the corrosion design for a large structure by simulation is the main goal of this study. Thus, not the localized corrosion which is focused on the micro-scale but the uniform corrosion in bulk structure was focused in this study.

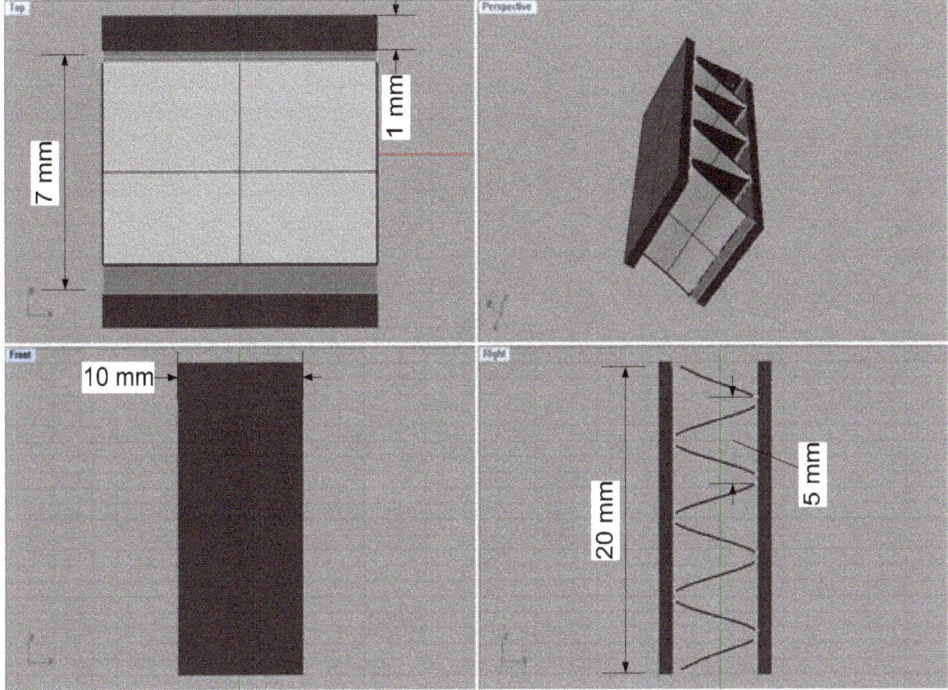

Figure 3. 3D models of the Al fin-tube heat exchanger with detailed dimensions (view of top, front, right and perspective).

3. Experimental Procedures

3.1. Materials and Solution

As mentioned above, AA 1100, AA 3003, and AA 4343 or modified AA 4343 (which was modified to include more Zn to decrease the corrosion potential), are used in the tube, fin, and filler metals, respectively. The chemical composition of each part is listed in Table 1. Cathodic and anodic joints indicate that the higher and lower corrosion potential than the tube materials. Fin material always has lower potential than tubes and joints for the corrosion protection of the tube. The test solution consisted of a seawater acidified solution which is a synthetic sea salt solution made with the addition of glacial acetic acid (pH 3.0), as described by ASTM G85. Test temperature of all the tests is 49 °C.

Table 1. Chemical composition of the parts (Al 1100, Al 3003, Al 4343 and modified Al 4343) of Al fin-tube heat exchanger.

Parts	Chemical Composition (wt%)				
	Cu	Fe	Si	Zn	Mn
Tube (AA 1100)	0.001	0.200	0.010	-	-
Fin (AA 3003)	0.002	0.210	0.220	0.150	0.640
Cathodic joint (AA 4343)	0.110	0.120	5.380	0.080	0.010
Anodic joint (modified AA 4343)	0.120	0.120	7.730	0.480	0.010

3.2. Potentiodynamic Tests

Polarization data for the different parts of the Al fin-tube heat exchanger (i.e., AA 1100, AA 3003, and AA 4343 or modified AA 4343) were needed to conduct simulations. Potentiodynamic polarization tests were conducted by a conventional three-electrode cell. A purified carbon rod was used as the counter electrode and a saturated calomel electrode (SCE) was used as the reference electrode. The tested specimen for carrying out the polarization test of each part was extracted from the Al fin-tube heat exchanger which was machined by a micro-cutting machine (The CUTLAM®micro 1.1, LAMPLAN, Gaillard, France). The working electrode was abraded by a series of abrasive papers (from 220 to 600 grit), rinsed ultrasonically with ethanol, and dried with nitrogen (N_2) gas. The specimen was covered with silicone rubber, leaving an area of 25 mm^2 unmasked. The prepared specimen was then exposed to the test solution for 1 h. Potentiodynamic polarization tests were carried out using a Bio-Logic VSP-300 potentiostat. The potential range was from -0.3 V_{SCE} *vs.* the open-circuit potential (OCP) to -0.4 V_{SCE} at a scan rate of 0.166 mV/s. All polarization tests were repeated a minimum of three times to ensure accuracy.

3.3. Sea Water Acetic Acid Test (SWAAT)

As a corrosion reliability test of fin-tube heat exchanger, SWAAT has been used in the field of Al fin-tube heat exchanger. Thus, to compare the corrosion reliability of Al fin-tube heat exchanger applied anodic and cathodic joint and validate the corrosion resistance of suggested joint materials, SWAAT was conducted in this study. SWAAT was produced based on ASTM G85 which utilizes the following cycle: 30-minute spray followed by a 90-minute soak at above 98% relative humidity and 49 °C. Whole Al fin-tube heat exchangers were tested to evaluate the tube leakage time caused by corrosion degradation. To record the leakage time caused by corrosion degradation, the tube was filled with air at the inlet part and the pressure gage was installed at the outlet part. Until the pressure gage was set to 5 MPa, the air was filled. The leakage of the Al fin-tube exchanger was determined from the decrease of a pressure gage during the SWAAT.

4. Results and Discussion

4.1. Polarization Curves

Figure 4 shows the polarization curves of the tube (AA 1100), fin (AA 3003), cathodic joint (AA 4343), and anodic joint (modified AA 4343) in the seawater acetic acid solution. Parameters such as the anodic and cathodic Tafel slopes (β_a and β_c), corrosion current density (i_{corr}), and corrosion potential (E_{corr}) were determined from the polarization curves and are listed in Table 2. Although i_{corr} is different for each part of the Al fin-tube heat exchanger due to the alloying and microstructure effects [16,17,22–24], the actual current density in the Al-fin tube heat exchanger should vary due to the galvanic corrosion between the three metals. E_{corr} exhibits the following decreasing order: cathodic joint > tube > anodic joint > fin. This means that the fin acts as a sacrificial anode for both of the assembled cases; cathodic joint > tube > fin and tube > anodic joint > fin. Thus, the effect of different tri-metal galvanic couplings on the corrosion reliability should be considered for designing Al fin-tube heat exchangers.

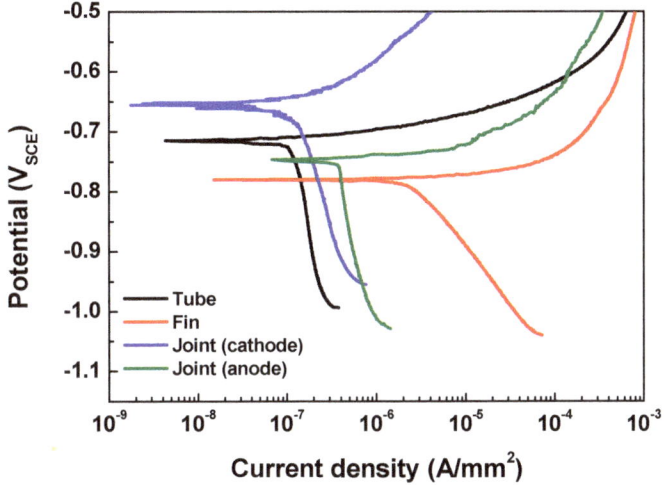

Figure 4. Polarization curves of each part of the Al fin-tube heat exchanger.

Table 2. Potentiodynamic polarization test results of different parts of the Al fin-tube heat exchanger (Al 1100, Al 3003, Al 4343, and modified Al 4343).

Parts	β_a (V/decade)	β_c (V/decade)	i_{corr} (μA/mm^2)	E_{corr} (V_{SCE})
Tube (AA 1100)	0.021 ± 0.003	0.886 ± 0.01	0.109 ± 0.05	−0.716 ± 0.02
Fin (AA 3003)	0.010 ± 0.005	0.223 ± 0.01	2.351 ± 0.25	−0.792 ± 0.02
Cathodic joint (AA 4343)	0.109 ± 0.03	0.748 ± 0.09	0.129 ± 0.06	−0.657 ± 0.01
Anodic joint (modified AA 4343)	0.018 ± 0.002	0.859 ± 0.05	0.349 ± 0.09	−0.746 ± 0.03

Generally, in tri-metal galvanic couples that include three metals (A_1, A_2, and A_3) with different corrosion potentials in the order of $E_{A1} > E_{A2} > E_{A3}$, the polarization of the middle potential metal (E_{A2}) can be defined by the mixed potential theory [25,26]. If the tri-metal galvanic coupled metals (A_1, A_2, and A_3) comply with the Tafel system (a linear system), the polarization state of the middle potential metal can be determined by the Tafel superposition relation. Thus, the tri-metal galvanic couple can be investigated by considering the A_1-A_2 and A_2-A_3 couples. Then, the net current (I_{net}) on the A_2 metal can be determined as $I_{A1-A2} - I_{A2-A3}$, where I_{A1-A2} and I_{A2-A3} are currents flowing from A_1 to A_2 and from A_2 to A_3, respectively. This means that A_2 acts as an anode in the case of a positive I_{net} and as a cathode in the case of a negative I_{net}.

Based on the above determination made via the polarization state method, the polarization states of the tube in the two assembled cases (case 1: fin-tube-cathodic joint and case 2: fin-tube-anodic joint) were determined. Schematic polarization curves of the tube, fin, and joint are shown in Figure 5, which reflect the area of the 3D model used to interpret the polarization state; this is done because the ratio of the cathode-to-anode surface areas is an important factor in galvanic corrosion. In case 1, the intersection between the Tafel slopes of the cathodic joint and the tube cannot occur. Thus, the galvanic currents on the anode parts, which are the anodic currents on the tube and fin (I_{T-CJ} and I_{F-T}), are not increased after galvanic coupling according to the mixed-potential theory. Alternatively, the galvanic currents on the cathodic parts, which are the cathodic currents on the cathodic joint and tube (I_{CJ-T} and I_{T-F}), are increased. Consequently, I_{net} of the tube in case 1 is negative, acting as a cathode. In case 2, the Tafel slopes of the anodic joint and tube make an intersection. The galvanic currents on the anode parts, which are the anodic currents on the anodic joint and fin ($I_{AJ\text{-}T}$ and $I_{F\text{-}T}$), are increased only in the anodic joint and tube couple. The galvanic currents on the cathode parts, which are the cathodic currents on the tube ($I_{T\text{-}AJ}$ and $I_{T\text{-}F}$), are increased in case 2. The value of I_{net} for the tube in case 2 is also negative, acting as a cathode. In short, the polarization state of the tube in both cases was cathodic; however, this determination did not consider the distance between the anode and cathode or the electrolyte conductivity. Thus, a more detailed determination is needed to design Al fin-tube heat exchangers adequately.

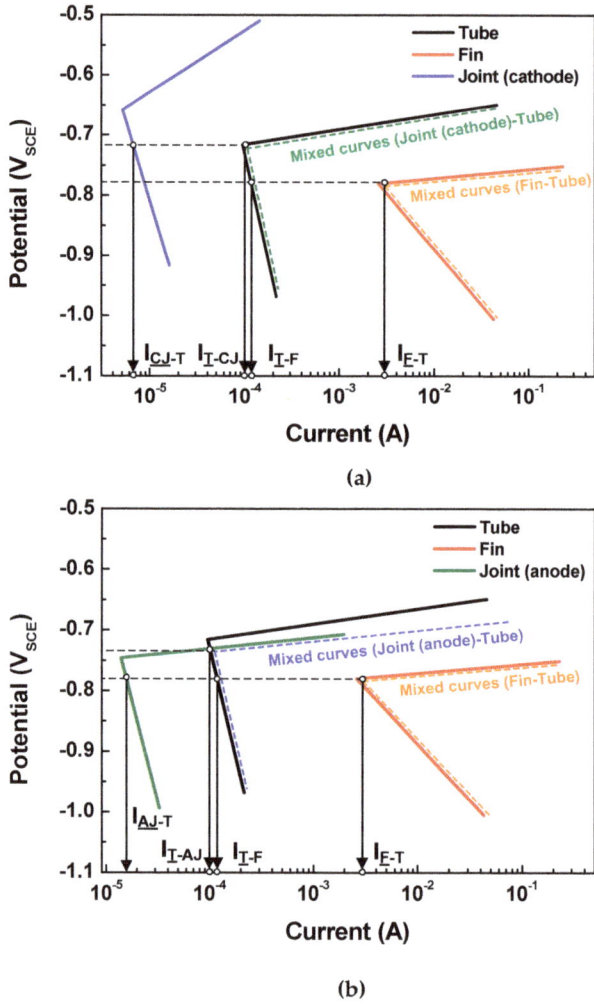

Figure 5. Tafel slopes of the polarization curves applied to the surface areas of the tube, fin, and joint in the 3D model and the calculated galvanic current based on mixed-potential theory: (**a**) case 1 and (**b**) case 2.

4.2. Corrosion Simulation

Figures 6 and 7 show the 3D models of the Al-fin tube heat exchanger with the cathodic and anodic joints, respectively, in high-conductivity electrolyte (0.4 S/m). The potential and current density ranges of the Al-fin tube heat exchanger assembled with a cathodic joint (Figure 6) were −775.2 mV$_{SCE}$

to -774.9 mV$_{SCE}$ and -2.21×10^{-7} A/mm^2 to 3.52×10^{-7} A/mm^2, respectively. While those of the Al-fin tube heat exchanger assembled with an anodic joint (Fig. 7) were -786.2 mV$_{SCE}$ to -786.1 mV$_{SCE}$ and -4.09×10^{-7} A/mm^2 to 1.37×10^{-7} A/mm^2, respectively. In both cases, the anodic current was observed only on the fin part and the current density increased near the tube and joint parts. This is due to the increased galvanic effect between the metals which had different corrosion potentials. The potential distribution was not broad (almost single potentials of -775 mV$_{SCE}$ and -786 mV$_{SCE}$) in the anodic potential range of the fin. This means that the fin part acted as a sacrificial anode to the tube and joint (anodic and cathodic joints) in the high-conductivity electrolyte, regardless of the polarization state of the joint. This result is similar to the determination of the polarization state using the polarization curves.

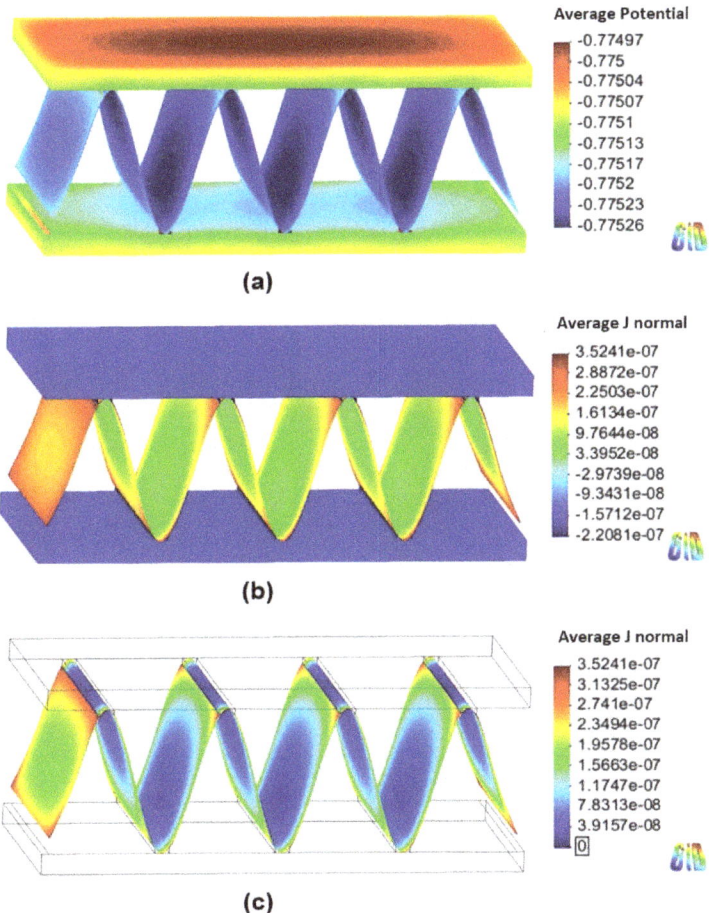

Figure 6. Simulation results of the Al fin-tube heat exchanger assembled with a cathodic joint in high-conductivity electrolyte: (**a**) potential, (**b**) current density, and (**c**) anodic current density distribution.

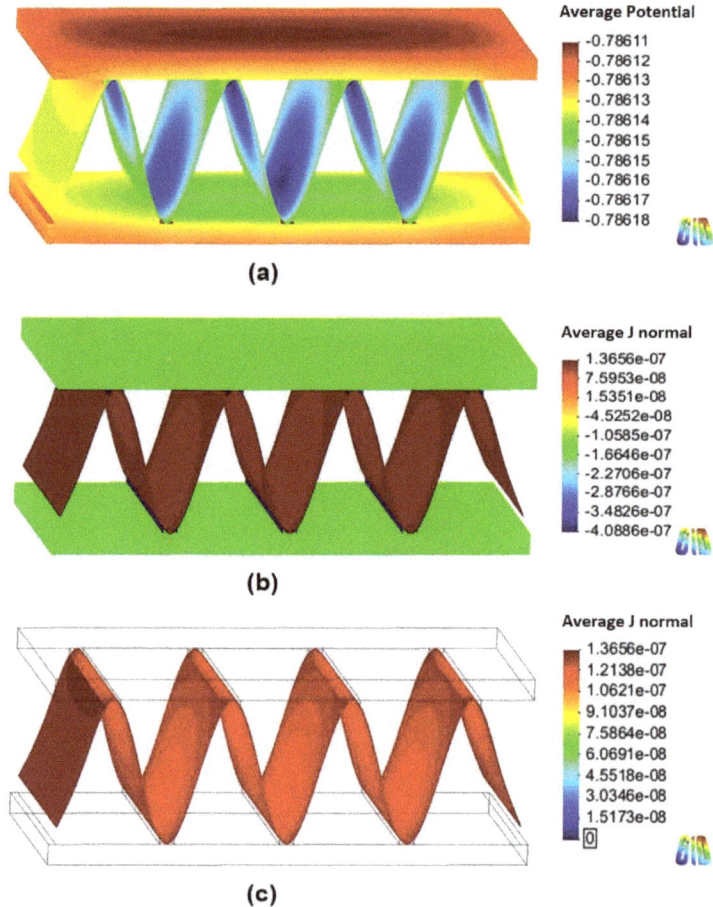

Figure 7. Simulation results of the Al fin-tube heat exchanger assembled with an anodic joint in high-conductivity electrolyte: (**a**) potential, (**b**) current density, and (**c**) anodic current density distribution.

The total net currents of each part of cathodic or anodic joints in high-conductivity electrolyte were -8.87×10^{-6} A (cathodic joint), -5.93×10^{-5} A (tube), 6.81×10^{-5} A (fin) and -1.09×10^{-6} A (anodic joint), -5.94×10^{-5} A (tube), 6.05×10^{-5} A (fin). The total net current is related to the amount of current flowing from the anode or cathode. A decrease in the total net current simply indicates a decrease of the corrosion rate (according to Faraday's law). Thus, the decrease of the total net current of the fin in the anodic joint implies increased corrosion life, which means that an anodic joint extends the corrosion life of the Al-fin tube heat exchanger compared to the cathodic joint in the high-conductivity electrolyte.

Figures 8 and 9 show the 3D models of the Al-fin tube heat exchanger using cathodic and anodic joints in low-conductivity electrolyte (0.00004 S/m), respectively. Contrary to the results of the high-conductivity electrolyte, the potential distribution was clearly separated in each part of the Al fin-tube heat exchanger in both cases (anodic and cathodic joints). This means that the galvanic effect among the tube, fin, and joint was significantly decreased due to the low conductivity [19]. Also, these results indicate that the cathodic protection effect between exchanger components was very low, so that sufficient protection could not be obtained at the regions that are far away from the junction. The cathodic current was observed in the case of the cathodic joint, while anodic and cathodic currents were found in the case of the anodic joint. In the tube and fin, only the anodic current was shown, which implies that the current flow is too low to change the polarization state from anodic to cathodic. However, the anodic current was slightly decreased near the anodic joint (Figure 9c) because of a small current effect from the anodic joint. Consequently, the total net current was slightly lower in the case of the anodic joint (6.205×10^{-3} A) than in the case of the cathodic joint (6.329×10^{-3} A). However, according to the simulation results, the anodic joint was more favorable for extending the corrosion life of the Al fin-tube heat exchanger (especially in the tube part) in both conductivity environments. This can be verified by practical corrosion testing.

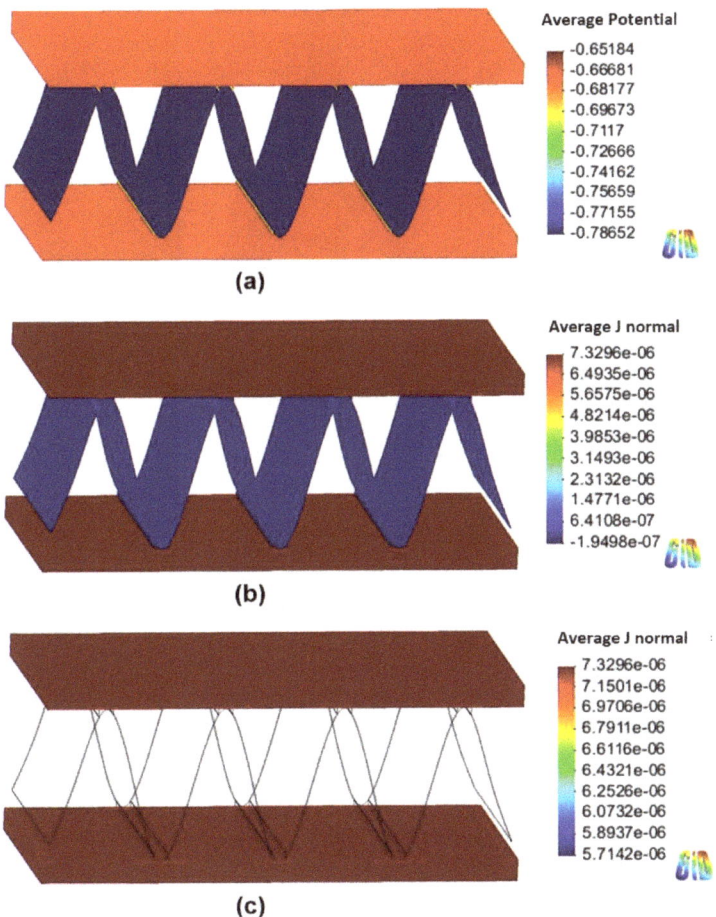

Figure 8. Simulation results of the Al fin-tube heat exchanger assembled with a cathodic joint in low-conductivity electrolyte: (**a**) potential distribution, (**b**) current density distribution, and (**c**) current density distribution at the tube.

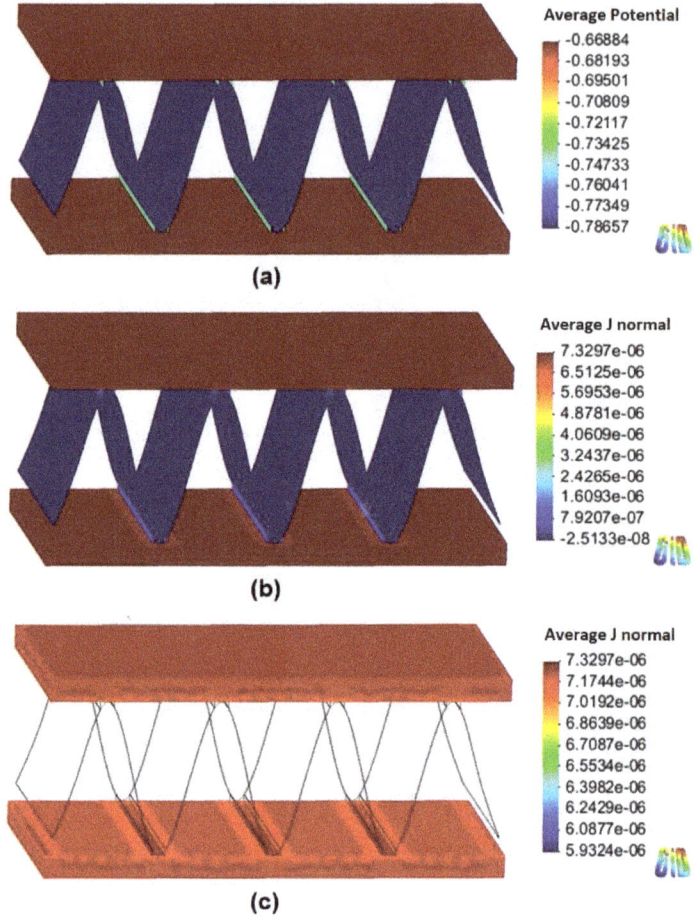

Figure 9. Simulation results of the Al fin-tube heat exchanger assembled with an anodic joint in low-conductivity electrolyte: (**a**) potential distribution, (**b**) current density distribution, and (**c**) current density distribution at the tube.

4.3. SWAAT Test

To compare the corrosion lifetimes of Al fin-tube heat exchangers assembled with cathodic or anodic joints, SWAAT was conducted until leakage of the tube occurred. The polarization states of the Al fin-tube heat exchangers assembled by cathodic or anodic joints and the corresponding leakage time are listed in Table 3. Although the polarization state was changed according to the electrolyte conductivity, the anodic joint was efficient in extending the corrosion life because the anodic joint either decreased the net current on the fin or it protected the tube by serving as a sacrificial anode. In short, based on the corrosion simulation results, it was revealed that the anodic joint can increase the corrosion life. In SWAAT, an increase in the corrosion life was observed on the Al fin-tube assembled with an anodic joint. The leakage time in the case of the anodic joint was increased by about 42% relative to the lifetime with the cathodic joint. This suggests that the numerical simulation results correlated well with the corrosion test of the product. Thus, corrosion simulations can be helpful for the design of Al fin-tube assembly in low and high-conductivity electrolytes.

Table 3. Sea water acetic acid test (SWAAT) results and polarization states of Al fin-tube heat exchangers assembled with cathodic or anodic joints.

	Polarization State FROM SIMULATION		Leakage Time (day)
	High conductivity	Low conductivity	
Cathodic joint	Cathode	Cathode	
Tube	Cathode	Anode	56 days
Fin	Anode	Anode	
Anodic joint	Cathode	Anode/Cathode	
Tube	Cathode	Anode	80 days
Fin	Anode	Anode	

5. Conclusions

In this study, the multi-galvanic effect in Al fin-tube heat exchangers assembled with cathodic or anodic joints was investigated using a polarization determination method and numerical simulation in the presence of electrolytes with high and low conductivities. SWAAT was also conducted to verify the simulation results. When determining the polarization state from the polarization curves, the tube and joint were found to be cathodes, while the fin acted as an anode, regardless of the corrosion potential of the joint. These results are in good agreement with the numerical simulation results in the high-conductivity environment. However, discrepancies were observed in the low-conductivity environment because the galvanic effect between the fin, tube, and joint was small. Although the polarization state changed according to the electrolyte conductivity, the total net current of the tube, which is related to the corrosion rate, was lower in the case of the anodic joint. Thus, the Al fin-tube heat exchanger assembled with the anodic joint was superior to the exchanger assembled with the cathodic joint. This result was verified by SWATT, and the leakage time of the Al fin-tube heat exchanger assembled with the anodic joint was 42% longer than that of the exchanger assembled with the cathodic joint. Due to the good correlation with the practical corrosion test results, numerical simulation of the multi-galvanic situation can be applied to improve the corrosion design of products.

Author Contributions: Conceptualization, Y.-S.K. and I.-J.P.; Software, Y.-S.K.; Validation, Y.-S.K. and I.-J.P.; Formal Analysis, Y.-S.K. and I.-J.P.; Investigation, Y.-S.K.; Resources, J.-G.K.; Data Curation, Y.-S.K.; Writing-Original Draft Preparation, Y.-S.K.; Visualization, J.-G.K.; Supervision, J.-G.K.; Project Administration, J.-G.K.; Funding Acquisition, J.-G.K.

Funding: This work was supported by the National Research Foundation of Korea (NRF) grant funded by the Korea Government (Ministry of Education, Science and Technology) (No.NRF-2016R1A2B4016027).

Conflicts of Interest: The authors declare no conflict of interest.

Nomenclature

i	**Current density, A/mm^2**
V	potential, V_{SCE}
i_{corr}	corrosion current density, $\mu A/mm^2$
E_{corr}	corrosion potential, V_{SCE}
β_a	anodic Tafel slope, V/decade
β_c	cathodic Tafel slope, V/decade
I_{net}	net current, A

Subscripts

Ω	electrolyte domain
Ω_{bulk}	bulk electrolyte domain
$\Omega_{thin\text{-}layer}$	thin-layer electrolyte domain
Γ_n	surface of the electrolyte
Γ_a	surface of the tube part
Γ_b	surface of the fin part
Γ_c	surface of the joint part
Φ	electrical potential
$f_a(\Phi_a)$	non-linear functions of the tube surface
$f_b(\Phi_b)$	non-linear functions of the tube fin
$f_c(\Phi_c)$	non-linear functions of the tube joint
A_1, A_2, and A_3	three metals
E_{A1}, E_{A2}, and E_{A3}	corrosion potential of three metals, V
$I_{\underline{A}\text{-}B}$	net current of A in galvanic condition between A and B

Abbreviation

Aluminum	Al
heating, ventilation, air conditioning, and refrigeration	HVACR
sulphur oxides	SO_x
sea water acetic acid test	SWAAT
saturated calomel electrode	SCE
3 dimension	3D
American Society for Testing and Materials	ASTM
nitrogen	N_2
open-circuit potential	OCP

References

1. Andreatta, F.; Anzytti, A.; Fedrizzi, L. Corrosion behaviour of AA 8xxx aluminium fins in heat exchanger. *Surf. Interface Anal.* **2016**, *48*, 789–797. [CrossRef]
2. Tierce, S.; Pébère, N.; Blanc, C.; Casenave, C.; Mankowski, G.; Robidou, H. Corrosion behaviour of brazing material AA 4343. *Electrochim. Acta* **2006**, *52*, 1092–1100. [CrossRef]
3. Adams, T.M.; Dowling, M.F.; Abdel-Khalik, S.I.; Jeter, S.M. Applicability of traditional turbulent single-phase forced convection correlations to noncircular microchannels. *Int. J. Heat Mass Trans.* **1999**, *42*, 4411–4415. [CrossRef]
4. Akers, W.W.; Rosson, H.F. Condensation inside a horizontal tube. *Chem. Eng. Prog. S Ser.* **1960**, *56*, 145–150.
5. Cavallini, A.; Col, D.D.; Doretti, L.; Matkovic, M.; Rosetto, L.; Zilio, C. Condensation in horizontal smooth tubes: A new heat transfer model for heat exchanger design. *Heat Transf. Eng.* **2006**, *27*, 31–38. [CrossRef]
6. Abdulstaar, M.; Mhaede, M.; Wagner, L.; Wollmann, M. Corrosion behaviour of Al 1050 severely deformed by rotary swaging. *Mater. Design* **2014**, *57*, 325–329. [CrossRef]
7. Hong, M.S.; Park, I.J.; Kim, J.G. Alloying effect of copper concentration on the localized corrosion of aluminum alloy for heat exchanger tube. *Met. Mater. Int.* **2017**, *23*, 708–714. [CrossRef]
8. Chen, G.; Chen, Q.; Wang, B.; Du, Z. Microstructure evolution and tensile mechanical properties of thixoformed high performance Al–Zn–Mg–Cu alloy. *Met. Mater. Int.* **2015**, *21*, 897–906. [CrossRef]
9. ASM International, Materials Park. *Metals Handbook: Welding and Brazing*, 8th ed.; ASM International, Materials Park: Novelty, OH, USA, 1971; Volume 6.
10. Tierce, S.; Pébère, N.; Blanc, C.; Mankowski, G.; Robidou, H.; Vaumousse, D.; Lacaze, J. Solidification and phase transformations in brazed aluminum alloys used in automotive heat exchangers. *Int. J. Cast. Met. Res.* **2005**, *18*, 370–376. [CrossRef]
11. Lacaze, J.; Tierce, S.; Lafont, M.C.; Thebault, Y.; Pébère, N.; Mankowski, G.; Blanc, C.; Robidou, H.; Vanmosse, D.; Daloz, D. Study of the microstructure resulting from brazed aluminium materials used in heat exchangers. *Mater. Sci. Eng. A* **2005**, *413*, 317–321. [CrossRef]

12. Sekulić, D.P. Molten aluminum equilibrium membrane formed during controlled atmosphere brazing. *Int. J. Eng. Sci.* **2001**, *39*, 229–241. [CrossRef]
13. Sekulic, D.P.; Zellmer, B.J.; Nigro, N. Influence of joint topology on the formation of brazed joints. *Model. Simul. Mater. Sci. Eng.* **2001**, *9*, 357–370. [CrossRef]
14. Zellmer, B.J.; Nigro, N.; Sekulic, D.P. Numerical modelling and experimental verification of the formation of 2D and 3D brazed joints. *Model. Simul. Mater. Sci. Eng.* **2001**, *9*, 339–355. [CrossRef]
15. Marshall, G.J.; Bolingbroke, R.K.; Gray, A. Microstructural control in an aluminum core alloy for brazing sheet applications. *Metall. Trans. A* **1993**, *24A*, 1935–1942. [CrossRef]
16. Al-Kharafi, F.M.; Badawy, W.A. Corrosion and passivation of Al and Al-Si alloys in nitric-acid solutions II-Effect of chloride ions. *Electrochim. Acta* **1995**, *40*, 1811–1817. [CrossRef]
17. Abdel Rehim, S.S.; Hassan, H.H.; Amin, M.A. Chronoamperometric studies of pitting corrosion of Al and (Al-Si) alloys by halide ions in neutral sulphate solution. *Corros. Sci.* **2004**, *46*, 1921–1938. [CrossRef]
18. Amaya, K.; Aoki, S. Effective boundary element methods in corrosion analysis. *Eng. Anal. Bound. Elem.* **2003**, *27*, 507–519. [CrossRef]
19. McCafferty, E. Distribution of potential and current in circular corrosion cells having unequal polarization parameters. *J. Electrochem. Soc.* **1997**, *124*, 1869–1878. [CrossRef]
20. Adey, R.A.; Niku, S.M. Computer modeling of galvanic corrosion. *ASTM Spec. Tech. Publ.* **1998**, *978*, 96–117.
21. Ridha, M.; Amaya, K.; Aoki, S. Boundary element simulation for identification of steel corrosion in concrete using magnetic field measurement. *Corrosion* **2005**, *61*, 784–791. [CrossRef]
22. Jafarian, F.; Umbrello, D.; Jabbaripour, B. Indentification of new material model for machining simulation of Inconel 718 ally and the effect of tool edge geometry on microstructure changes. *Simul. Model. Pract. Theory* **2016**, *66*, 273–284. [CrossRef]
23. Gastli, A.; Metwally, I.A. Computation of eddy-current density on ESP motor and well casings under different operating conditions. *Simul. Model. Pract. Theory* **2008**, *16*, 483–493. [CrossRef]
24. Nisançıoğlu, K. Electrochemical behavior of aluminum-base intermetallics containing iron. *J. Electrochem. Soc.* **1990**, *137*, 69–77. [CrossRef]
25. Jones, D.A. *Principles and Prevention of Corrosion*, 2nd ed.; Prentice Hall: Upper Saddle River, NJ, USA, 1996.
26. Hu, Q.F.; Zhang, T.; Geng, S.J.; Wang, F.H. A method for determining the polarization state of the metal with the middle potential in order to calculate the corrosion rate of multi-metals complicated galvanic couple. *Mater. Corros.* **2017**, *68*, 935–942. [CrossRef]

© 2019 by the authors. Licensee MDPI, Basel, Switzerland. This article is an open access article distributed under the terms and conditions of the Creative Commons Attribution (CC BY) license (http://creativecommons.org/licenses/by/4.0/).

Article

Resistance of Thermally Aged DSS 2304 against Localized Corrosion Attack

Federica Zanotto [1,2,*], **Vincenzo Grassi** [1], **Andrea Balbo** [1], **Cecilia Monticelli** [1] and **Fabrizio Zucchi** [1]

1. "Aldo Daccò" Corrosion and Metallurgy Study Centre, University of Ferrara, Via G. Saragat 4A, 44122 Ferrara, Italy; vincenzo.grassi@unife.it (V.G.); andrea.balbo@unife.it (A.B.); mtc@unife.it (C.M.); zhf@unife.it (F.Z.)
2. Terra & AcquaTech Laboratory, University of Ferrara, Via Saragat 1, 44122 Ferrara, Italy
* Correspondence: zntfrc@unife.it; Tel.: +39-053-245-5195

Received: 31 October 2018; Accepted: 4 December 2018; Published: 5 December 2018

Abstract: In this paper, the effects of thermal aging in the 650–850 °C range on the localized corrosion behaviour of duplex stainless steel (DSS) 2304 was investigated. Pitting corrosion resistance was assessed by pitting potential (E_{pitt}) and critical pitting temperature (CPT) determination, while the degree of sensitisation (DOS) to intergranular corrosion (IGC) was evaluated by double loop electrochemical potentiokinetic reactivation (DL-EPR). The susceptibility to stress corrosion cracking (SCC), evaluated in standard NACE TM-0177 solution at pH 2.7 and 25 °C, with the addition of $S_2O_3^{2-}$ at 10^{-3} M, resulted in general good agreement with pitting and IGC behaviour. In fact, as-received DSS 2304 aged for 5 min at 650 °C or 750 °C presented a high resistance to localized corrosion. The alloy corrosion behaviour was severely impaired with an aging time of 60 min at 650 °C and of 10 or 60 min at 750 °C, due to the precipitation of finely distributed $M_{23}C_6$-type chromium carbides at ferrite/austenite interphases, which determined the formation of chromium and molybdenum depleted areas. The behaviour of samples aged at 850 °C also depended on the aging time, but, at 60 min, the rediffusion of passivating elements produced a recovery of the alloy resistance to pitting, IGC and SCC.

Keywords: duplex stainless steel; pitting corrosion; intergranular corrosion; stress corrosion cracking; CPT; DL-EPR

1. Introduction

The significant mechanical and corrosion resistance performances of duplex stainless steels (DSS) are essentially related to their biphasic microstructure, characterized by an austenite/ferrite ratio close to 1 and by their chemical composition, which includes elements like chromium, molybdenum and nitrogen. However, these elements can cause the formation of deleterious secondary phases, such as χ (chi) and σ (sigma) phases and chromium nitrides and/or carbides, if DSS are held, also for brief times, within the 550–950 °C temperature range [1,2]. The σ phase is an intermetallic compound, containing about 30% Cr, 4% Ni and 7% Mo, that nucleates preferentially at the ferrite and ferrite/austenite grain boundaries and then grows into the ferritic grains in DSS aged between 700 and 950 °C [3]. The χ-phase, containing about 25% Cr, 3% Ni and 14% Mo, also forms within 700–900 °C, but in smaller amounts than σ [3]. Escriba et al. [4] observed that this phase was metastable in DSS 2205 grade and was consumed by the σ-phase precipitation. The precipitation of Cr_2N and $M_{23}C_6$ type carbides is observed to occur simultaneously in the 550–1000 °C temperature range, but their formation is quicker at 700–900 °C [5]. Both the χ and/or σ phase and the chromium nitride and/or carbide precipitation can determine a Cr-depletion in adjacent areas [5–9], with consequent impairment of corrosion resistance. Moreover, Cr-depleted ferrite phase with high Ni content becomes unstable and eventually transforms into secondary austenite (γ_2) [10–13], which is characterized by a significantly lower chromium concentration [5].

DSS 2304 (UNS 32304) was the first lean duplex stainless steel (LDSS) to be commercialized. It is a low-cost grade, due to Mo savings (Mo mass% around 0.3%), but also as a σ-free grade [14]. In fact, as already observed in a preceding study on LDSS 2101 [15], which also contains a Mo mass% around 0.3%, the kinetics of precipitation of χ and σ phase mainly depend on Mo content [3]. In particular, in [15] it was observed that brief aging (from 5 to 60 min) between 650 and 850 °C on LDSS 2101 did not cause any σ phase formation, but the precipitation of chromium carbides and nitrides occurred at the ferrite (α) and austenite (γ) grain boundaries or phase interfaces [16,17]. These precipitates were found to adversely affect the localized corrosion resistance of LDSS [12,15–19].

Few studies in the literature deal with the corrosion resistance of DSS 2304 after microstructural modifications. Some concern the influence of welding conditions [20–22], while others the effect of thermal aging on DSS 2304 corrosion behaviour [23–25]. Thus, a wider characterization of pitting and intergranular corrosion behaviour of this grade after thermal aging within a temperature range in which deleterious secondary phases can form is evidently an interesting topic. Moreover, the susceptibility to stress corrosion cracking (SCC) in critical environments, such as in media containing chlorides and hydrogen sulphide (H_2S) as encountered by components of oil and gas extraction equipment, can be correlated to pitting and intergranular corrosion (IGC) resistance, thus giving information about the microstructural modifications which can cause early failures in field conditions.

With this aim, the pitting corrosion resistance of the alloy was evaluated by both anodic polarization curves recording in 1 M NaCl solution at 20 °C and critical pitting temperature (CPT) measurements in 0.1 M NaCl solution. The IGC susceptibility was studied by double loop electrochemical potentiokinetic reactivation (DL-EPR) technique carried out in 33% H_2SO_4 solution with different amounts of HCl as depassivator. These electrochemical tests were performed under conditions specifically chosen to differentiate the response of the material after different thermal treatments in the 650–850 °C range, in order to evidence any corrosion resistance weakening. The SCC susceptibility of DSS 2304 was evaluated in standard NACE TM-0177 solution at pH 2.7 and 25 °C, in the presence of $S_2O_3^{2-}$ at 10^{-3} M, by slow strain rate tests (SSRT). The use of $S_2O_3^{2-}$ ions in replacement of H_2S was discussed in our preceding papers [26–28]. Depending on the alloy potential and solution pH, $S_2O_3^{2-}$ allows the assessment of the influence of H_2S on SCC resistance under health-safe testing conditions and at lower costs of experimental setup [29]. Finally, the correlation between the microstructural modifications produced by thermal aging and the electrochemical test and SSRT results was studied by observations under optical microscope (OM) and scanning electron microscope equipped with energy dispersion spectroscopy (SEM-EDS).

2. Materials and Methods

Annealed DSS 2304 alloy was provided by Outokumpu Company. The stainless steel chemical composition (wt%), evaluated by optical emission spectroscopy (OES), is shown in Table 1, where the alloy pitting resistance equivalent number (PRE_N) is also reported.

Table 1. Chemical composition (wt%) of DSS 2304.

DSS	C	Mn	Cr	Ni	Mo	N	Si	Cu	V	S	P	Fe	PRE_N *
DSS 2304	0.03	1.34	23.55	4.88	0.38	0.1	0.41	0.25	0.1	0.012	0.021	Bal.	26

* PRE_N (pitting resistance equivalent number) = % Cr + 3.3% Mo + 16% N.

15 mm × 15 mm specimens were cut from a 1.5-mm thick steel sheet. The specimens were heat treated for 5, 10 and 60 min at 650, 750 and 850 °C and then cooled in air. The microstructures obtained were observed by Zeiss EVO MA15 SEM (Oberkochen, Germany), coupled to an Oxford Aztec EDS system (Oxford, United Kingdom).

2.1. Electrochemical Measurements

The electrochemical tests were performed on electrodes obtained by embedding DSS 2304 samples in an epoxy resin (exposed area about 0.45 cm^2). The exposed surface was prepared with emery papers down to 2500 grit, polished by a diamond colloidal suspension (from 6 to 1 µm), rinsed by deionized water and finally degreased by acetone.

The anodic polarization curves were recorded after 1 h immersion in 1 M NaCl solution at 20 °C, starting from the open circuit potential (E_{OCP}) and with a scan rate of 0.1 mV/s. The pitting potentials (E_{pitt}) were determined from these curves and corresponded to the potential values at which the current density rapidly exceeded 1 µA·cm^{-2}. Each E_{pitt} average value was obtained from triplicate tests.

CPT tests were carried out in 0.1 M NaCl solution by employing a potentiostatic polarisation method. The solution was thermostated at 5 °C before the working electrode immersion. First, the electrode was cathodically polarized at -0.9 V_{SCE} for 5 min in order to de-oxidize the surface and improve the test reproducibility [30]. Then, it was allowed to stabilize at E_{OCP} for 30 min. The CPT was determined by raising the electrolyte temperature by 1 °C/min [31], at an applied potential of $+0.75$ V_{SCE}. CPT was defined as the temperature at which the current exceeded 100 µA for at least 60 s. Each CPT average value was obtained from triplicate tests.

DL-EPR measurements were conducted in 33% H_2SO_4 solution, at 20 °C, with controlled addition of HCl (0.3, 0.45 and 0.6%) acting as depassivator [32]. The DSS samples were cathodically polarized at -0.6 V_{SCE} for 3 min in order to improve the reproducibility. After 10 min stabilization under free corrosion conditions, the potential was cycled from E_{OCP} to $+0.3$ V_{SCE} and then to E_{OCP} again, under a scan rate of 2.5 mV/s. According to the standard [33], the active dissolution in the depleted zones is proportional to the ratio Ir/Ia, where Ir is the peak current in the reverse scan (peak reactivation current) and Ia is the peak current in the anodic scan (peak activation current). The degree of susceptibility (DOS) to intergranular corrosion was estimated by the percent ratio (Ir/Ia) × 100 [34]. Below a (Ir/Ia) × 100 value of about 1, the corrosion rate calculated by weight loss is reported to be negligible, while above the ratio of 1 the change in the DOS is strongly reflected in the weight loss values [35]. Moreover, if (Ir/Ia) × 100 is higher than 5, the samples could fail the Streicher, Strauss and Huey tests [33]. After the tests, the IGC attack morphology was observed by both OM and SEM.

2.2. Slow Strain Rate Test in NACE TM-0177 Solution in the Presence of $S_2O_3^{2-}$

Slow strain rate tests were carried out on annealed and heat-treated samples in order to investigate their susceptibility to SCC. The SSRT were performed with a strain rate of 10^{-6} s^{-1} on samples (length 23 cm, gauge portion 20 × 5 × 1.5 mm^3) prepared by EDM (electrical discharge machining). Heat treatments of 5, 10 and 60 min at 650, 750 °C and 850 °C were applied, followed by air cooling. Before the SSRT, the sample surfaces were polished with abrasive paper (down to 800 grit) and protected with an epoxy resin, with the exception of the gauge portion to be exposed to the aggressive environment. The test environment consisted of a de-aerated and thermostated (T = 25 °C) solution containing 5% NaCl + 0.5% CH_3COOH (the basic standard solution NACE TM-0177 [36] but free of H_2S) with the addition of 10^{-3} M $Na_2S_2O_3$ (final pH = 2.7). Both open circuit potential values, E_{OCP}, and stress-strain curve were recorded during the tests. A saturated calomel electrode (SCE) was used as a reference electrode. Three tests were conducted for each adopted condition (annealed and heat treated samples, in solution and in air tests). The R ratios between the fracture strain percentage measured in the test solution and that in air were calculated for all heat treatment conditions in order to evaluate the corresponding susceptibility to SCC; in particular, R values ≥ 0.8 were considered as an indication of immunity to the SCC [37]. After the tests, the gauge portion of each sample was cut, polished and etched with Beraha's reagent to characterize the crack initiation and propagation morphology.

3. Results

3.1. Microstructure

Figure 1 presents the microstructure of DSS 2304 aged for 5 to 60 min in the 650–850 °C range. Through observation with SEM in back-scattered electron (BSE), the austenitic phase (elongated in the rolling direction) was brighter than the ferritic phase, because of the higher nickel content of the former [16,38]. Due to the low molybdenum content of this alloy, no χ and σ intermetallic phases were identified after the heat treatments adopted [17]. In the sample aged for 5 min at 650 °C (Figure 1b), no precipitates were observed at the grain boundaries. By prolonging the aging time to 60 min at 650 °C, very small black precipitates were detected at the α/γ interphases (indicated by arrows, Figure 1c). After 10 and 60 min at 750 °C (Figure 1d,e), a more important precipitation of these intermetallic phases was observed and a brighter phase was observable between the precipitates and the ferrite matrix. This phase was likely the so-called secondary austenite (γ_2) [17,38]. The thermal treatment performed for 60 min at 850 °C (Figure 1f) led to the formation of more voluminous black particles. In Figure 2 the results of SEM-EDS analysis, performed at several points along the yellow line, which passes through the black particles and the γ_2 phase, are reported. The increase in wt% of chromium, molybdenum and carbon evidenced that these precipitates were essentially $M_{23}C_6$-type carbides, which are generally observed in DSS [5,39,40].

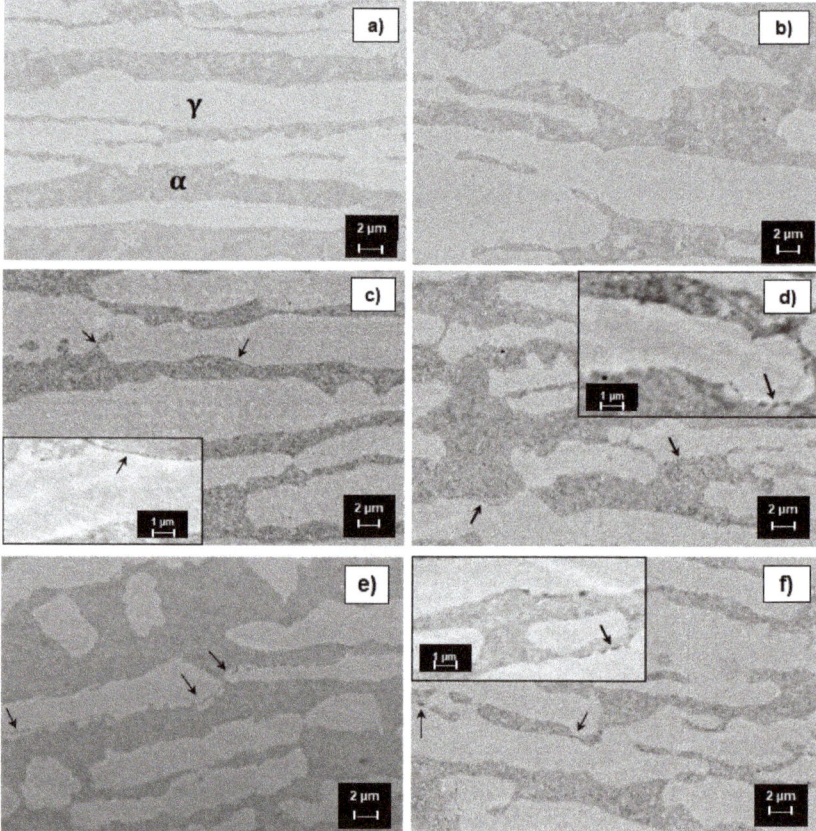

Figure 1. SEM-BSD micrographs of both duplex stainless steel (DSS) 2304 as-received (**a**) and aged for: 5 (**b**) and 60 min (**c**) at 650 °C, 10 (**d**) and 60 min (**e**) at 750 °C and 60 min at 850 °C (**f**).

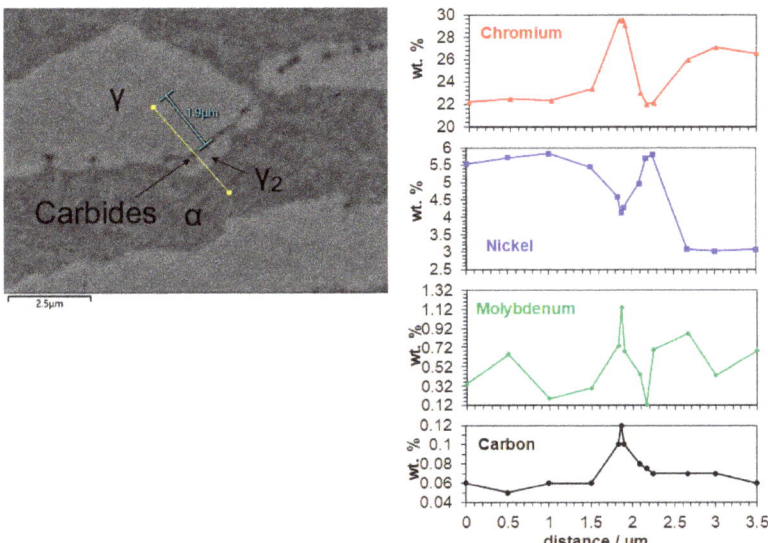

Figure 2. SEM-BSD image of DSS 2304 sample aged for 60 min at 750 °C. Quantitative profile line analysis of: chromium (**red**), nickel (**blue**), molybdenum (**green**) and carbon (**black**).

Figure 2 also clearly confirms the growth of γ_2 due to chromium and molybdenum depletion and nickel enrichment in the original ferrite phase adjacent to the chromium carbides.

3.2. Pitting Potential Measurements

Table 2 collects the average E_{pitt} values obtained by the anodic polarization curves in 1 M NaCl solution at 20 °C on DSS 2304 electrodes aged at the three investigated temperatures.

Table 2. Average E_{pitt} values with standard deviation evaluated in 1 M NaCl solution at 20 °C on both as-received and aged DSS 2304 electrodes.

E_{pitt} (V_{SCE})									
As-Received	650 °C			750 °C			850 °C		
0 min	5 min	10 min	60 min	5 min	10 min	60 min	5 min	10 min	60 min
0.918 * ± 0.03	0.884 * ± 0.05	0.750 ± 0.08	0.481 ± 0.04	0.890 * ± 0.01	0.615 ± 0.01	0.336 ± 0.03	0.635 ± 0.06	0.640 ± 0.07	0.813 ± 0.04

* Transpassive potential range.

As an example, in Figure 3 the anodic polarization curves recorded on samples, both as-received and aged at 750 °C for different times, are shown. For the as-received sample and for those aged for 5 min at 650 and 750 °C, a passive state was detected at E_{OCP} and at higher potentials up to about 0.8 V_{SCE}, after which a transpassive behaviour occurred. After a 10-min treatment at these temperatures, the 2304 grade became moderately susceptible to pitting corrosion and E_{pitt} values of 0.750 (at 650 °C) and 0.615 (at 750 °C) V_{SCE} were measured, but the passive current density (i_{pass}) remained constant. After 60 min aging at 650 and 750 °C, a further drop in E_{pitt} was observed and values of 0.481 and 0.336 V_{SCE} were obtained, respectively. A permanence of only 5 or 10 min at 850 °C produced a certain tendency to pitting corrosion at potentials higher than about 0.64 V_{SCE}, but, in this case, a recovery of the corrosion resistance was observed by increasing the aging time to 60 min (E_{pitt} = 0.813 V_{SCE}), most likely due to the replenishment of chromium by diffusion from the grain cores to the impoverished zones [11,15].

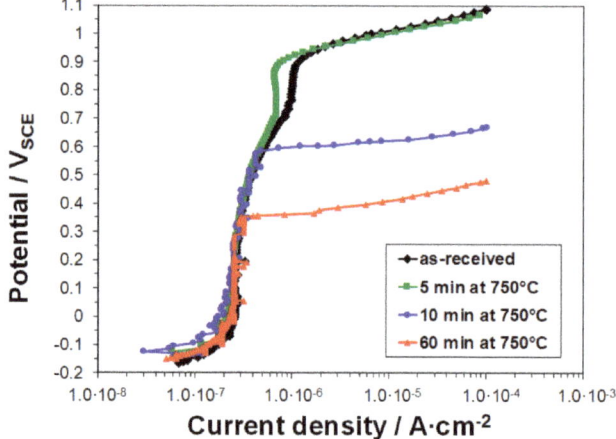

Figure 3. Anodic polarization curves recorded in 1 M NaCl solution at 20 °C on both as-received and aged DSS 2304.

3.3. CPT Test

Figure 4 shows the current density/temperature curves obtained for as-received and DSS 2304 electrodes aged at 750 °C during polarization at 0.75 V_{SCE} in 0.1 M NaCl solution. In agreement with the potentiodynamic tests, the as-received and 5-min aged samples maintained very low current densities up to temperatures over 32 °C. Then, an abrupt current increase was observed and current values exceeded 100 µA/cm^2 at temperatures of 35.3 ± 1.1 and 35.2 ± 1.3 °C, respectively. A moderate decrease in CPT (about 6 °C) was detected for the sample aged at 750 °C for 10 min and, by extending the treatment time to 60 min, CPT decreased to 21.2 ± 0.6 °C.

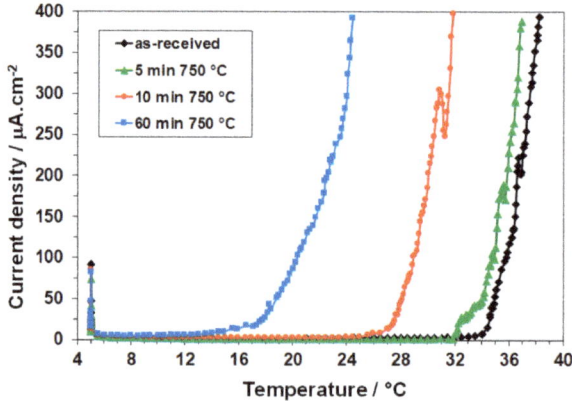

Figure 4. Current density vs. temperature curves obtained in 0.1 M NaCl solution on both as-received and aged DSS 2304.

Figure 5 reports the histogram that collects average CPT values in 0.1 M NaCl solution on DSS 2304 electrodes for all the tested aging conditions. The CPT values were in good agreement with the results of the potentiodynamic test. Both these techniques showed a significant reduction in pitting corrosion resistance after 60 min of aging at 650 °C, in comparison to that of as-received samples. At 750 °C, a 10-min aging was sufficient to have a relevant decrease in CPT and E_{pitt}, and a further strong reduction was observed after 60 min. A 5- and 10-min aging at 850 °C also determined a significant tendency to pitting corrosion in 0.1 M NaCl solution. However, a recovery was detected after 60-min aging, with CPT and E_{pitt} only slightly lower than that obtained on as-received specimens.

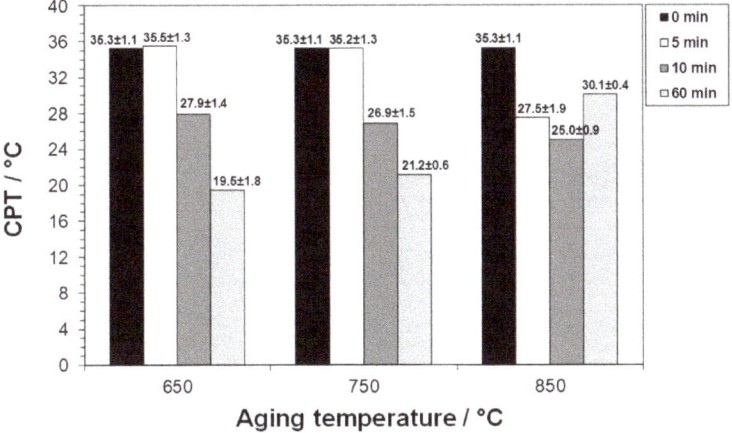

Figure 5. Average critical pitting temperature (CPT) values and standard deviations determined in 0.1 M NaCl solution for both as-received and aged DSS 2304.

Figure 6 shows a micrograph acquired on the surface of DSS 2304 electrode aged for 10 min at 750 °C after CPT test in 0.1 M NaCl solution. In agreement with the results of other authors [23,30], the localized corrosion attack appeared to start at the γ phase boundary, most likely in correspondence of Cr and Mo depleted areas around precipitates, then propagated in the ferrite phase, where relatively large pits formed.

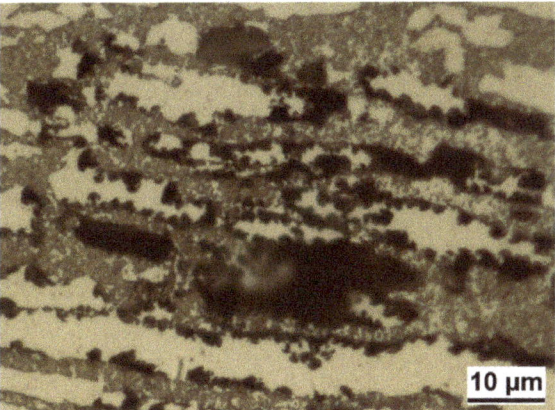

Figure 6. Micrograph of DSS 2304 electrode aged for 10 min at 750 °C after CPT test in 0.1 M NaCl solution (etching with Beraha's reagent).

3.4. DL-EPR Test

In order to determine the sensitivity of the alloys to IGC, the optimal concentration of the depassivator (HCl) to be added to the sulphuric acid solution was evaluated, prior to the extensive application of the DL-EPR technique. With this aim, preliminary tests were performed on both as-received and 60-min aged DSS 2304, with HCl concentrations of 0.3, 0.45 and 0.6%. In Table 3, the Ir/Ia% ratios obtained under the different conditions are reported. A 0.3% HCl content was not enough to differentiate the DOS to IGC of the samples aged for 60 min at 750 and 850 °C. Moreover, with a HCl concentration of 0.6%, the as-received specimen showed Ir/Ia% values exceeding 1%, evidencing that a significant generalized corrosion had occurred, together with IGC [41]. In the presence of a 0.45% HCl, the as-received specimen maintained an Ir/Ia% value lower than 1%, while a good Ir/Ia% variation was detectable for the heat-treated samples, indicating that the attack selectively occurred at the grain boundary areas depleted in passivating elements.

Table 3. Ir/Ia% ratios and standard deviations in 33% H_2SO_4 solution with different HCl concentrations, at 20°C. Double loop electrochemical potentiokinetic reactivation (DL-EPR) tests performed on DSS 2304, both as-received and 60-min aged at 650, 750 and 850 °C.

HCl Concentration (%)	Ir/Ia%			
	As-Received	60 min 650 °C	60 min 750 °C	60 min 850 °C
0.3	0.02 ± 0.002	4.4 ± 0.05	0.2 ± 0.01	0.03 ± 0.003
0.45	0.08 ± 0.003	5.9 ± 0.07	1.4 ± 0.04	0.2 ± 0.02
0.6	1.7 ± 0.02	9.2 ± 0.11	5.2 ± 0.08	3.0 ± 0.05

Therefore, 0.45% HCl was chosen as the correct depassivator concentration to be used for all DL-EPR tests, and the DOS values obtained in 33% H_2SO_4 + 0.45% HCl are presented as a histogram in Figure 7. Aging times of 5 and 10 min at 650 °C did not affect the resistance to IGC, whereas, 60-min aging at 650 °C determined a very high increase in Ir/Ia% parameter (significantly higher than that obtained with a treatment of 10 min). An aging of 10 min at 750 °C was sufficient to cause a high susceptibility to IGC (Ir/Ia% = 1.7%), maintained also on the sample aged for 60 min at the same temperature. The temperature of 850 °C did not determine susceptibility to IGC (Ir/Ia% < 1%), in spite of the formation of relatively large precipitates (Figure 1) [42].

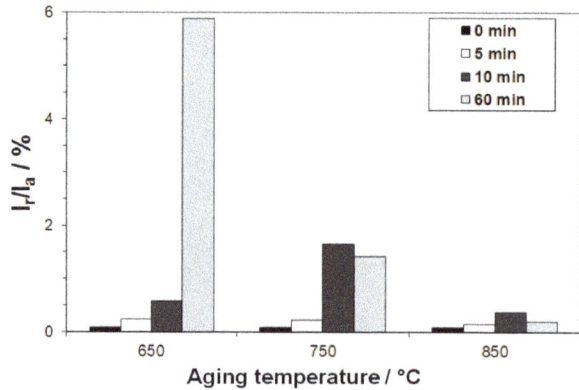

Figure 7. Ir/Ia% determined for as-received and aged DSS 2304, by DL-EPR tests in 33% H_2SO_4 + 0.45% HCl solution, at 20 °C.

In Figure 8, the micrographs acquired by OM after DL-EPR tests on DSS 2304 electrodes heat treated for 60 min at different temperatures are reported. Confirmation of a marked IGC attack was detected in the 650 °C aged sample (DOS value of 5.9%). Instead, the IGC attack became less evident with the rise in aging temperature to 750 and 850 °C (DOS values of 1.4 and 0.2%, respectively).

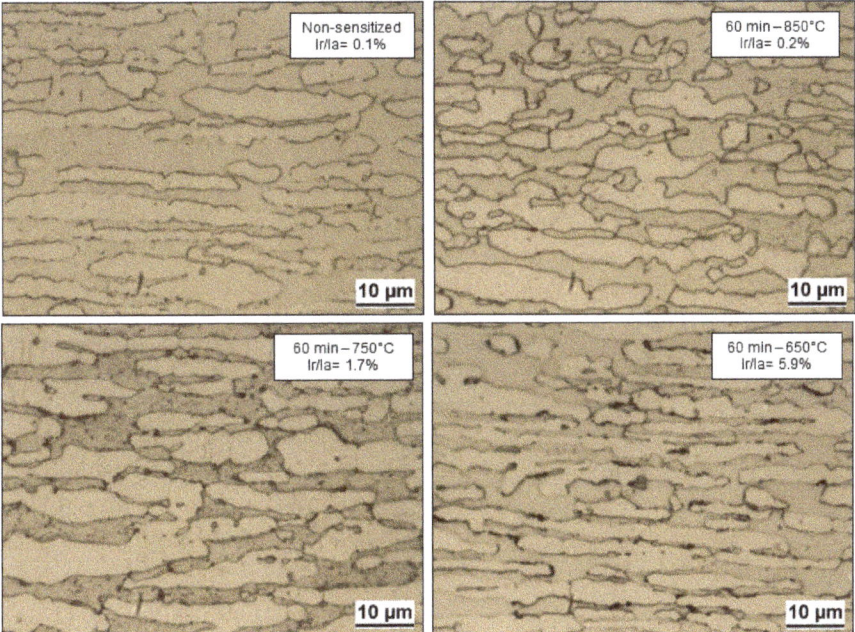

Figure 8. Optical microscope micrographs acquired on DSS 2304 electrodes after DL-EPR tests in 33% H2SO4 solution, at 20 °C, with 0.45% HCl addition.

The DL-EPR results obtained for the 650 and 750 °C aged samples are in agreement with CPT values, suggesting that the thermal aging that were critical for pitting corrosion (i.e., 60 min at 650 °C and 10 and 60 min at 750 °C), also determined a very high susceptibility to IGC. The samples aged at 850 °C had very low DOS values even if a moderate decrease in pitting corrosion resistance was observed. Figure 9 reports some representative results of SEM-EDS analysis acquired after DL-EPR tests performed on DSS 2304 electrodes. The images of samples aged at 650 and 750 °C for 60 min show that the attack mainly occurred around the precipitates at the α/γ interfaces, that is, on γ and likely on γ_2 phases, due to passivating element depletion. The line elemental analysis performed by EDS across the precipitates revealed peaks related to chromium, molybdenum and carbon, confirming the presence of chromium and molybdenum carbides ($M_{23}C_6$-type precipitates [5]), while the small peak related to nitrogen suggests the likely concomitant formation of some nitrides [23,39].

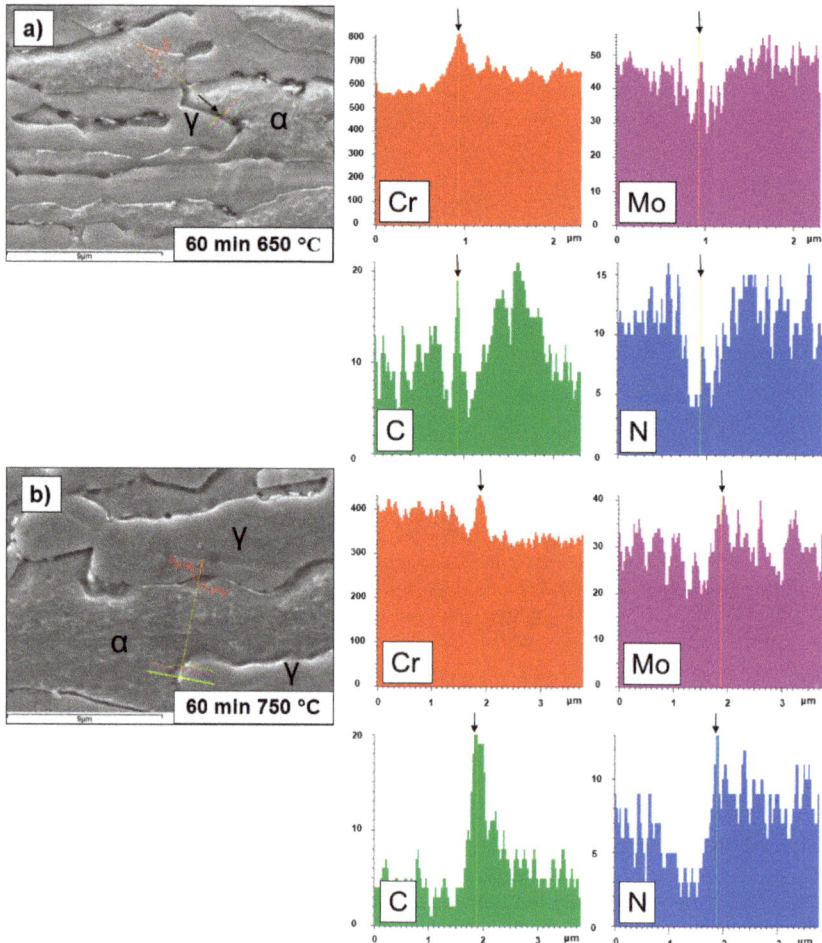

Figure 9. SEM-BSD micrographs of DSS 2304 aged for 60 min at 650 (**a**) and 750 °C (**b**), after DL-EPR test. Scanning electron microscope–energy dispersion spectroscopy (SEM-EDS) line profile analysis of: chromium (red), molybdenum (green), carbon (blue) and nitrogen (purple).

3.5. SSRT

Table 4 collects the average $\varepsilon_f\%$ values obtained with SSRT carried out in air at 25 °C and in NACE TM-0177 in the presence of 10^{-3} M $S_2O_3^{2-}$, before and after thermal aging.

Table 4. Average values and standard deviations of $\varepsilon_f\%$ from slow strain rate tests (SSRT) performed in air at 25 °C and in NACE TM-0177 in the presence of 10^{-3} M $S_2O_3^{2-}$ on DSS 2304 as-received and thermally aged.

$\varepsilon_f\%$	As Received	650 °C			750 °C			850 °C		
		5 min	10 min	60 min	5 min	10 min	60 min	5 min	10 min	60 min
Air at 25 °C	52 ± 1	52 ± 3	52 ± 2	50 ± 1	51 ± 2	52 ± 3	56 ± 2	51 ± 1	53 ± 2	57 ± 1
NACE TM-0177 with 10^{-3} M $S_2O_3^{2-}$	51 ± 2	51 ± 1	43 ± 1	18 ± 3	50 ± 3	22 ± 1	25 ± 3	51 ± 3	22 ± 2	39 ± 2

In air at 25 °C, only the samples aged for 60 min at 750 and 850 °C showed a moderately higher ductility (increase in $\varepsilon_f\%$ of about 10%, Table 4), due to the subtraction of interstitial atoms from solid solutions, after the precipitation of chromium carbides and nitrides at the grain boundaries (Figure 2). All other aged samples presented $\varepsilon_f\%$ values close to that of the as-received one. As an example, Figure 10a shows the behaviour of the as-received sample and the samples aged at 850 °C for different times. A similar behaviour was observed on the alloy LDSS 2101, with a nitrogen content of 0.22 wt%, where an increment in ductility of about 40% after 30 min aging at 850 °C was detected, essentially due to a large chromium nitride precipitation [28]. In DSS 2304, the observed formation of ductile γ_2 phase under prolonged aging could also have contributed to the $\varepsilon_f\%$ increase [43].

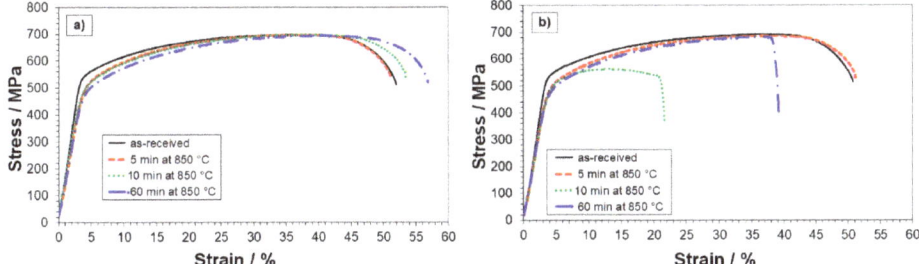

Figure 10. Stress—strain curves obtained with SSRT performed in air at 25 °C (**a**) and in NACE TM-0177 in the presence of 10^{-3} M $S_2O_3^{2-}$ (**b**) on DSS 2304, both as-received and aged at 850 °C for 5, 10 and 60 min.

In NACE TM-0177 containing 10^{-3} M $S_2O_3^{2-}$, all samples aged for 5 min maintained the same ductility as the as-received sample. On the contrary, at increasing aging time and temperature, a general reduction in $\varepsilon_f\%$ was observed, while only the sample aged for 60 min at 850 °C exhibited a partial recovery in SCC resistance (Table 4). In fact, as highlighted in Figure 10b, the sample aged for 5 min at 850 °C maintained its ductility, an aging of 10 min at 850 °C determined a relevant decrease in $\varepsilon_f\%$ from 53 to 22% and a recovery of ductility was observed by extending the aging time up to 60 min ($\varepsilon_f\% = 39\%$).

Figure 11 compares the extent of necking and the surface aspect of both the as-received sample (Figure 11a) and the samples aged at 850 °C (Figure 11b–d), at the end of SSRT. The as-received sample and that aged for 5 min showed a ductile-type fracture and a light general corrosion attack only on the latter. A brittle-type fracture occurred in DSS 2304 aged 10 min at 850 °C (Figure 11c), with evidence of numerous secondary cracks and presence of areas in which the corrosion attack became strongly localized (dark stains). For the sample aged at 850 °C, for the longest time (Figure 11d), the fracture was still of brittle-type, with presence of secondary crack propagation, but again a more generalized corrosion attack, instead of a localized one, was visible. The samples aged for 10 and 60 min at 750 °C and that aged for 60 min at 650 °C (not shown) presented a fracture morphology and a surface attack similar to the severe one exhibited in Figure 11c. Instead, DSS 2304 aged for only 10 min at 650 °C (not shown) behaved similarly to that aged 60 min at 850 °C. The behaviour of the other samples was similar to that of the as-received one.

The morphology of the corrosion attack observed in section on samples aged at 850 °C is shown in Figure 12. The sample in Figure 11b, aged for 5 min, clearly evidences the formation of several small pits which did not initiate cracks (Figure 12a). Differently, the sections of the sample in Figure 11c, aged for 10 min (Figure 12b), and that in Figure 11d, aged for 60 min (picture not reported), show that pits triggered many cracks, indicating SCC failure. These cracks mainly propagated in the ferrite phase or followed austenite/ferrite interface.

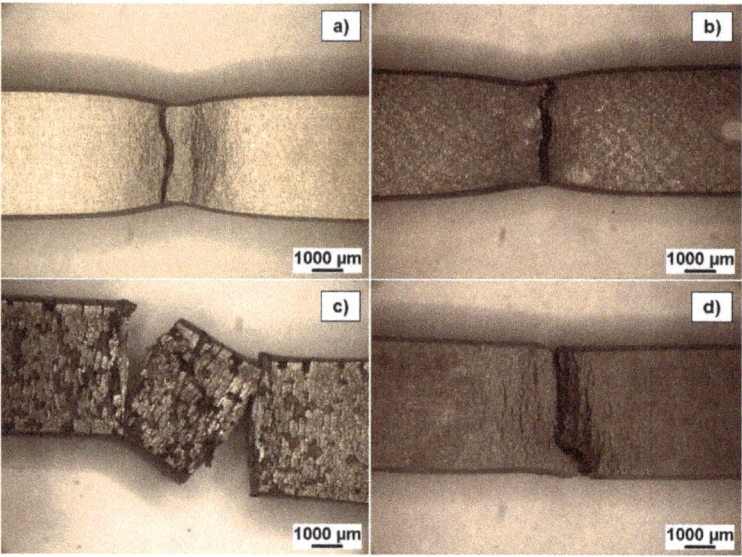

Figure 11. Macrograph acquired on the gauge length of DSS 2304 as-received (**a**) and aged at 850 °C for 5 (**b**), 10 (**c**) and 60 min (**d**) after SSRT in NACE TM-0177 in the presence of 10^{-3} M $S_2O_3^{2-}$.

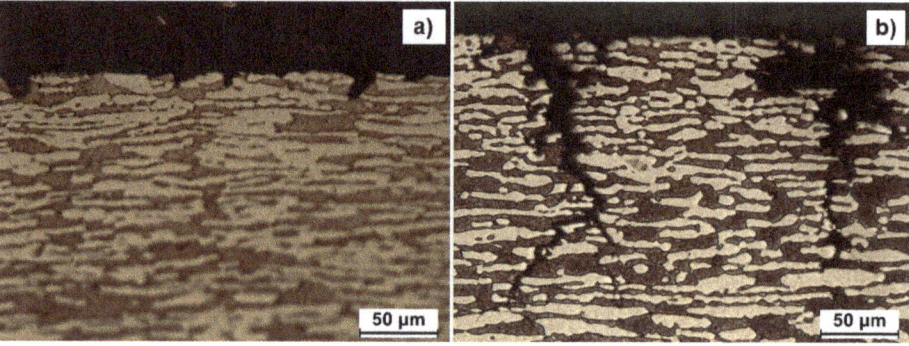

Figure 12. Micrographs of DSS 2304 aged for 5 (**a**) and 10 min (**b**) at 850 °C after SSRT in NACE solution containing 10^{-3} M $S_2O_3^{2-}$ (long transverse sections, parallel to load direction).

These differences may be rationalized by comparing the different potential trends shown during SSRT by the as-received sample and by the samples aged at 850 °C for different times (Figure 13). The E_{OCP} values of the as-received sample remained rather noble (around -0.25 V_{SCE}) throughout the test, indicating passive conditions [27,44], as confirmed by the visual aspect of the sample at the end of the exposure (Figure 11a). A significant initial E_{OCP} drop towards negative values was detected for DSS 2304 aged for 5 min at 850 °C. However, after a strain of about 6%, the sample recovered a passive state. Conversely, the E_{OCP} values of the sample aged for 10 min at 850 °C decreased rapidly and reached values close to -0.55 V_{SCE}. The same E_{OCP} trend was shown by the other samples showing significant ductility reduction (aging for 60 min at 650 and for 10 and 60 min at 750 °C). The sample aged for 60 min at 850 °C, characterized by a moderate ductility decrease, reached quite negative E_{OCP} values of about -0.5 V_{SCE}, but at a much slower rate. This also occurred for the sample aged at 650 °C for 10 min, exhibiting comparable ductility behaviour.

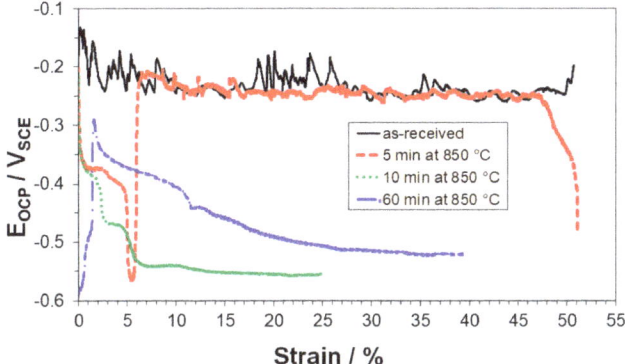

Figure 13. E_{OCP}—strain curves obtained with SSRT performed in NACE TM-0177 in the presence of 10^{-3} M $S_2O_3^{2-}$ on DSS 2304, both as-received and aged at 850 °C for 5, 10 and 60 min.

The potential-pH diagrams calculated for elementary sulphur and oxygen adsorbed on metals, such as Fe, Ni or Cr, and those calculated for the S-Fe (or Ni or Cr)-water systems at low S molality (10^{-4} mole/kg), at the temperature of 25 °C [44–46], suggest that at pH 2.7 all these E_{OCP} values in the range $-0.25/-0.55$ V_{SCE} are compatible with $S_2O_3^{2-}$ ion reduction to adsorbed sulphur and H_2S. The amounts of these reduced species, capable of impairing the alloy passivity and favouring hydrogen penetration in the alloy, reasonably increased at decreasing E_{OCP} values, thus justifying the higher SCC susceptibility the longer the time persistence at quite negative E_{OCP} values.

The histogram reported in Figure 14 collects the R values obtained from SSRT results. The as-received sample and those aged for 5 min at the three different temperatures were not susceptible to SCC in the presence of $S_2O_3^{2-}$, as their R values were close to 1. By prolonging the aging time up to 10 min, the alloy became moderately susceptible to SCC already at 650 °C and much more susceptible at 750 and 850 °C. With an aging time of 60 min, the susceptibility was high at 650 °C and gradually decreased by raising the treatment temperature, showing a restoring of SCC resistance at 850 °C, as previously observed for E_{pitt} and CPT results.

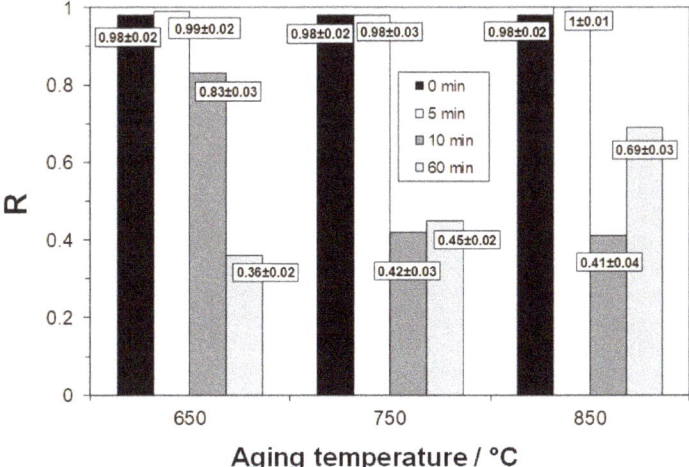

Figure 14. SCC susceptibility (R index) of DSS 2304, before and after thermal aging, obtained by SSRT in NACE solution containing 10^{-3} M $S_2O_3^{2-}$.

4. Discussion

This research shows that, in agreement with the literature [2,5,10], aging DSS 2304 samples between 650 and 850 °C induces the precipitation of chromium and molybdenum carbides (Figure 2), and perhaps nitrides (Figure 9). The nucleation of these precipitates occurs essentially at the α/γ interface, acting as a connection between Cr-rich ferrite (which is also relatively rich in Mo) and C-rich austenite [10,47,48]. Then, carbides grow into the ferrite phase, because of the higher diffusivity of Cr and Mo in this phase [49]. As a consequence of carbide precipitation, conversion of ferrite phase close to carbides into γ_2 phase (having a lower Cr and Mo content in comparison to γ) occurs to a certain extent (Figure 2), because of ferrite instability after depletion in ferrite-stabilizing elements [4,40]. γ_2 phase and, likely, the narrow zone close to carbides in the original γ phase (depleted in passivating elements, like γ_2) become the preferential sites for localized corrosion and IGC attack (Figures 6 and 9).

The 60 min treatments at 650 and 750 °C are the most critical for pitting corrosion resistance and IGC susceptibility, as they significantly worsen the alloy behaviour against these corrosion forms. The thermal aging at 850 °C did not affect IGC sensitization, but produced a moderate worsening in pitting resistance already after 5 min of permanence, which persisted after 10 min. A recovery of localized corrosion resistance was observed after the longest aging time, likely due to the high-temperature rediffusion of the key passivating alloying elements (Cr, Mo, N) from the phase cores towards the phase boundaries [11,15,50,51]. In agreement with the findings of other authors [1,52,53], at low aging temperatures (i.e., 650 and 750 °C) the precipitated carbides were small and finely distributed, while at 850 °C more discontinuous larger precipitates were detected as a consequence of the higher diffusion rates (Figure 1). Thus, the higher sensitization to localized corrosion attack obtained after long aging treatments at the two lower temperatures could also be due to the higher width and continuity of the Cr- and Mo-depleted areas.

The SSRT results on SCC susceptibility of as-received and thermally aged DSS 2304 in NACE solution containing 10^{-3} M $S_2O_3^{2-}$ are in good general agreement with pitting corrosion and IGC resistance. In particular, the conditions determining the best alloy behaviour towards pitting and IGC corrosion (as-received, 5 min at 650 °C and 5 min at 750 °C) are also quite resistant to SCC. Moreover, the aging conditions determining the lowest pitting resistance and highest sensitization to IGC (at 650 °C for 60 min and at 750 °C for 10 and 60 min) also induce the highest susceptibility to SCC in DSS 2304. In fact, the synergistic effect of Cl^- and $S_2O_3^{2-}$ [54] leads to a much easier pit development in the areas depleted in passivating elements and finally to an earlier SCC failure, due to crack growth at the bottom of pits.

During the elastic deformation step in SSRT, the most SCC susceptible samples quickly reached quite negative E_{OCP} values of about -0.55 V_{SCE}, which were maintained for long times. Under these potential/pH conditions, a significant conversion of $S_2O_3^{2-}$ into adsorbed sulphur and H_2S is expected [55], suggesting that, beside an active path mechanism, hydrogen penetration may also contribute to SCC failure [29,45]. As for pitting corrosion and IGC, also in the case of SCC, a long aging at 850 °C determines a recovery of DSS 2304 performance, most likely due to Cr- and Mo-rediffusion and to the formation of less continuous Cr- and Mo-depleted areas.

Some discrepancies among the relative resistance to different forms of localized corrosion refer to specimens subjected to thermal aging of intermediate severity. The differences detected can be ascribed to the different nature of the localized corrosion forms addressed and to the different conditions applied during the tests. As an example, both specimens aged for 5 and 10 min at 850 °C show some tendency to pitting corrosion in neutral chloride solution, due the presence of Cr- and Mo-depleted regions, even if the DOS values to IGC are low ($I_r/I_a\% < 1$). It is likely that the impoverished areas are relatively narrow and discontinuous at this high aging temperature thus determining a limited IGC attack. The susceptibility to SCC of these same specimens is certainly influenced by the presence of regions depleted in passivating elements but, during the dynamic conditions applied by SSRT, cracks only develop if the metal repassivation rate decreases and becomes at least comparable to the rate of formation of new bare metal surfaces at the pit bottom. This is likely to occur in the case of the

10-min aged specimens that show a low R value. Instead, in the case of the 5-min aged specimens the repassivation rate likely remains too high for severe crack development and R values of about 1 are obtained.

5. Conclusions

1. The heat treatment in the 650–850 °C range on DSS 2304 determined the formation of chromium carbides at α/γ interphase and, under some conditions, the growth of Cr- and Mo-depleted γ_2 phase.
2. These microstructural modifications affected the localized corrosion performances of this alloy. Pitting and IGC mainly initiated in Cr- and Mo-depleted regions near to precipitates inside the γ_2 and likely also the γ phases, then propagated in the ferrite matrix.
3. E_{pitt} and CPT values indicated a decrease in pitting resistance of DSS 2304 after 10 min aging at 650 and 750 °C, and the pitting behaviour worsened after longer aging time at these temperatures. At 850 °C, a 5-min aging was sufficient to markedly decrease the pitting resistance, but a recovery was observed after 60 min of aging.
4. Similarly, DL-EPR results evidenced a significant IGC sensitization of DSS 2304 after 10 min at 750 °C and 60-min aging at both 650 and 750 °C but no aging treatment at 850 °C was detrimental to the alloy IGC resistance.
5. As for pitting corrosion, SCC susceptibility in NACE solution containing 10^{-3} M $S_2O_3^{2-}$ was also detected after 10 min aging at 650 and 750 °C and increased after longer aging time. SCC also occurred on the sample aged for 10 min at 850 °C. A longer heat treatment at this high temperature ensured a recovery of SCC resistance.
6. SCC failure initiated at the bottom of pits and was likely stimulated by hydrogen penetration.

Author Contributions: F.Z. and C.M. designed and planned the experiments; F.Z. and V.G. carried out all the tests; F.Z. and Andrea Balbo examined the obtained data; F.Z., C.M. and Andrea Balbo wrote the paper.

Funding: This research received no external funding.

Acknowledgments: The authors wish to thank Mattia Merlin for heat treatments of the studied materials and Marco Frigo (Customer Service Manager of Outokumpu S.p.A.) for supplying DSS 2304 samples.

Conflicts of Interest: The authors declare no conflict of interest.

References

1. Pohl, M.; Storz, O.; Glogowski, T. Effect of intermetallic precipitations on the properties of duplex stainless steel. *Mat. Charact.* **2007**, *58*, 65–71. [CrossRef]
2. Badji, R.; Kherrouba, N.; Mehdi, B.; Cheniti, B.; Bouabdallah, M.; Kahloun, C.; Bacroix, B. Precipitation kinetics and mechanical behavior in a solution treated and aged dual phase stainless steel. *Mater. Chem. Phys.* **2014**, *148*, 664–672. [CrossRef]
3. Wang Chan, K.; Chin Tjong, S. Effect of secondary phase precipitation on the corrosion behavior of Duplex Stainless Steels. *Materials* **2014**, *7*, 5268–5304. [CrossRef] [PubMed]
4. Escriba, D.M.; Materna-Morris, E.; Plaut, R.L.; Padilha, A.F. Chi-phase precipitation in a duplex stainless steel. *Mater. Charact.* **2009**, *60*, 1214–1219. [CrossRef]
5. Knyazeva, M.; Pohl, M. Duplex Steels. Part II: Carbides and Nitrides. *Metallogr. Microstruct. Anal.* **2013**, *2*, 343–351. [CrossRef]
6. Moura, V.S.; Lima, L.D.; Pardal, J.M.; Kina, A.Y.; Corte, R.R.A.; Tavares, S.S.M. Influence of microstructure on the corrosion resistance of the duplex stainless steel UNS S31803. *Mat. Charact.* **2008**, *59*, 1127–1132. [CrossRef]
7. Lopez, N.; Cid, M.; Puigalli, M. Infuence of σ-phase on mechanical properties and corrosion resistance of duplex stainless steels. *Corros. Sci.* **1999**, *41*, 1615–1631. [CrossRef]

8. Lopez, N.; Cid, M.; Puigalli, M.; Azkarate, I.; Pelayo, A. Application of double loop electrochemical potentiodynamic reactivation test to austenitic and duplex stainless steels. *Mater. Sci. Eng. A* **1997**, *A229*, 123–128. [CrossRef]
9. Chen, T.H.; Yang, J.R. Effect of solution treatment and continuous cooling on σ-phase precipitation in a 2205 duplex stainless steel. *Mater. Sci. Eng. A* **2001**, *A311*, 28–41. [CrossRef]
10. Kashiwar, A.; Phani Vennela, N.; Kamath, S.L.; Khatirkar, R.K. Effect of solution annealing temperature on precipitation in 2205 duplex stainless steel. *Mat. Charact.* **2012**, *74*, 55–63. [CrossRef]
11. Sathirachinda, N.; Petterson, R.; Wessman, S.; Pan, J. Study of nobility of chromium nitrides in isothermally aged duplex stainless steels by using SKPFM and SEM/EDS. *Corros. Sci.* **2010**, *52*, 179–186. [CrossRef]
12. Zhang, L.; Jiang, Y.; Deng, B.; Zhang, W.; Xu, J.; Li, J. Effect of aging on the corrosion of 2101 lean duplex stainless steel. *Mat. Charact.* **2009**, *60*, 1522–1528. [CrossRef]
13. Ramirez, A.J.; Lippold, J.C.; Brandi, S.D. The relationship between Chromium Nitride and Secondary Austenite precipitation in Duplex Stainless Steels. *Metallurg. Mater. Trans. A* **2003**, *34A*, 1575–1596. [CrossRef]
14. Charles, J.; Chemelle, P. The history of duplex developments, nowadays DSS properties and duplex market future trends. In Proceedings of the 8th Duplex Stainless Steels Conference, Beaune, France, 13–15 October 2010.
15. Zanotto, F.; Grassi, V.; Merlin, M.; Balbo, A.; Zucchi, F. Effect of brief heat treatments performer between 650 and 850 °C on corrosion behaviour of a lean duplex stainless steel. *Corros. Sci.* **2015**, *94*, 38–47. [CrossRef]
16. Berner, M.; Liu, H.P.; Olsson, C.O.A. Estimating localized corrosion resistance of low alloy stainless steels: comparison of pitting potentials and critical pitting temperatures measured on lean duplex stainless steel LDX 2101 after sensitization. *Corros. Eng. Sci. Techn.* **2008**, *43*, 111–116. [CrossRef]
17. Liu, H.; Johansson, P.; Liljas, M. Structural evolution of LDX 2101® (EN 1.4162) during isothermal ageing at 600–850°C. In Proceedings of the 6th European Stainless Steel Conference, Science and Market, Helsinki, Finland, 10–13 June 2008; pp. 555–560.
18. Sun, K.; Zeng, M.; Shi, Y.; Hu, Y.; Shen, X. Microstructure and corrosion behavior of S32101 stainless steel underwater dry and wet welded joints. *J. Mater. Process. Tech.* **2018**, *256*, 190–201. [CrossRef]
19. Zhang, Z.; Zhao, H.; Zhang, H.; Yu, Z.; Hu, J.; He, L.; Li, J. Effect of isothermal aging on the pitting corrosion resistance of UNS S82441 duplex stainless steel based on electrochemical detection. *Corros. Sci.* **2015**, *93*, 120–125. [CrossRef]
20. Tan, H.; Wang, Z.; Jiang, Y.; Yang, Y.; Deng, B.; Song, H.; Li, J. Influence of welding thermal cycles on microstructure and pitting corrosion resistance of 2304 duplex stainless steels. *Corros. Sci.* **2012**, *55*, 368–377. [CrossRef]
21. Chen, L.; Tan, H.; Wang, Z.; Li, J.; Jiang, Y. Influence of cooling rate on microstructure evolution and pitting corrosion resistance in the simulated heat-affected zone of 2304 duplex stainless steels. *Corros. Sci.* **2012**, *58*, 168–174. [CrossRef]
22. Jiang, Y.; Tan, H.; Wang, Z.; Hong, J.; Jiang, L.; Li, J. Influence of Creq/Nieq on pitting corrosion resistance and mechanical properties of UNS S32304 duplex stainless steel welded joints. *Corros. Sci.* **2013**, *70*, 252–259. [CrossRef]
23. Pezzato, L.; Lago, M.; Brunelli, K.; Breda, M.; Piva, E.; Calliari, I. Effect of Secondary Phases Precipitation on Corrosion Resistance of Duplex Stainless Steels. *Mater. Sci. Forum* **2016**, *879*, 1495–1500. [CrossRef]
24. Guo, L.; Li, X.; Sun, T.; Xu, J.; Li, J.; Jiang, Y. The influence of sensitive temperature on the localized corrosion resistance of duplex stainless steel SAF2304. *Acta Metall. Sin.* **2012**, *48*, 1503–1509. [CrossRef]
25. Zhang, Z.; Han, D.; Jiang, Y.; Shi, C.; Li, J. Microstructural evolution and pitting resistance of annealed lean duplex stainless steel UNS S32304. *Nuclear Eng. Design* **2012**, *243*, 56–62. [CrossRef]
26. Zanotto, F.; Grassi, V.; Balbo, A.; Monticelli, C.; Zucchi, F. Stress corrosion cracking of LDX 2101® duplex stainless steel in chloride solutions in the presence of thiosulphate. *Corros. Sci.* **2014**, *80*, 205–212. [CrossRef]
27. Zanotto, F.; Grassi, V.; Balbo, A.; Monticelli, C.; Zucchi, F. Stress-Corrosion Cracking Behaviour of Lean-Duplex Stainless Steels in Chloride/Thiosulphate Environments. *Metals* **2018**, *8*, 237. [CrossRef]
28. Zanotto, F.; Grassi, V.; Balbo, A.; Monticelli, C.; Melandri, C.; Zucchi, F. Effect of brief thermal aging on stress corrosion cracking susceptibility of LDSS 2101 in the presence of chloride and thiosulphate ions. *Corros. Sci.* **2018**, *130*, 22–30. [CrossRef]

29. Tsujikawa, S.; Miyasaka, A.; Ueda, M.; Ando, S.; Shibata, T.; Haruna, T.; Katahira, M.; Yamane, Y.; Aoki, T.; Yamada, T. Alternative for evaluating sour gas resistance of low-alloy steels and corrosion-resistant alloys. *Corrosion* **1993**, *49*, 409–419. [CrossRef]
30. He, L.; Guo, Y.-J.; Wu, X.-Y.; Jiang, Y.-M.; Li, J. Effect of Solution Annealing Temperature on Pitting Behavior of Duplex Stainless Steel 2204 in Chloride Solutions. *J. Iron Steel Res. Int.* **2016**, *23*, 357–363. [CrossRef]
31. International Organization for Standardization. *UNI EN ISO 17864:2005: Corrosion of Metals and Alloys—Determination of the Critical Pitting Temperature under Potientiostatic Control*; International Organization for Standardization: Geneva, Switzerland, 2005.
32. ASTM International. *ASTM A262 Standard Practices for Detecting Susceptibility to Intergranular Attack in Austenitic Stainless Steel*; ASTM International: West Conshohocken, PA, USA, 2015.
33. International Organization for Standardization. *ISO 12732:2006 Corrosion of Metals and Alloys—Electrochemical Potentiokinetic Reactivation Measurement Using the Double Loop Method (Based on Cihal's Method)*; International Organization for Standardization: Geneva, Switzerland, 2006.
34. Hong, J.; Han, D.; Tan, H.; Li, J.; Jiang, Y. Evaluation of aged duplex stainless steel UNS S32750 susceptibility to intergranular corrosion by optimized double loop electrochemical potentiokinetic reactivation method. *Corros. Sci.* **2013**, *68*, 249–255. [CrossRef]
35. Wasnik, D.N.; Kain, V.; Samajdar, I.; Verlinden, B.; De, P.K. Resistance to sensitization and intergranular corrosion through extreme randomization of grain boundaries. *Acta Mater.* **2002**, *50*, 4587–4601. [CrossRef]
36. *NACE standard TM-0177-90 Standard Test Method Laboratory Testing of Metals for Resistance to Sulfide Stress Cracking in H2S Environments*; NACE International: Huston, TX, USA, 1990.
37. Barteri, M.; De Cristofaro, N.; Scoppio, L.; Cumino, G.; Della Pina, G. Corrosion resistance of martensitic stainless steels in moderately sour oilfield environments. In Proceedings of the Corrosion '95, NACE, Houston, TX, USA, 26–31 March 1995; p. 76.
38. Garzon, C.M.; Ramirez, A.J. Growth kinetics of secondary austenite in the welding microstructure of a UNS S32304 duplex stainless steel. *Acta Mater.* **2006**, *54*, 3321–3331. [CrossRef]
39. Calliari, I.; Pellizzari, M.; Baldo, S.; Zanellato, M.; Ramous, E. Analysis of phase stability in Cr-Ni and Cr-Mn DSS. In Proceedings of the 8th Duplex Stainless Steels Conference, Beaune, France, 10–13 June 2008.
40. Maetz, J.-Y.; Douillard, T.; Cazottes, S.; Verdu, C.; Kléber, X. $M_{23}C_6$ carbides and Cr_2N nitrides in aged duplex stainless steel: A SEM, TEM and FIB tomography investigation. *Micron* **2016**, *84*, 43–53. [CrossRef]
41. Gong, J.; Jiang, Y.M.; Deng, B.; Xu, J.L.; Hu, J.P.; Li, J. Evaluation of intergranular corrosion susceptibility of UNS S31803 duplex stainless steel with an optimized double loop electrochemical potentiokinetic reactivation method. *Electrochem. Acta* **2010**, *55*, 5077–5083. [CrossRef]
42. Amadou, T.; Braham, C.; Sidhom, H. Double Loop Electrochemical Potentiokinetic Reactivation Test Optimization in Checking of Duplex Stainless Steel Intergranular Corrosion Susceptibility. *Metall. Mater. Tran. A* **2004**, *35A*, 3499–3513. [CrossRef]
43. Fang, Y.L.; Liu, Z.Y.; Xue, W.Y.; Song, H.M.; Jiang, L.Z. Precipitation of secondary phases in lean duplex stainless steel 2101 during isothermal ageing. *ISIJ Int.* **2010**, *50*, 286–293. [CrossRef]
44. Marcus, P.; Oudar, J. *Corrosion Mechanism in Theory and Practice*, 2nd ed.; Marcel Dekker Inc.: New York, NY, USA, 1995; pp. 240–248. ISBN 0-8247-0666-8.
45. Choudhary, L.; Macdonald, D.D.; Alfantazi, A.A. Role of thiosulfate in the corrosion of steels: A review. *Corrosion* **2015**, *71*, 1147–1168. [CrossRef]
46. Marcus, P.; Protopopoff, E. Thermodynamics of thiosulphate reduction on surfaces of iron, nickel and chromium in water at 25 and 300 °C. *Corros. Sci.* **1997**, *39*, 1741–1752. [CrossRef]
47. Lee, K.M.; Cho, H.S.; Choi, D.C. Effect of isothermal treatment of SAF 2205 duplex stainless steel on migration of interface boundary and growth of austenite. *J. Comp.* **1999**, *285*, 156–161. [CrossRef]
48. Cheng, X.; Wang, Y.; Lia, X.; Dong, C. Interaction between austein-ferrite phases on passive performance of 2205 duplex stainless steel. *J. Mater. Sci. Technol.* **2018**, *34*, 2140–2148. [CrossRef]
49. Devine, T.M. Kinetics of Sensitization and De-Sensitization on Duplex 308 Stainless Steel. *Acta Metall.* **1988**, *36*, 1491–1501. [CrossRef]
50. Aydoğdu, G.H.; Aydinol, M.K. Determination of susceptibility to intergranular corrosion and electrochemical reactivation behaviour of AISI 316L type stainless steel. *Corros. Sci.* **2006**, *48*, 3565–3583.
51. Mehrer, H. *Diffusion in Solids Fundamentals, Methods, Materials, Diffusion-Controlled Processes*; Springer: Berlin, Germany, 2007; pp. 127–130.

52. Deng, B.; Jiang, Y.; Gong, J.; Zhong, C.; Gao, J.; Li, J. Critical pitting and repassivation temperatures for duplex stainless steel in chloride solutions. *Electrochim. Acta* **2008**, *53*, 5220–5225. [CrossRef]
53. Bettini, E.; Kivisäkk, U.; Leygraf, C.; Pan, J. Study of corrosion behavior of a 22% Cr duplex stainless steel: Influence of nano-sized chromium nitrides and exposure temperature. *Electrochim. Acta* **2013**, *113*, 280–289. [CrossRef]
54. Newman, R.C.; Isaacs, H.S.; Alman, B. Effect of sulfur compounds on the pitting behavior of type 304 stainless steel in near-neutral chloride solutions. *Corrosion* **1982**, *38*, 261–264. [CrossRef]
55. Rhodes, P.R.; Welch, G.A.; Abrego, L. Stress corrosion cracking susceptibility of duplex stainless steels in sour gas environments. *J. Mater. Energy Syst.* **1983**, *5*, 3–18. [CrossRef]

© 2018 by the authors. Licensee MDPI, Basel, Switzerland. This article is an open access article distributed under the terms and conditions of the Creative Commons Attribution (CC BY) license (http://creativecommons.org/licenses/by/4.0/).

Article

Initial Stages of AZ31B Magnesium Alloy Degradation in Ringer's Solution: Interpretation of EIS, Mass Loss, Hydrogen Evolution Data and Scanning Electron Microscopy Observations

Lucien Veleva [1], Mareny Guadalupe Fernández-Olaya [1] and Sebastián Feliu Jr. [2,*]

[1] Applied Physics Department, Center for Investigation and Advanced Study (CINVESTAV-IPN), Unidad Merida, 97310 Merida, Mexico; veleva@cinvestav.mx (L.V.); marfer.005@gmail.com (M.G.F.-O.)
[2] Centro Nacional de Investigaciones Metalúrgicas CSIC, 28040 Madrid, Spain
* Correspondence: sfeliu@cenim.csic.es; Tel.: +34-915-538-900

Received: 14 October 2018; Accepted: 9 November 2018; Published: 12 November 2018

Abstract: The initial stages of corrosion of AZ31B magnesium alloy, immersed in Ringer's solution at 37 °C body temperature for four days, have been evaluated by independent gravimetric and chemical methods and through electrochemical impedance spectroscopy (EIS) measurements. The corrosion current densities estimated by hydrogen evolution are in good agreement with the time-integrated reciprocal charge transfer resistance values estimated by EIS. The change in the inductive behavior has been correlated with difference in the chemical composition of corrosion layers. At the shorter immersion of 2 days, EDS analysis of cross section of the uniform corrosion layer detected Cl and Al elements, perhaps as formed aluminum oxychlorides salts.

Keywords: alloy; magnesium; SEM-EDS; EIS; mass loss; corrosion layers

1. Introduction

Magnesium (Mg) alloys are being suggested as biodegradable implant materials for clinical applications [1–5], because Mg is non-toxic, biocompatible and beneficial for bone growth and metabolic processes in the human body [6–9]. As a consequence, these alloys are widely used as materials for biomedical applications [10–14]. Among the AZ series, AZ31 (Mg-3%Al-1%Zn) is considered to be suitable as biodegradable material for biomedical applications, having several advantages including: reduced aluminum content, microstructure refinement, low fatigue, corrosion resistance similar to other Mg alloys [7,15–18], not harmful to tissue [19,20] and promoter of new bone cell formation [7,16]. Cathodically active intermetallic Al-Mn particles located within the grains of α-Mg matrix are the main second phase constituents of AZ31B microstructure, such as Al_8Mn_5, ε-AlMn, $Al_{11}Mn_4$ and β-Mn(Al) [21–25]. The corrosion mechanism directly depends on their distribution and the solution composition [26–32]. As test media in this work was selected Ringer's solution, which is an isotonic (physiological) aqueous solution of NaCl with additional compounds, as found in human body fluids (blood serum).

Magnesium is highly active, presenting rapid and continuing dissolution in aqueous solutions. During its degradation hydrogen gas is produced, which is a problem for cardiovascular stents, or for temporary orthopedic implants and could cause a complete failure of the medical device before the bone is healed. Assessing the corrosion rate of Mg-based implant is a critical issue and different alternative techniques could be used, in order to prevent some method limitations [33–36]. Besides, Mg corrosion process near the surface-electrolyte is very dynamic, which is altered with time and therefore, the instantaneous test results and those of long-term methods do not conform well [37].

Widely used technique for corrosion tests in simulated body fluids (SBF) is electrochemical impedance spectroscopy (EIS), for in vitro investigation of the corrosion behavior of metals and alloys [3,4,38–40]. The properties of the electrode surface are not altered after an EIS measurement, since only a small amplitude (~10 mV) AC signal is applied. The high accuracy and reproducibility of the results are well-known advantages of this technique, applied in many material systems and applications [41–44]. The nondestructive character of the impedance technique allows derivation of in situ changes in Mg-based implant material degradation, which in clinical applications is a rather long-term dynamic process [3,4,38]. Some research effort has been focused on the relationship between EIS parameters and independently-measured corrosion rate data, such as average of corrosion mass loss [34,45–49], solution analysis by atomic absorption spectroscopy [50], or hydrogen gas volume evolution [34,36,46–48,51]. Good agreement between corrosion rates of bulk magnesium and several alloys calculated from charge transfer resistance values, R_t, obtained from impedance diagrams and atomic absorption or gravimetric measurements have been observed by Makar and Kruger [45] and Pebere et al. [50]. Recently, King et al. [34] and Bland et al. [46,47] have obtained excellent correlation between the values of the corrosion rate determined by weight loss, hydrogen gas collection and EIS measurements, extrapolating the inductive loop of the impedance diagrams to zero frequency, defined as the polarization resistance (R_p), rather than R_t. Most of the studies are conducted at the room temperature, using rather simple electrolytes, for example, NaCl, $Na_2B_4O_7$ and Na_2CO_3, Na_2SO_4, ammonium/carbonate solutions. Likewise, there is a lack of similar investigations on the use of EIS as a method to determine corrosion rates of Mg-based materials for clinical implants, exposed to simulated body fluids at a body temperature of 37 °C. Xin et al. [52] have reported degradation rates of Mg immersed in SBF having different concentrations of HCO_3^- and correlated the EIS data with hydrogen evolution values. Recently, Liu et al. [49] have established a good agreement between EIS-estimated R_t values, hydrogen evolution measurements and mass loss, performed on a Mg-1Ca alloy exposed to SBF solution.

In the low frequency (LF) region, the impedance diagrams of Mg usually are characterized by a well-marked inductive loop [34,53], while the high frequency loop is considered to be a consequence of the formed corrosion film and its influence on the charge transfer process [54]. The inductive behavior could be associated with the occurrence of pitting corrosion, accompanied by the absorption of $Mg(OH)^+_{ads}$ or $Mg(OH)_2$ species [50,54], or because of accelerated anodic dissolution [34]. The inductive loop could disappear when corrosion protective layer is formed on the surface [55,56]. However, the effect of the precise chemical species, which are responsible for the characteristic inductive loop, is still unclear [34] and its interpretation remains controversial [51].

The aim of this study is to follow the evolution of AZ31B magnesium alloy surface activity during the initial stages of corrosion in Ringer's solution, maintained at a body temperature of 37 °C. One goal is to explore the EIS capabilities for reliable corrosion rate determination. A key objective is to determine which resistance, as obtained from the impedance diagrams, has a stronger correlation with the calculated corrosion rate, based on volume of hydrogen evolution. In this investigation, the EIS technique is used in combination with the SEM-EDS analysis applied on the cross-sections of the test samples, in order to provide information on the grown corrosion films on AZ31B and establish a relationship between their EIS-inductive behavior in artificial physiological environment.

2. Materials and Methods

2.1. Sample Preparation

The composition of the rolled AZ31B Mg-alloy sheet (Magnesium Elektron Ltd., Manchester, UK) is given in Table 1. Square coupon specimens of dimensions $20 \times 20 \times 3$ mm^3 and $50 \times 50 \times 3$ mm^3 were used. Before tests all samples were abraded with 2000 grit SiC paper and mirror polished with 1-µm diamond paste, using ethanol as lubricant, then they were sonicated in ethanol and dried in warm air flow.

Table 1. Chemical composition of AZ31B magnesium alloy (X-ray fluorescence analysis).

Element	Mg	Al	Zn	Mn
Wt%	95.8	3.0	1.0	0.2

2.2. Immersion Test

The isotonic Ringer' solution is a physiological solution, which is an aqueous solution of NaCl with additional compounds, as constituents of the human body fluids (blood serum). It was prepared as described elsewhere [57] with deionized water (18.2 MΩ·cm) and analytical grade reagents (NaCl, KCl, CaCl$_2$) supplied from Sigma-Aldrich, St. Louis, MO, USA. During the test the temperature of the solution was kept stable thermostatically at 37 °C ± 1 °C, the body temperature, using a Digital Circulating Water Bath (Ultrasons Medi-II, J.P. Selecta).

2.3. Microstructure Characterization

Frontal and cross-sectional SEM (Jeol JSM 6500F, Jeol Ltd., Tokyo, Japan) and optical microscopy (Olympus BX-51, Olympus, Tokyo, Japan) images of AZ31B surface were used to characterize the morphological and microstructural changes occurring during the corrosion process.

The corrosion product's elemental composition was characterized through SEM-EDS (EDS, Oxford instrument, Oxford, Oxfordshire, UK) and the phase composition by low-angle X-ray diffraction instrument (XRD, Bruker AXS D8 diffractometer, Bruker AXS, Karlsruhe, Germany) with CuKα radiation.

2.4. Hydrogen Evolution Measurement

Immediately after immersion of the AZ31B samples in the Ringer's solution, hydrogen gas evolution is observed (Equation (1)), produced by magnesium corrosion [1,2,26]:

$$Mg + 2H_2O \rightarrow Mg(OH)_2 + H_2 \qquad (1)$$

Hydrogen collection was performed by placing the entire specimen surface (10.4 cm^2) into the Ringer's solution under an inverted burette system [58]. The height of the solution level in the burette was recorded.

The relationship between the measured volume of hydrogen gas (V_H, mL) and the corrosion current density (i_{H_2}, mA cm^{-2}), is determined via a combination of Faraday's and ideal gas laws:

$$i_{H_2} = 0.091 \frac{V_H}{A \cdot t} \qquad (2)$$

where A is the surface area (cm^2) and t is the time (days) of exposure.

2.5. Electrochemical Measurements

Electrochemical impedance (EIS) measurements were carried out using a potentiostat with a frequency response analyzer (Autolab PGSTAT30, Metrohm, Herisau, Switzerland). The three electrode cell consisted of AZ31B working electrode (9 cm^2 of area), Pt spiral auxiliary electrode and saturated Ag/AgCl (sat. KCl) reference electrode. In order to establish a relatively stable open circuit potential (OCP), the electrochemical measurements were performed after 30 min of immersion of the AZ31B in the Ringer's solution. The tests were carried out varying the exposure time of AZ31B from 1 h to 4 days. The EIS measurements were conducted at OCP (open circuit potential) conditions, applying 10 mV sinusoidal signal amplitude with frequencies ranging from 100 kHz to 1 mHz. The EIS data were numerically fitted with equivalent circuits using Zview software (3.0a Scribner Associates, Inc., Southern Pines, NC, USA).

The corrosion layers of the EIS tested samples were removed after a 10 min cleaning procedure with a solution of 10 g/L AgNO$_3$ (Sigma-Aldrich, St. Louis, MO, USA) and 200 g/L CrO$_3$ (Sigma-Aldrich, St. Louis, MO, USA), at room temperature (21–22 °C), rinsed with distilled water and ethanol, then dried in warm air flow before determining the weight loss. All tests and measurements reported in this study were triplicated to ensure repeatability.

3. Results

3.1. Surface Morphology and Cross-Sectional Analysis of Corrosion Layers

The morphologies of AZ31B surfaces, after immersion in Ringer's solutions at 37 °C for different times, are shown in Figure 1. After 2 days of immersion, a thick layer of corrosion completely covered the AZ31B alloy surface (Figure 1a). The surface morphology after 4 days immersion shows severe pitting corrosion, which could reach a diameter of ~200 µm (white arrows in Figure 1b). The cross sectional morphology images (BSE) of the corrosion layers formed after 2 and 4 days of immersion (Figure 2a,d) and their EDS quantitative analysis are compared in Figure 2b,c and Figure 2e,f, respectively. The EDS reveals that at 2 days (Figure 2b) the corrosion layer is mostly composed of Mg and O. An enrichment in Al and high Cl contents are also detected (Figure 2c). Average atomic composition obtained by EDS and the respective standard deviations from twenty positions across the corrosion layer are shown in Table 2. Within the limits of the EDS analysis, the atomic ratio between Cl and Al is approximately 2.0 ± 0.4, an evidence for possible formation of Al(OH)Cl$_2$ and Al(OH)$_2$Cl aluminum oxychlorides, as previously suggested by Wang et al. [59]. In contrast, no significant amount of Cl was detected in the uniform corrosion layer formed after 4 days of immersion (Figure 2d,f and Table 2), where only Mg and O were observed (Figure 2e).

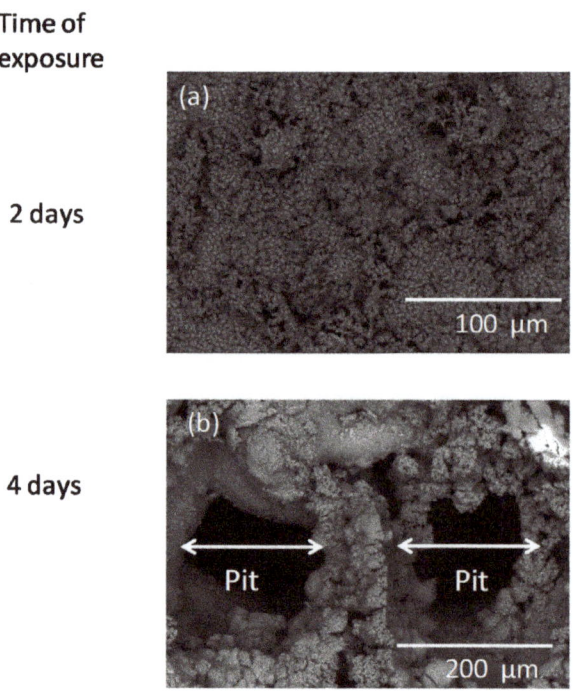

Figure 1. (**a**) BSE images showing AZ31B surface after immersion in Ringer's solution (37 °C) for 2 days; (**b**) 4 days.

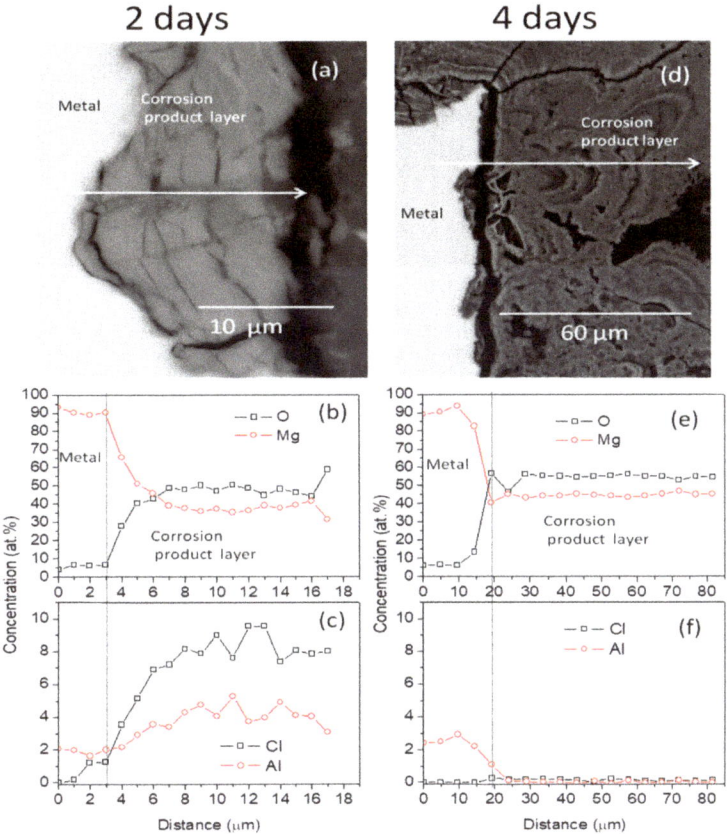

Figure 2. (**a**,**d**) BSE images and (**b**,**c**,**e**,**f**) EDS quantitative analysis of the cross-section of the uniform corrosion layer formed on AZ3B1 surface after immersion in Ringer's solution (37 °C) for (**a**–**c**) 2 days and (**d**–**f**) 4 days.

Table 2. Atomic composition (EDS) of the cross-section of uniform corrosion layer formed on AZ31B surface after immersion in Ringer's solution (37 °C) for 2 and 4 days. Average values from 20 spectra.

Exposure Time (days)	O (at%)	Mg (at%)	Al (at%)	Cl (at%)	Zn (at%)	Cl/Al at.ratio
2	45 ± 6	43 ± 8	3.5 ± 1	7.0 ± 2	1.5 ± 0.5	2.0 ± 0.4
4	55 ± 1	45 ± 1	0.0 ± 0.1	0.1 ± 0.1	0.1 ± 0.1	0.0 ± 0.1

3.2. XRD Analysis of Corrosion Layers

The XRD spectra (Figure 3) of AZ31B surfaces, after their exposure in the Ringer's solution for varying immersion periods, indicated that *brucite* $Mg(OH)_2$ is the primary phase of the corrosion layer. In contrast with the EDS data, no diffraction peaks associated with Al-oxychlorides were registered, suggesting that these compounds appear in an amorphous form [60], or are below the detection limit for second phases using XRD technique (>5% usually).

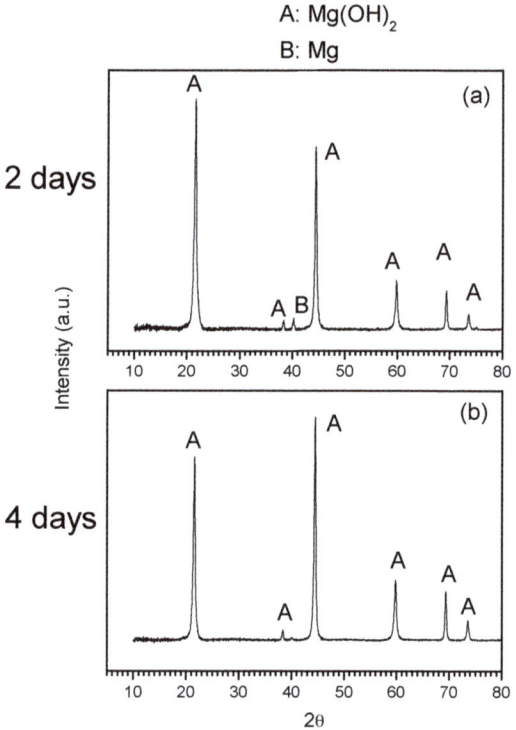

Figure 3. Low-angle XRD spectra of AZ31B surface after immersion in Ringer's solution (37 °C) for 2 and 4 days.

3.3. Hydrogen Evolution Measurement

The volume of hydrogen gas (Figure 4) for the AZ31B samples immersed in Ringer's solution for up to 4 days slowly increased during the first 24 h. After three days of immersion, the volume of hydrogen evolution increased markedly, indicating the formation of non-protective corrosion layer [61].

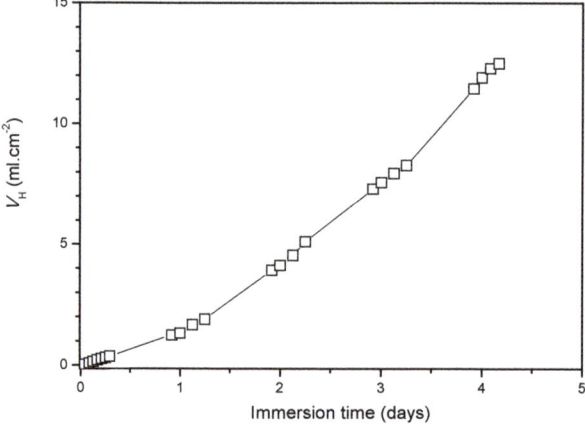

Figure 4. Volume of hydrogen evolution from AZ31B after immersion in Ringer's solution (37 °C) for 4 days.

3.4. EIS Diagrams

Nyquist diagrams of AZ31B samples immersed in the Ringer's solution (37 °C) display one semi-circle (Figure 5). The diameter decreases with time and at lower frequencies an inductive loop is visible, which decreases at the end of the experiment (enlarged impedance spectra in the top-left corner of Figure 5). The electrical equivalent circuit (EEC), used to fit the EIS diagrams is shown in Figure 6. Because of the depressed Nyquist diagram in the center (Figure 5), a constant phase element (CPE_1) was introduced instead of the capacitor, as a characteristic of the electrical double layer in addition to the solution resistance (R_s) and the charge transfer resistance (R_t). An inductor (L) and a resistance (R_1) have also been included to represent the inductive response appearing at the low frequency [38,62]. By using this EEC, a good fit is obtained with an average value of χ^2 around 10^{-3}.

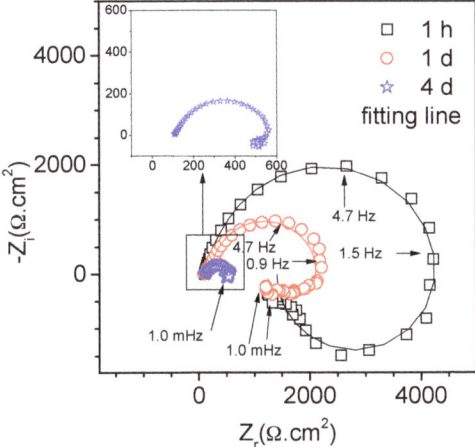

Figure 5. Examples of Nyquist diagrams with respective fitting line for AZ31B samples after 1 h, 1 and 4 days of immersion at open circuit potential in Ringer's solution (37 °C).

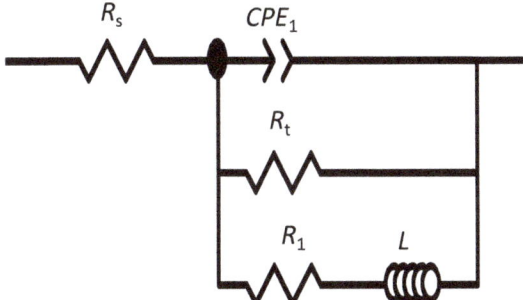

Figure 6. The equivalent circuit used for fitting experimental EIS spectra of AZ31B immersed in Ringer's solution (37 °C).

The polarization resistance (R_p) values, corresponding to the equivalent circuit (Figure 6), were calculated using the following equation [34,46,47]:

$$\frac{1}{R_p} = \frac{1}{R_t} + \frac{1}{R_1} \qquad (3)$$

Table 3 shows the results of the fitting of the electrochemical parameters with respect to the considered equivalent circuit. After the first hour of immersion, the values of the modulus of CPE_1 (Q_1) increased (Table 3), consistent with the changes occurring on the metal surface with the advance of the corrosion process. The exponent of the constant phase element (n_1) decreased with immersion time (Table 3), being 0.78 after four days, significantly lower than 1, suggesting a behavior far from ideal, because of more pronounced heterogeneity and growth of large pits on AZ31B surface, as shown in Figure 1b. Besides, after the first hour of immersion of AZ31B, the R_t value was twice lower (Table 3), indicating decreased protective ability of the formed corrosion layer and as a consequence of the activation of the cathodic process [47]. No significant difference was observed in R_1 values during the test period.

Table 3. Fitting parameters obtained from the EIS measurements of AZ31 immersed in Ringer's solution (37 °C) up to 4 days.

Immersion Time	R_S (KΩ·cm^2)	Q_1 (s^n/Ω·cm^2)	n_1	R_t (KΩ·cm^2)	R_1 (KΩ·cm^2)	L (H·cm^2)	R_p (KΩ·cm^2)
1h	0.1	9.1 × 10^{-6}	0.92	4.3	1.9	3.0	1.3
1d	0.1	2.2 × 10^{-5}	0.87	2.2	2.6	4.2	1.2
4d	0.1	3.8 × 10^{-5}	0.78	0.5	2.2	0.5	0.4

Figure 7 presents the time evolution of the R_t and R_p values extracted from the numerical fitting. Since the first day (24 h) the change in the R_p value (Figure 7b) generally followed the variation in the R_t value (Figure 7a).

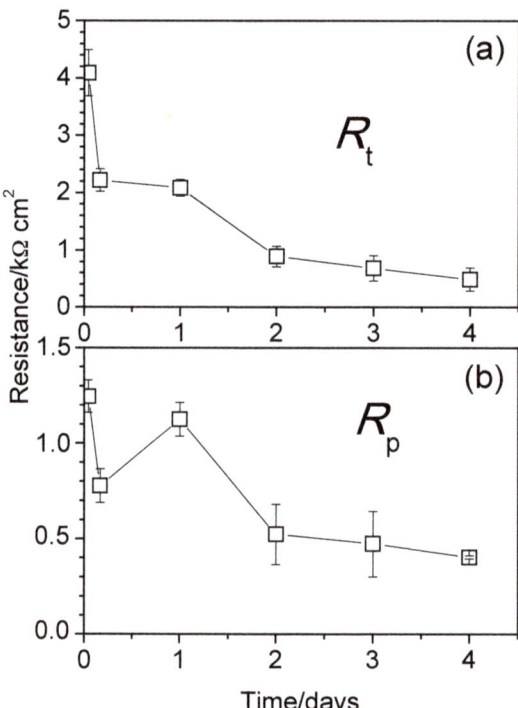

Figure 7. Variations in the resistances obtained from fitting of the EIS spectra of AZ31B, as a function of immersion time in Ringer's solution (37 °C). Scatter bands are the standard deviation of 3 measurements.

Similar to other electrochemical methods, the value of the EIS-estimated resistance (R) is converted into a value of corrosion current density (j_{corr}), using the Stern-Geary equation [63]:

$$j_{corr} = B/R \tag{4}$$

where B is the Stern Geary coefficient, a function of the anodic and cathodic Tafel slopes (βa and βc) [43,44]. Because there is no reasonable linear regions in the anodic branches of Mg and its alloys, the accuracy of the value of B obtained from the polarization curves is questionable [44,48,64]. The aforementioned problem made it necessary to use apparent Stern Geary coefficients B' estimated from the calibration relationship between EIS and hydrogen evolution or mass loss measurements [48,49,51,65].

It has been demonstrated that the anodic charge obtained from the integration of the j_{corr} values, as found from the EIS measurements varying immersion periods (Equation (5)), is similar to the consumed anodic charge Q_a^{WL} found on the mass loss [34,46,47]:

$$Q_a^{WL} = j_{WL} \cdot t = \int j_{corr} \cdot dt = \int \frac{B'}{R} \cdot dt \tag{5}$$

where j_{WL} is the corrosion current found on the mass loss measurement. Assuming a constant value (with time > 3 days) of the 'apparent' Stern–Geary coefficient (B') for a given metal/environment system [66], it follows:

$$B' = \frac{Q_a^{WL}}{\int \left(\frac{1}{R}\right) \cdot dt} \tag{6}$$

Table 4 lists the B' values, empirically calculated from a comparison between the EIS data (R_t and R_p resistances) and mass loss measurement (Q_a^{WL}) of AZ31 tested samples after removal of the corrosion products.

Table 4. 'Apparent' Stern–Geary coefficient values B', calculated from EIS-estimated R_t or R_p values for AZ31B magnesium alloy immersed in Ringer's solution (37 °C) up to 4 days and anodic charge Q_a^{WL} determined by mass loss of the EIS tested sample.

Exposure Time (days)	Mass Loss (mg·cm^{-2}·day)	Q_a^{WL} (mg·cm^{-2}·day)	$\int (1/R_t)\,dt$ (Ω^{-1}·cm^{-2}·day)	B' Estimated Using R_t Values (mV)	$\int (1/R_p)\,dt$ (Ω^{-1}·cm^{-2}·day)	B' Estimated Using R_p Values (mV)
4	3.4	1.24	3.92×10^{-3}	317	6.07×10^{-3}	204

The change in the corrosion current density (Equation (4)) as a function of the immersion period of AZ31B in Ringer's solution (at 37 °C) is shown in Figure 8. For the current calculation the 'apparent' Stern–Geary coefficients B' and EIS-estimated R_t and R_p values (Table 4) were used, as well the hydrogen evolution measurements (Figure 4). It can be seen that there is a good agreement between the corrosion current densities from EIS-estimated R_t values and hydrogen measurements, however, an overestimation of the corrosion current density was observed when considering EIS-estimated R_p values.

Similar to our previous studies [48,67], the contribution of the inductive response (difference between R_p and R_t [34]) has been quantified by using the ratio δ, obtained from the diameters of the inductive loop R_t–R_p and the overall capacitive loop R_t. The evolution of the δ values with increasing immersion time is displayed in Figure 9. During the first hours the δ values are 0.5–0.7 but after four days they decrease significantly to approximately 0.1–0.2.

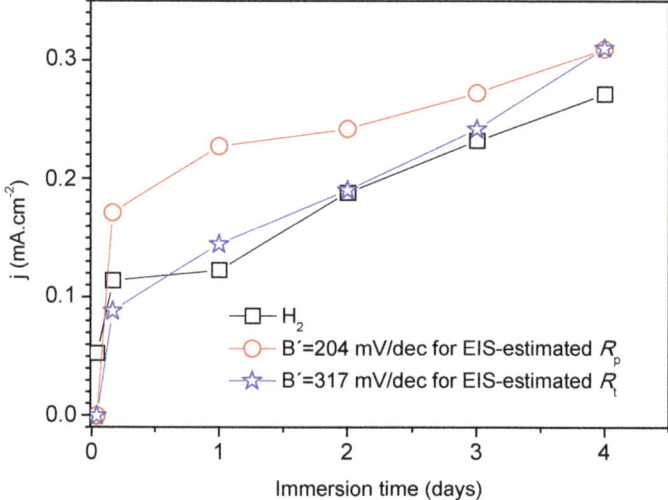

Figure 8. Variation in corrosion current density (mA cm^{-2}) as a function of immersion time, obtained from EIS and hydrogen evolution measurements during immersion of AZ31B alloy in the Ringer's solution for 4 days (37 °C).

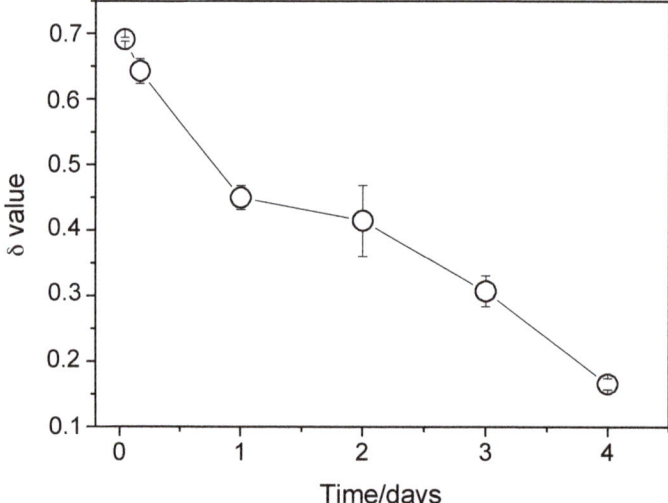

Figure 9. Evolution of the δ values with immersion time of AZ31B in Ringer's solution (37 °C). Scatter bands are the standard deviation of 3 measurements.

To illustrate the relationship between the localized pitting corrosion of AZ31B and the inductive behavior, Figure 10 presents a comparison between the EIS spectra after two (Figure 10a) and four (Figure 10b) days of immersion in the Ringer's solution (37 °C) and the optical cross-sectional images of the tested samples (Figure 10c,d). It can be observed that the inductive loop size is independent of the severity of the pitting corrosion damage.

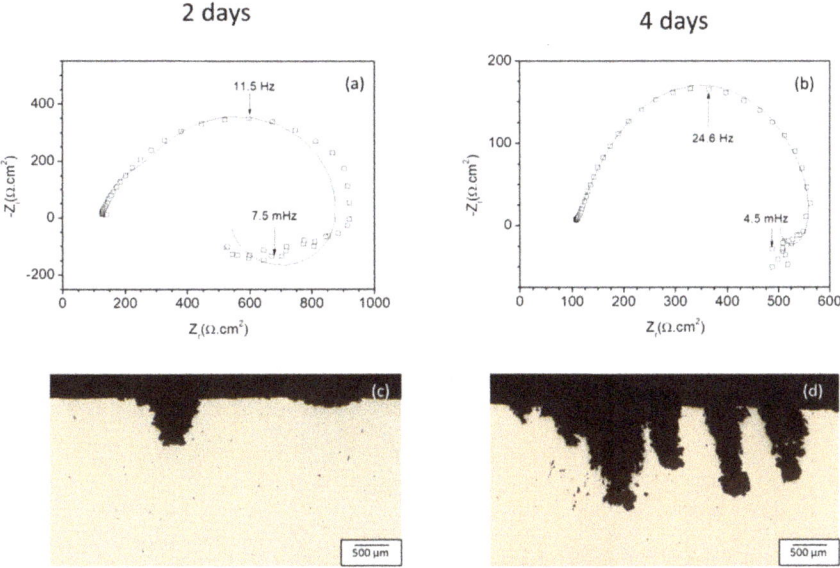

Figure 10. Comparison between electrochemical impedance spectra obtained from AZ31B exposed in Ringer's solution (37 °C) for 2 and 4 days (**a**,**b**) and optical cross-sectional images of samples tested under similar conditions (**c**,**d**).

4. Discussion

4.1. Changes in the Chemistry of the Uniform Corrosion Layer Formed on AZ31B Mg-Alloy Surface in Ringer's Solution and Correlation with EIS Results

Because of the high chloride concentration of Ringer's solution (about 0.15 M), the formed insoluble magnesium hydroxide (Equation (1)) on the AZ31B Mg-alloy surface is transformed to a highly soluble MgCl$_2$ (Equation (7)), which is a source of released magnesium ions (Equation (8)) and an increase of the local pH [3,38]:

$$Mg(OH)_2 + 2Cl^- \rightarrow MgCl_2 + 2OH^- \tag{7}$$

$$MgCl_2 \rightarrow Mg^{2+} + 2Cl^- \tag{8}$$

Instead of MgCl$_2$, the results of this study may be suggestive for chloride association with aluminum, as evidenced by the EDS analysis of the corrosion layer (Figure 2c and Table 2). EDS results of Beldjoudi et al. [68] have shown that the corrosion products of AZ91 alloy exposed to 5% NaCl (saturated with Mg(OH)$_2$), were enriched with Al but no Cl was detected. To our knowledge, there are no studies, which have established the presence of significant amounts of aluminum oxychlorides salts in the uniform corrosion layers formed on the magnesium-aluminum alloy surfaces, after being exposed to simulated physiological solutions. However, reported results for Al immersed in 1 M NaCl at pH = 11 suggest that because the concentration of Cl$^-$ ions within the pits is much higher than that of OH$^-$ ions, it may be expected that the hydrolyzed aluminum cations would complex with the chloride ions to form basic salts [69]. The low pH maintained inside the pit was also considered to be a consequence of the presence of aluminum salts in the interior of pit [70]. Although the reasons are still unclear, the accumulation of metal chloride at the metal interface should be considered as a promoter of the oxide film rupture and nucleation-propagation of growing pits [71]. Besides, the thickening of the formed corrosion layer could bring to the rupture in the surface layer (unfavorable Pilling Bedwoth ratio < 1).Thus, one could speculate that the formation of corrosion products with poorer protection properties and the increase in the tendency for pitting (Figure 10c,d) are promoted by the significant content of Cl in the form of aluminum oxychlorides salts, observed across the uniform corrosion layer grown on AZ31B after 2 days immersion in the Ringer's solution (37 °C). For longer exposure times (4 days) the Cl disappears from the uniform corrosion layer. As the local pH value becomes quite alkaline, due to the creation of OH^- ions (Equation (7)), this fact may contribute to the increase of dissolution rate of the aluminum oxychlorides present in the uniform corrosion layers at the beginning of immersion test.

The shrinking of the EIS inductive loop with immersion time (Figures 5 and 9) could be considered to be a consequence of the important decrease in the aluminum salts content observed in the uniform corrosion layer commented previously. Reports concerning the aluminum corrosion suggest that EIS diagrams present the appearance of low frequency pseudo-inductive loop when Cl^- ions are chemisorbed on aluminum-oxide surface and oxide-chloride complex [72], or aluminum salt film [73] is formed.

4.2. Correlation between Corrosion Current Densities Estimated by EIS and Hydrogen Evolution Volume

In our work integration over the exposure period was applied to correlate corrosion current values, estimated by EIS, hydrogen collected gas and gravimetric measurements. It should be noted that the corrosion current densities determined from parameter R_t quantitatively agree with the corrosion current densities calculated using independent hydrogen evolution measurement, over the whole duration of the measurement (Figure 8). Our suggestion is that R should be closely related only to the activation (charge-transfer) controlled cathodic process of hydrogen evolution, which occurs at the interface metal-electrolyte, in the absence of such control in the anodic process. With this assumption, R_t could be inserted in the Stern-Geary equation to estimate the corrosion current densities. If the parameter R_p and value of $B' = 204$ mV are used (Table 4), the estimated corrosion current yielded 20–50% greater values than those based on the hydrogen evolution volumes during the first two days of AZ31B immersion in the Ringer's solution at 37 °C (Figure 8). By prolonging immersion time, a significant decrease occurs in the differences between the corrosion current values derived from EIS-estimated R_p measurements and those derived from the hydrogen evolution. The reason is that R_p characterizes the overall corrosion process (cathodic and anodic, besides the surface changes), having a meaning more complex than that of the R_t values. We hypothesize that R_p may be composed of other resistances (in addition to R_t), such as diffusion of the reacting species, including passivation, adsorption and salt film Ohmic resistance. The electrochemical overestimation of the corrosion current, using the EIS-R_p values, may be assigned to a possible superposition, due to the initial formation of a salt film on the uniform corrosion layer and its subsequent dissolution during immersion of AZ31B in the Ringer's solution.

With regard to the 'apparent' Stern–Geary coefficients B', based on EIS-estimated R_t or R_p values (Table 4), the recent work reported by Curioni et al. [49,51] could serve as a reference, despite the use of the reciprocal R_t values. In our study, the estimated B' is $\sim$317 mV for the AZ31B magnesium alloy exposed to the Ringer's solution (at 37 °C), when the time-integrated reciprocal R_t resistance values were used. When AZ31B was exposed to 0.6 M NaCl (at 21 °C), the value was $\sim$97 mV [48]. This difference should be attributed to the variance in the chemical composition of both tested solutions: The Ringer's solution contains four times lower chloride concentration (0.15 M NaCl and with additional compounds) and the experiments were performed at higher temperature (37 °C). The nature of the metal or alloys and the specific corrosive environment could influence the value of B [66], which correlates to the R_p and the instantaneous corrosion rate. The results of our study suggest that the chloride concentration in the test solution is a critical factor for the Stern–Geary coefficient B'.

5. Conclusions

(1) The use of mass loss data and time-integrated reciprocal values of the charge transfer resistance R_t estimated by the EIS diagrams, allows to follow the variation in the corrosion current

densities of AZ31B Mg-alloy as a function of the immersion time up to four days in Ringer's solution (at 37 °C), similar to those measured by an independent chemical (I.e., non-electrochemical) method of hydrogen evolution.

(2) The value of the 'apparent' Stern–Geary coefficients B' ($\approx$ 300 mV) was empirically obtained.

(3) The marked decrease in the EIS inductive loop, with increased immersion time in the Ringer's solution, tends to reflect the dissolution of aluminum chloride salt, which is probably formed across the uniform corrosion layer during the initial stages, as suggested by the EDS analysis.

(4) The formation and dissolution of metallic salt, as a part of the corrosion layer formed on AZ31 in the Ringer's solution at 37 °C, seems to be responsible for the decrease in the accuracy of the corrosion current densities derived from EIS estimated R_p.

Author Contributions: M.G.F.-O. performed the preparation of samples and their corrosion tests. L.V. and S.F.Jr. discussed the results and wrote the manuscript with contributions from all authors. L.V. and S.F.Jr. conceived and supervised the project. All correspondence should be addressed to S.F.

Funding: This research was funded by the Mexican National Council for Science and Technology (CONACYT) and the Spanish Ministry of Economy and Competitiveness (project MAT2015-65445-C2-1-R) for their financial support. Mareny Fernández-Olaya is grateful to CONACYT for her scholarship as an M.Sci. student at CINVESTAV-IPN and for the research stay at CENIM/CSIC (the National Centre for Metallurgical Research of Madrid, Spain).

Conflicts of Interest: The authors declare no conflict of interest.

References

1. Song, G.; Song, S. A possible biodegradable magnesium implant material. *Adv. Eng. Mater.* **2007**, *9*, 298–302. [CrossRef]
2. Witte, F.; Hort, N.; Vogt, C.; Cohen, S.; Kainer, K.U.; Willumeit, R.; Feyerabend, F. Degradable biomaterials based on magnesium corrosion. *Curr. Opin. Solid State Mater. Sci.* **2008**, *2*, 63–72. [CrossRef]
3. Ascencio, M.; Pekguleryuz, M.; Omanovic, S. An investigation of the corrosion mechanisms of WE43 Mg alloy in a modified simulated body fluid solution: The influence of immersion time. *Corros. Sci.* **2014**, *87*, 489–503. [CrossRef]
4. Ascencio, M.; Pekguleryuz, M.; Omanovic, S. An investigation of the corrosion mechanisms of WE43 Mg alloy in a modified simulated body fluid solution: The effect of electrolyte renewal. *Corros. Sci.* **2015**, *91*, 297–310. [CrossRef]
5. Mao, L.; Shen, L.; Chen, J.H.; Zhang, X.B.; Kwak, M.; Wu, Y.; Fan, R.; Zhang, L.; Pei, J.; Yuan, G.Y.; et al. A promising biodegradable magnesium alloy suitable for clinical vascular stent application. *Sci. Rep.* **2017**, *7*, 46343. [CrossRef] [PubMed]
6. Seiler, H.G.; Sigel, H.; Sigel, A. *Handbook on Toxicity of Inorganic Compounds*; Marcel Dekker Inc.: New York, NY, USA, 1988.
7. Witte, F.; Kaese, V.; Haferkamp, H.; Switzer, E.; Meyer-Lindenberg, A.; Wirth, C.J.; Windhagen, H. In vivo corrosion of four magnesium alloys and the associated bone response. *Biomaterials* **2005**, *26*, 3557–3563. [CrossRef] [PubMed]
8. Zhang, E.; Xu, L.P.; Yang, K. Formation by ion plating of Ti-coating on pure Mg for biomedical applications. *Scr. Mater.* **2005**, *53*, 523–527. [CrossRef]
9. Xiong, H.Q.; Liang, Z.F.; Wang, Z.F.; Qin, C.L.; Zhao, W.M.; Yu, H. Mechanical properties and degradation behavior of Mg(100-7x)Zn6xYx(x = 0.2, 0.4, 0.6, 0.8) alloys. *Metals* **2018**, *8*, 261. [CrossRef]
10. Staiger, M.P.; Pietak, A.M.; Huadmai, J.; Dias, G. Magnesium and its alloys as orthopedic biomaterials: A review. *Biomaterials* **2006**, *27*, 1728–1734. [CrossRef] [PubMed]
11. Witte, F.; Fischer, J.; Nellesen, J.; Crostack, H.-A.; Kaese, V.; Pisch, A.; Beckmann, F.; Windhagen, H. In vitro and in vivo corrosion measurements of magnesium alloys. *Biomaterials* **2006**, *27*, 1013–1018. [CrossRef] [PubMed]
12. Keim, S.; Brunner, J.G.; Fabry, B.; Virtanen, S. Control of magnesium corrosion and biocompatibility with biomimetic coatings. *J. Biomed. Mater. Res. B* **2011**, *96*, 84–90. [CrossRef] [PubMed]

13. Waizy, H.; Seitz, J.M.; Reifenrath, J.; Weizbauer, A.; Bach, F.W.; Meyer-Lindenberg, A.; Denkena, B.; Windhagen, H. Biodegradable magnesium implants for orthopedic applications. *J. Mater. Sci.* **2013**, *48*, 39–50. [CrossRef]
14. Hiromoto, S.; Inoue, M.; Taguchi, T.; Yamane, M.; Ohtsu, N. In vitro and in vivo biocompatibility and corrosion behavior of a bioabsorbable magnesium alloy coated with octacalcium phosphate and hydroxyapatite. *Acta Biomater.* **2015**, *11*, 520–530. [CrossRef] [PubMed]
15. Gray-Munro, J.E.; Seguin, C.; Strong, M. Influence of surface modification on the in vitro corrosion rate of magnesium alloy AZ31. *J. Biomed. Mater. Res. A* **2009**, *91*, 221–230. [CrossRef] [PubMed]
16. Duygulu, O.; Kaya, R.A.; Oktay, G.; Kaya, A.A. Investigation on the potential of Magnesium alloy AZ31 as a bone implant. *Mater. Sci. Forum* **2007**, *546*, 421–424. [CrossRef]
17. Zhang, L.; Zhang, J.; Chen, C.F.; Gu, Y. Advances in microarc oxidation coated AZ31 Mg alloys for biomedical applications. *Corros. Sci.* **2015**, *91*, 7–28. [CrossRef]
18. Srinivasan, A.; Shin, K.S.; Rajendran, N. Influence of bicarbonate concentration on the conversion layer formation onto AZ31 magnesium alloy and its electrochemical corrosion behavior in simulated body fluid. *RSC Adv.* **2016**, *6*, 49910–49922. [CrossRef]
19. Witte, F.; Abeln, I.; Switzer, E.; Kaese, V.; Meyer-Lindenberg, A.; Windhagen, H. Evaluation of the skin sensitizing potential of biodegradable magnesium alloys. *J. Biomed. Mater. Res. A* **2008**, *86*, 1041–1047. [CrossRef] [PubMed]
20. Xue, D.C.; Yun, Y.H.; Tan, Z.Q.; Dong, Z.Y.; Schulz, M.J. In Vivo and in vitro degradation behavior of Magnesium alloys as biomaterials. *J. Mater. Sci. Technol.* **2012**, *28*, 261–267. [CrossRef]
21. Wei, L.Y.; Westengen, H.; Aune, T.K.; Albright, D. Characterisation of manganese-containing intermetallic particles and corrosion behavior of die cast Mg-Al-based alloys. In *Magnesium Technology 2000*; Kaplan, H.L., Hryn, J.N., Eds.; The Minerals, Metals and Materials Society (TMS): Nashville, TN, USA, 2000; pp. 153–160.
22. Cao, P.; StJohn, D.H.; Qian, M. The effect of manganese on the grain size of commercial AZ31 alloy. *Mater. Sci. Forum* **2005**, *488*, 139–142. [CrossRef]
23. Cheng, Y.L.; Qin, T.W.; Wang, H.M.; Zhang, Z. Comparison of corrosion behaviors of AZ31, AZ91, AM60 and ZK60 magnesium alloys. *Trans. Nonferr. Met. Soc. China* **2009**, *19*, 517–524. [CrossRef]
24. Liu, F.; Song, Y.; Shan, D.; Han, E. Corrosion behavior of AZ31 magnesium alloy in simulated acid rain solution. *Trans. Nonferr. Met. Soc. China* **2010**, *20*, 638–642. [CrossRef]
25. Pawar, S.; Zhou, X.; Thompson, G.E.; Scamans, G.; Fan, Z. The role of intermetallics on the corrosion initiation of twin roll cast AZ31 Mg alloy. *J. Electrochem. Soc.* **2015**, *162*, C442–C448. [CrossRef]
26. Song, G.; Atrens, A. Understanding magnesium corrosion—A framework for improved alloy performance. *Adv. Eng. Mater.* **2003**, *5*, 837–858. [CrossRef]
27. Eliezer, D.; Uzan, P.; Aghion, E. Effect of second phases on the corrosion behavior of magnesium alloys. *Mater. Sci. Forum* **2003**, *419*, 857–866. [CrossRef]
28. Zeng, R.-C.; Zhang, J.; Huang, W.-J.; Dietzel, W.; Kainer, K.; Blawert, C.; Wei, K. Review of studies on corrosion of magnesium alloys. *Trans. Nonferr. Met. Soc. China* **2006**, *16*, s763–s771. [CrossRef]
29. Xin, Y.; Hu, T.; Chu, P.K. In vitro studies of biomedical magnesium alloys in a simulated physiological environment: A review. *Acta Biomater.* **2011**, *7*, 1452–1459. [CrossRef] [PubMed]
30. Mena-Morcillo, E.; Veleva, L.; Wipf, D.O. In situ investigation of the initial stages of AZ91D Magnesium alloy biodegradation in simulated body fluid. *Int. J. Electrochem. Sci.* **2018**, *13*, 5141–5150. [CrossRef]
31. Mena-Morcillo, E.; Veleva, L.; Wipf, D.O. Multi-scale monitoring the first stages of electrochemical behavior of AZ31B magnesium alloy in simulated body fluid. *J. Electrochem. Soc.* **2018**, *165*, C749–C755. [CrossRef]
32. Saikrishna, N.; Reddy, G.P.K.; Munirathinam, B.; Sunil, B.R. Influence of bimodal grain size distribution on the corrosion behavior of friction stir processed biodegradable AZ31 magnesium alloy. *J. Magn. Alloys* **2016**, *4*, 68–76. [CrossRef]
33. Kirkland, N.T.; Birbilis, N.; Staiger, M. Assessing the corrosion of biodegradable magnesium implants: A critical review of current methodologies and their limitations. *Acta Biomater.* **2012**, *8*, 925–936. [CrossRef] [PubMed]
34. King, A.D.; Birbilis, N.; Scully, J.R. Accurate electrochemical measurement of magnesium corrosion rates; a combined impedance, mass-loss and hydrogen collection study. *Electrochim. Acta* **2014**, *121*, 394–406. [CrossRef]

35. Tkacz, J.; Minda, J.; Fintová, S.; Wasserbauer, J. Comparison of electrochemical methods for the evaluation of cast AZ91 Magnesium alloy. *Materials* **2016**, *9*, 925. [CrossRef] [PubMed]
36. Yang, Y.; Scenini, F.; Curioni, M. A study on magnesium corrosion by real-time imaging and electrochemical methods: Relationship between local processes and hydrogen evolution. *Electrochim. Acta* **2016**, *198*, 174–184. [CrossRef]
37. Cao, F.; Shi, Z.; Hofstetter, J.; Uggowitzed, P.J.; Song, G.; Liu, M.; Atrens, A. Corrosion of ultra-high-purity Mg in 3.5% NaCl solution saturated with Mg(OH)$_2$. *Corros. Sci.* **2013**, *75*, 78–99. [CrossRef]
38. Jamesh, M.; Kumar, S.; Sankara Narayanan, T.S.N. Corrosion behavior of commercially pure Mg and ZM21 Mg alloy in Ringer's solution—Long term evaluation by EIS. *Corros. Sci.* **2011**, *53*, 645–654. [CrossRef]
39. Jamesh, M.I.; Wu, G.S.; Zhao, Y.; McKenzie, D.R.; Bilek, M.M.M.; Chu, P.K. Effects of zirconium and oxygen plasma ion implantation on the corrosion behavior of ZK60 Mg alloy in simulated body fluids. *Corros. Sci.* **2014**, *82*, 7–26. [CrossRef]
40. Tkacz, J.; Sloukova, K.; Minda, J.; Drabikova, J.; Fintova, S.; Dolezal, P.; Wasserbauer, J. Influence of the composition of the Hank's balanced salt solution on the corrosion behavior of AZ31 and AZ61 magnesium alloys. *Metals* **2017**, *7*, 465. [CrossRef]
41. Lorenz, W.J.; Mansfeld, F. Determination of corrosion rates by electrochemical DC and AC methods. *Corros. Sci.* **1981**, *21*, 647–672. [CrossRef]
42. Song, G.; Shi, Z. Corrosion mechanism and evaluation of anodized magnesium alloys. *Corros. Sci.* **2014**, *85*, 126–140. [CrossRef]
43. Feliu, S., Jr.; Galván, J.C.; Pardo, A.; Merino, M.C. Estimation of the corrosion rate in circumstances of difficult implementation of the common methods for electrochemical measurements. In *Applied Electrochemistry (Chemistry Research and Applications) 2009*; Singh, V.G., Ed.; Nova Science Publishers, Inc.: Hauppauge, NY, USA, 2009; pp. 387–403.
44. Feliu, S., Jr.; Garcia-Galvan, F.R.; Llorente, I.; Diaz, L.; Simancas, J. Influence of hydrogen bubbles adhering to the exposed surface on the corrosion rate of magnesium alloys AZ31 and AZ61 in sodium chloride solution. *Mater. Corros.* **2017**, *68*, 651–663. [CrossRef]
45. Makar, G.L.; Kruger, J. Corrosion studies of rapidly solidified Magnesium alloys. *J. Electrochem. Soc.* **1990**, *137*, 414–421. [CrossRef]
46. Bland, L.G.; King, A.D.; Birbilis, N.; Scully, J.R. Assessing the corrosion of commercially pure magnesium and commercial AZ31B by electrochemical impedance, mass-loss, hydrogen collection and inductively coupled plasma optical emission spectrometry solution analysis. *Corrosion* **2015**, *71*, 128–145. [CrossRef]
47. Bland, L.G.; Scully, L.C.; Scully, J.R. Assessing the corrosion of multi-phase Mg-Al alloys with high Al content by electrochemical impedance, mass loss, hydrogen collection and inductively coupled plasma optical emission spectrometry solution analysis. *Corrosion* **2017**, *73*, 526–543. [CrossRef]
48. Delgado, M.C.; Garcia-Galvan, F.R.; Barranco, V.; Feliu, S., Jr. A measuring approach to assess the corrosion rate of magnesium alloys using electrochemical impedance measurements. In *Magnesium Alloys*; Aliofkhazraei, M., Ed.; Intech: Rijeka, Croatia, 2017; pp. 129–159.
49. Liu, Y.X.; Curioni, M.; Liu, Z. Correlation between electrochemical impedance measurements and corrosion rates of Mg-1Ca alloy in simulated body fluid. *Electrochim. Acta* **2018**, *264*, 101–108. [CrossRef]
50. Pebere, N.; Riera, C.; Dabosi, F. Investigation of magnesium corrosion in aerated sodium-sulfate solution by electrochemical impedance spectroscopy. *Electrochim. Acta* **1990**, *35*, 555–561. [CrossRef]
51. Curioni, M.; Scenini, F.; Monetta, T.; Bellucci, F. Correlation between electrochemical impedance measurements and corrosion rate of magnesium investigated by real-time hydrogen measurement and optical imaging. *Electrochim. Acta* **2015**, *166*, 372–384. [CrossRef]
52. Xin, Y.; Hu, T.; Chu, P.K. Degradation behavior of pure magnesium in simulated body fluids with different concentrations of HCO_3^-. *Corros. Sci.* **2011**, *53*, 1522–1528. [CrossRef]
53. Mathieu, S.; Rapin, C.; Hazan, J.; Steinmetz, P. Corrosion behavior of high pressure die-cast and semi-solid cast AZ91D alloys. *Corros. Sci.* **2002**, *44*, 2737–2756. [CrossRef]
54. Baril, G.; Blanc, C.; Keddam, M.; Pebere, N. Local electrochemical impedance spectroscopy applied to the corrosion behavior of an AZ91 magnesium alloy. *J. Electrochem. Soc.* **2003**, *150*, B488–B493. [CrossRef]
55. Baril, G.; Blanc, C.; Pebere, N. AC impedance spectroscopy in characterizing time-dependent corrosion of AZ91 and AM50 magnesium alloys—Characterization with respect to their microstructures. *J. Electrochem. Soc.* **2001**, *148*, B489–B496. [CrossRef]

56. Xin, Y.; Huo, K.; Tao, H.; Tang, G.; Chu, P.K. Influence of aggressive ions on the degradation behavior of biomedical magnesium alloy in physiological environment. *Acta Biomater.* **2008**, *4*, 2008–2015. [CrossRef] [PubMed]
57. ISO 16428, Implants for Surgery—Test Solutions and Environmental Conditions for Static and Dynamic Corrosion Tests on Implantable Materials and Medical Devices. 2005. Available online: https://www.iso.org/standard/30280.html (accessed on 9 November 2018).
58. Song, G.; Atrens, A.; Stjohn, D. An hydrogen evolution method for the estimation of the corrosion rate of Magnesium alloys. In *Essential Readings in Magnesium Technology*; Suveen, N.M., Alan, A.L., Neale, N.R., Eric, A.N., Wim, H.S., Eds.; John Wiley & Sons, Inc.: Hoboken, NJ, USA, 2014; pp. 565–572.
59. Wang, B.; Zhang, L.W.; Su, Y.; Xiao, Y.; Liu, J. Corrosion behavior of 5A05 aluminum alloy in NaCl solution. *Acta Metall. Sin. (Engl. Lett.)* **2013**, *26*, 581–587. [CrossRef]
60. Lucuta, P.G.; Halliday, J.D.; Christian, B. Phase evolution in Al_2O_3 fibre prepared from an oxychloride precursor. *J. Mater. Sci.* **1992**, *27*, 6053–6061. [CrossRef]
61. Zhang, J.; Yang, D.H.; Ou, X.B. Microstructures and properties of aluminum film and its effect on corrosion resistance of AZ31B substrate. *Trans. Nonferr. Metals Soc. China* **2008**, *18*, s312–s317. [CrossRef]
62. Li, Z.C.; Song, G.L.; Song, S.Z. Effect of bicarbonate on biodegradation behavior of pure magnesium in a simulated body fluid. *Electrochim. Acta* **2014**, *115*, 56–65. [CrossRef]
63. Stern, M.; Geary, A.L. Electrochemical polarization-1. A theoretical analysis of the shape of polarization curves. *J. Electrochem. Soc.* **1957**, *104*, 56–63. [CrossRef]
64. McCafferty, E. Validation of corrosion rates measured by the Tafel extrapolation method. *Corros. Sci.* **2005**, *47*, 3202–3215. [CrossRef]
65. Feliu, S., Jr.; Maffiotte, C.; Samaniego, A.; Galván, J.C.; Barranco, V. Effect of the chemistry and structure of the native oxide surface film on the corrosion properties of commercial AZ31 and AZ61 alloys. *Appl. Surf. Sci.* **2011**, *257*, 8558–8568. [CrossRef]
66. Hsieh, M.K.; Dzombak, D.A.; Vidic, R.D. Bridging gravimetric and electrochemical approaches to determine the corrosion rate of metals and metal alloys in cooling systems: Bench scale evaluation method. *Ind. Eng. Chem. Res.* **2010**, *49*, 9117–9123. [CrossRef]
67. Feliu, S., Jr.; Maffiotte, C.; Samaniego, A.; Galvan, J.C.; Barranco, V. Effect of naturally formed oxide films and other variables in the early stages of Mg-alloy corrosion in NaCl solution. *Electrochim. Acta* **2011**, *56*, 4454–4565. [CrossRef]
68. Beldjoudi, T.; Fiaud, C.; Robbiola, L. Influence of homogenization and artificial aging heat-treatments on corrosion behavior of Mg–Al alloys. *Corrosion* **1993**, *49*, 738–745. [CrossRef]
69. Wong, K.P.; Alkire, R.C. Local chemistry and growth of single corrosion pits in aluminum. *J. Electrochem. Soc.* **1990**, *137*, 3010–3015. [CrossRef]
70. Silva, F.S.D.; Bedoya, J.; Dosta, S.; Cinca, N.; Cano, I.G.; Guilemany, J.M.; Benedetti, A.V. Corrosion characteristics of cold gas spray coatings of reinforced aluminum deposited onto carbon steel. *Corros. Sci.* **2017**, *114*, 57–71. [CrossRef]
71. Burstein, G.T.; Liu, C.; Souto, R.M. The effect of temperature on the nucleation of corrosion pits on titanium in Ringer's physiological solution. *Biomaterials* **2005**, *26*, 245–256. [CrossRef] [PubMed]
72. Garrigues, L.; Pebere, N.; Dabosi, F. An investigation of the corrosion inhibition of pure aluminum in neutral and acidic chloride solutions. *Electrochim. Acta* **1996**, *41*, 1209–1215. [CrossRef]
73. Lee, E.J.; Pyun, S.I. The effect of oxide chemistry on the passivity of aluminum surfaces. *Corros. Sci.* **1995**, *37*, 157–168. [CrossRef]

© 2018 by the authors. Licensee MDPI, Basel, Switzerland. This article is an open access article distributed under the terms and conditions of the Creative Commons Attribution (CC BY) license (http://creativecommons.org/licenses/by/4.0/).

Article

Passivity of Spring Steels with Compressive Residual Stress

Kyu-Hyuk Lee [1], Seung-Ho Ahn [2], Ji-Won Seo [2] and HeeJin Jang [3,*]

[1] Department of Advanced Materials Engineering, Chosun University, 309 Pilmundaero, Dong-gu, Gwangju 61452, Korea; lgh900916@gmail.com
[2] Accelerated Durability Development Team, Hyundai Motor R&D Center, 772-1 Jangdeok-dong, Hwaseong-si, Gyeonggi-do 18280, Korea; scoupeman@hyundai.com (S.-H.A.); bamtor@hyundai.com (J.-W.S.)
[3] Department of Materials Science and Engineering, Chosun University, 309 Pilmundaero, Dong-gu, Gwangju 61452, Korea
* Correspondence: heejin@chosun.ac.kr; Tel.: +82-62-230-7196

Received: 24 August 2018; Accepted: 25 September 2018; Published: 2 October 2018

Abstract: The electrochemical corrosion behavior and the semiconducting properties of the passive film formed on a coil spring steel were investigated in a buffer solution at pH 9. The anodic dissolution was mitigated, and the passive film grew faster for the spring steel with compressive residual stress than the one without stress. The passive films had an n-type semiconducting property with a high density of oxygen vacancy, and the defect density was lower for the specimens with compressive stress. The passive current density of the specimens with stress was higher and showed fluctuation. These characteristics imply that the growth mechanism of passive film and the transport of vacancies in the film on metals and alloys depend on the residual stress on the metallic surface.

Keywords: corrosion; spring steel; shot peening; Mott–Schottky analysis; point defect; passive film

1. Introduction

Coil springs for automobile suspension usually suffer corrosion fatigue [1–7] or fatigue [8–15]. Shot peening, proper design of alloy composition, and heat treatment are known to improve the resistance to fatigue [1–16]. Commercial coil springs are usually shot-peened. Shot peening provides compressive residual stress on the surface, and hence suppresses the crack growth under tensile stress during operation [1,2,4,6,7,11]. The effects of residual stress on the mechanical properties, which include fatigue and corrosion fatigue, have been extensively studied by many researchers [1–16]. However, little is known about the corrosion behavior of surfaces with residual stress.

Corrosion resistance of metallic materials is largely determined by its passivity. Passivity refers to the phenomenon whereby a metal or alloy shows a very low corrosion rate in spite of its thermodynamically high activity in a corrosion environment. This originates in the passive film, which is an oxide film formed naturally on the surface with a thickness of several nanometers. The passive film protects the metal from severe corrosion and can decelerate the corrosion rate drastically. The Point Defect Model [17] of passivity suggests that the growth and breakdown of passive films are controlled by the generation, annihilation, and transport of point defects such as vacancies and interstitials. The type and concentration of point defects are obtained from the semiconducting properties of the passive film, according to many corrosion researchers who have investigated them by using the Mott–Schottky analysis [18–48].

In this study, the authors aim to examine the effect of compressive residual stress on the corrosion and passivity of spring steels. Electrochemical polarization tests were used to evaluate the corrosion behavior, and a Mott–Schottky analysis was performed to determine the type and concentration of point defects in the passive film.

2. Experimental Procedures

A coil spring part from an automobile provided by Hyundai Motors Company (Seoul, Korea) was used as the specimen. The spring was made of JIS SUP-10 steel (Table 1) and was shot-peened. The samples were taken from an undamaged part of the front left-hand coil spring after a proving ground test. The paint was removed by immersing the samples into a methyl ethyl ketone solution for 10 min. The samples were categorized into two groups, the S and M groups. The specimens with residual stress, designated as the S group, were prepared by slightly polishing the surface of the spring to make a flat area of about 0.1 cm^2. The depth of polishing was not the same for all specimens, so it was expected that the residual stress of each S specimen would be different. The specimens exposing the cross-section of the spring were also prepared by cutting the center of the spring for a comparative study and were designated as the M group. Multiple numbers of specimens were prepared for both groups, and each specimen was designated as S1, S2, M1, M2, and M3. The surface of the samples was finished with a 0.05-μm alumina paste.

Table 1. Chemical composition of SUP-10 steel.

Element	C	Si	Mn	P	S	Cr	V
Composition (wt%)	0.45–0.55	0.15–0.35	0.65–0.95	<0.035	<0.035	0.80–1.10	0.15–0.25

The microstructure of specimens was observed using a field emission scanning electron microscope (FE-SEM) (Jeol, Tokyo, Japan) after etching with a 3% nital solution.

The residual stress of the specimens was measured by instrumented indentation testing (IIT) (Frontics, Seoul, Korea). IIT requires a reference sample without stress in order to measure the residual stress quantitatively, and the authors presumed that the residual stress of an M specimen was 0. The stress was measured 10 times for each S specimen and the results were then averaged.

The hardness was measured using a Vickers hardness tester (Future-Tech, Kanagawa, Japan) the test was repeated five times.

Potentiodynamic polarization, potentiostatic polarization, and Mott–Schottky analysis were performed sequentially for each specimen, using a potentiostat/galvanostat (Ivium Technologies, AJ Eindhoven, The Netherlands). The surface of the specimens was sealed with a silicone sealant, leaving an exposed area of 0.04–0.1 cm^2. A three-electrode electrochemical cell was made up of a working electrode (i.e., the specimen), a counter electrode made of Pt wire, and a saturated calomel electrode (SCE) reference electrode.

A buffer solution at pH 9 was used as the electrolyte in order to establish a stable passivity as suggested by the E-pH diagram of iron (Figure 1) [49]. The solution was made of $H_3BO_3 + C_6H_9O_7 \cdot H_2O + Na_3PO_4 \cdot 12H_2O$ at ambient temperature. The solution was purged with 99.999% N_2 gas during tests. The working electrode was cathodically cleaned at -1 V_{SCE} for 30 min. The open circuit potential was monitored for 30 min. The potential was scanned from -0.3 V with respect to the open circuit potential to 0.5 V_{SCE} at a rate of 1 mV/s. Subsequently, the specimen was passivated by a potentiostatic polarization at 0.5 V_{SCE} for 24 h. Finally, the capacitance was measured at potentials from 0.5 V_{SCE} to -0.7 V_{SCE} with the potential sweep rate of -10 mV/s for the Mott–Schottky analysis. The frequency of AC was 1 kHz [50] and the amplitude was 0.01 V.

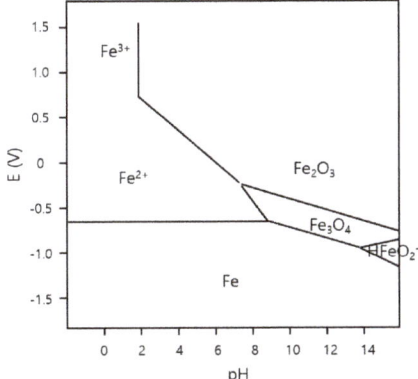

Figure 1. E-pH diagram of Fe in water at 25 °C, assuming $[Fe^{2+}] = 10^{-6}$.

3. Results and Discussion

Experimental Results

Figure 2 shows the SEM images of the specimens. Lath martensite, which is usually found in spring steels with a carbon content of less than 0.6 wt% [51], is observed for both S and M group samples. Some distortion of grains due to the pressure involved by shot peening is seen from the surface to a depth of about 10–20 µm. Cracks were found in the corrosion product layer between the alloy and the coating. During the proving ground test, the paint coating was degraded and water penetrated into the coating–metal interface. Corrosion products formed on the metal below the coating and caused the detachment of the coating layer. Wet–dry cycles, temperature variation, and impingement of sand or fine gravel led to the cracking of the coating and corrosion products.

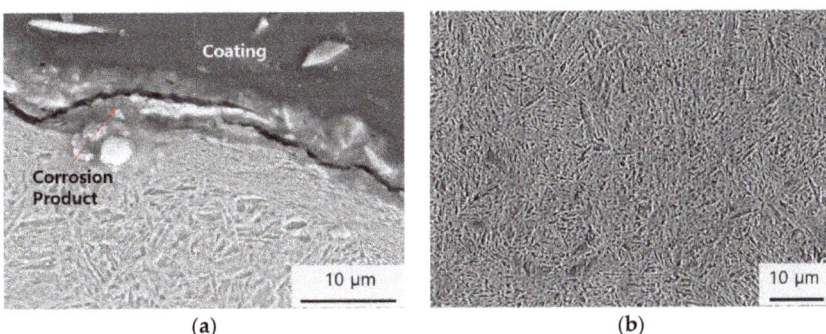

Figure 2. Microstructure of cross sections (**a**) near the surface and (**b**) at the middle of the spring.

The residual stress of S group specimens was measured by IIT and is shown in Figure 3. The compressive stress of the two specimens, S1 and S2, are 155 and 116 MPa on average, respectively.

Figure 4 shows the hardness of the S and M specimens. The hardness of the S1 and S2 samples was Hv 617 and Hv 682, respectively. The hardness of the M specimens was approximately Hv 590. The hardness of the alloy surface was increased due to the increase of the dislocation density by compressive stress during shot peening [52]. The deviation between the data for the M specimens was much lower than that for the S specimens. The variations in the residual stress and hardness were large between the S specimens and also between the repetitive measurements for a given S specimen. The different depth of grinding between the S1 and S2 specimens caused a different compressive stress and hardness because the compressive stress had a gradual increase and decrease profile with depth.

The deviation of data for a specimen was presumed to be due to the shot peening, which created locally irregular stress on the surface from the random overlapping of impingements.

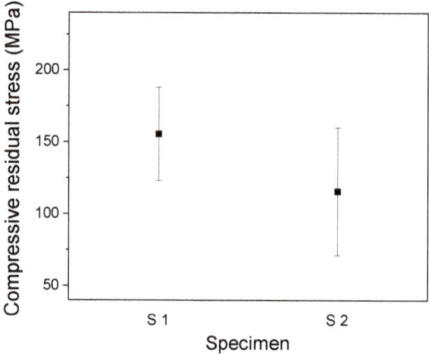

Figure 3. Residual stress of the S specimens measured by instrumented indentation testing (IIT).

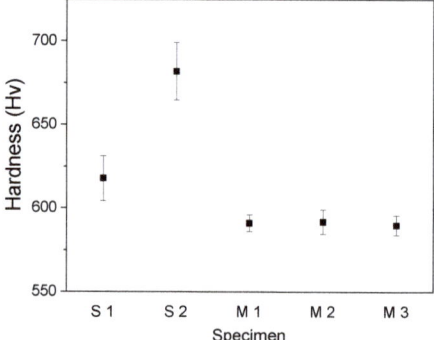

Figure 4. Hardness of the coil spring specimens.

The potentiodynamic curves of the spring steels are shown in Figure 5. The corrosion potential of the specimens was about -0.75 V_{SCE} and the critical anodic current density was measured to be 1.7×10^{-5}–6.6×10^{-5} A/cm^2 at approximately -0.63 V_{SCE}. The current density began to decrease at -0.63 V_{SCE} and the passive current density was between 5×10^{-6}–1.4×10^{-5} A/cm^2 at potentials below 0.5 V_{SCE}.

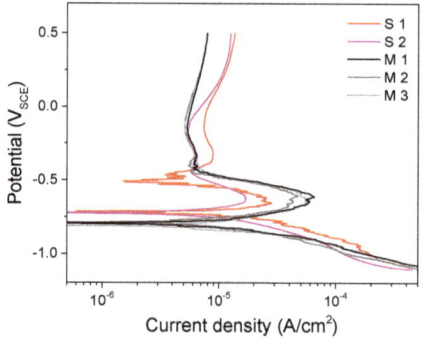

Figure 5. Potentiodynamic polarization curves.

The corrosion potential (E_{corr}), the corrosion rate (i_{corr}), and the passive current density (i_{pass}) at 0.5 V_{SCE} are plotted in Figure 6. The corrosion potential and the passive current density at 0.5 V_{SCE} of the S group were a little higher than those of the M group. The corrosion rate did not appear to show a dependence on the specimens. The critical anodic current density was measured to be 1.6×10^{-5}–2.6×10^{-5} A/cm^2 for the S group and 4.6×10^{-5}–6.4×10^{-5} A/cm^2 for the M group at -0.65–-0.60 V_{SCE}, indicating that the maximum dissolution rate of the S group was lower than that of the M group. It is remarkable that the S specimens, which have residual stress, dissolved less before passivation but that the passive current density of the S specimens was higher than the M specimens without residual stress.

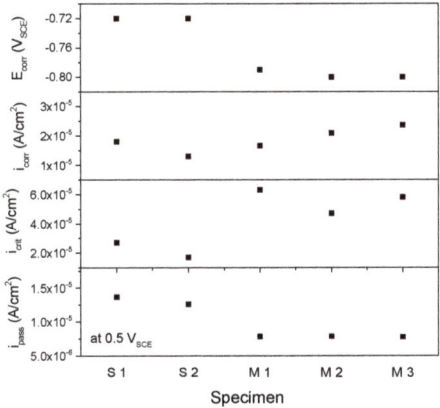

Figure 6. Corrosion parameters determined from the potentiodynamic polarization curves.

The potentiostatic polarization curves presented in Figure 7 also show a different passivation behavior for the S and M specimens. The passive current density of S1 and S2 decreased very rapidly and reached a minimum after about 1.5 h. Their current density showed slight increases and decreases during potentiostatic polarization. On the other hand, the current density of the M specimens decreased slowly during 16–17 h, and then showed a little increase. The passive current density of the M specimens became lower than that of the S specimens after 5–11 h.

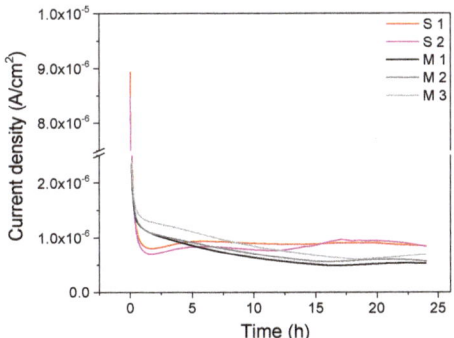

Figure 7. Current transients during potentiostatic passivation at 0.5 V_{SCE} for 24 h.

The passive current density reached 8.4×10^{-7}–8.5×10^{-7} A/cm^2 for the S specimens and 5.3×10^{-7}–6.9×10^{-7} A/cm^2 for the M specimens after 24 h of passivation (Figure 8). The S specimens had a higher current density than that of the M specimens, although the ranking of the passive current density of each specimen shown in the potentiostatic polarization (Figure 8) was different from that shown by the i_pass at 0.5 V_SCE from the potentiodynamic polarization (Figure 6).

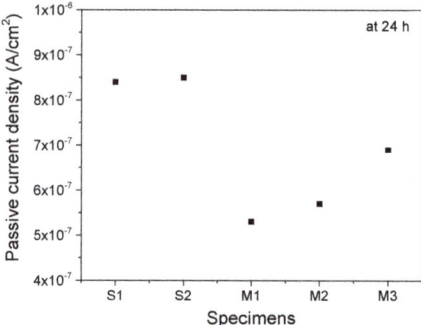

Figure 8. Passive current density after 24 h.

The difference in the potentiodynamic and potentiostatic polarization behaviors means that the electrochemical dissolution and passivation mechanisms of spring steels can be different under the influence of residual stress. It seems that the steel surface with compressive stress is less sensitive to an anodic dissolution and is rapidly passivated, but that its passive film is less stable than the surface without stress.

The type and concentration of point defects in the passive film were examined using a Mott–Schottky analysis. The authors of this study point out that the Mott–Schottky analysis has limitations as applied to passive films, because a passive film is not a well-defined semiconductor. Therefore, several assumptions are employed commonly and the results of the analysis are interpreted with care [53–55]. In particular, the quantitative property derived from the Mott–Schottky analysis (i.e., the donor density in this study) cannot be directly compared with that of other alloys. Nevertheless, the relative value of the donor density within the boundary of this study can be discussed.

The Mott–Schottky plots (Figure 9) of all specimens showed similar behavior, indicating a linear region with a positive slope between −0.3–0.3 V_SCE. This means that these specimens had passive films with an n-type semiconductivity. The donor density, which implies the concentration of oxygen vacancy in the passive film [17], was determined by the Mott–Schottky relationship, as shown in Equation (1):

$$\frac{1}{C^2} = \frac{2}{\varepsilon \varepsilon_0 e N_D}\left(E_\text{app} - E_\text{FB} - \frac{kT}{e}\right) \quad (1)$$

where C is the capacitance of the space charge layer, ε is the dielectric constant of the passive film, ε_0 is the permittivity of the vacuum, e is the charge of an electron, N_D is the donor density, E_app is the applied potential, E_FB is the flat band potential, k is the Boltzmann constant, and T is the temperature. ε_0 of the passive films on the specimens in this study was presumed to be 15.6, as accepted usually for the passive film of steels [20,46–48].

The donor density and the flat band potential of the passive films formed on the S and M specimens are shown in Figure 10. The donor density of the S1 and S2 specimens was measured to be 2.01×10^{19} cm^{-3} and 1.37×10^{19} cm^{-3}, respectively, whereas that of the M specimens was 3.03×10^{19}–4.03×10^{19} cm^{-3}. The density of point defect, which is thought commonly to be oxygen vacancy for the n-type passive film [50], was found to be higher in the passive film of the steels without residual stress than in the film on the specimens with stress. The flat band potential of passive films on the M specimens was −0.427–−0.434 V_SCE and a little higher than that (−0.448–0.463 V_SCE) of the film on the S specimens.

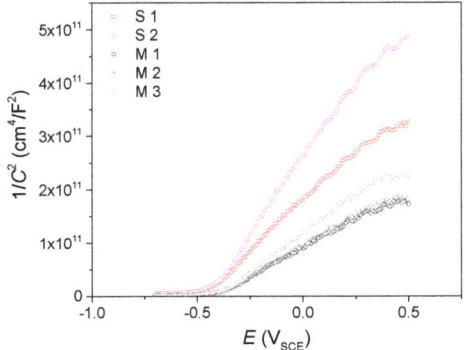

Figure 9. Mott–Schottky plots.

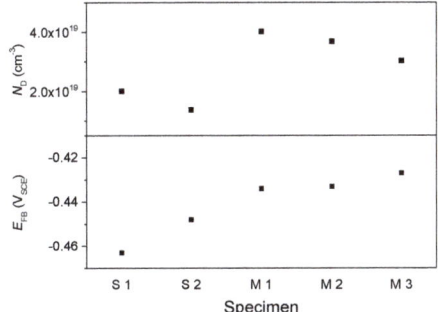

Figure 10. Donor density and flat band potential determined by Mott–Schottky analysis.

Figure 11 shows the i_{pass} at 24 h vs. the N_D plot, based on the data presented in Figures 8 and 10. The passive current density decreased with an increase in the defect density in the passive film generally, although such dependence was not explicit for S2. The authors could anticipate easily that a low density of point defects leads to a slow mass transport and hence a low passive current density. However, the results of this study presented the opposite, in that the passive current density for the M specimens was lower than that for the S specimens although the donor density of the M specimens was higher.

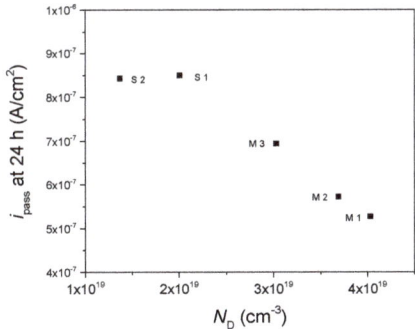

Figure 11. Relationship between donor density and steady state passive current density.

Previous reports about the relationship between the corrosion resistance and the point defect density fall into three categories. One group has suggested that the low density of point defect causes high corrosion resistance. Extensive experimental results apparently agree with this opinion [18–37]. Other researchers have reported a high corrosion rate or a low corrosion resistance for the metals or alloys with passive films with low point defect density [38–42], corresponding to the results drawn from this work. Another group has proposed no clear dependence of corrosion resistance on the point defect density [43–45]. These studies dealt with passivity or corrosion behaviors of various metals and alloys in various aqueous environments. However, the authors of this study could not find any relationship between the three different suggestions and the experimental conditions or semiconducting types from the previous works. Therefore, the origin of such discordance is not yet known. Many of the reports included a proposition that the concentration of point defect would affect the stability of the passive film and hence the corrosion resistance but did not provide a mechanism supported theoretically [18,19,21–38].

Only a few researchers, represented by Ahn et al. [44] and Park et al. [20], presented the theoretical mechanism of degradation of corrosion resistance with an increase of point defect density validated by the experimental phenomena, based on the Point Defect Model. Ahn et al. [44] and Park et al. [20] commonly conducted studies on the passive film of Ni, which is a p-type semiconductor. It was suggested that a high concentration of cation vacancy in the passive film created voids at the metal–film interface and promoted a breakdown of the passive film. Nevertheless, Ahn et al. also showed that the passive current density of Fe, which is an n-type semiconductor, did not depend on the point defect density, that is, the concentration of oxygen vacancy [44]. Their work implies that the relationship between the defect density and the corrosion resistance should be discussed while considering the type of the defect, although adequate evidence is not yet established.

Figure 12 shows the i_{pass} at 24 h, the N_D, and the hardness vs. the residual stress plots. The hardness of the S specimens was higher than that of the M specimens, but not increased with an increase in the residual stress. The passive current density was generally higher for the specimens with stress but did not appear to depend on the residual stress in the S group. The donor density was lower for the S specimens, but it had rather increased with an increase in the compressive stress in the S group. It is not clear whether the compressive residual stress involves any relationship which can be described as a linear or a high-order function from this work. However, it was noted that the bulk specimen without residual stress (i.e., the M group) and the sub-surface specimen with residual stress (i.e., the S group) had a different corrosion and passivity behavior.

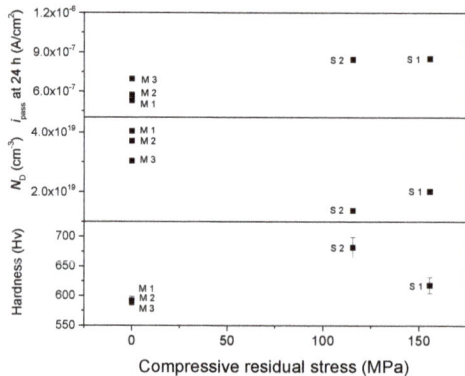

Figure 12. Effects of compressive residual stress on steady state passive current density, donor density, and hardness.

Several reports suggest that the corrosion resistance is improved with compressive residual stress [56–59]. On the contrary, other studies present that shot peening lowers the corrosion resistance. Irregularity and micro-cracks due to immoderate shot peening sometimes accelerate corrosion [52]. There was a case where the corrosion rate of an SAE 5155 steel after shot peening was higher than the specimen without peening in a relatively short test, although the results were inversed after a prolonged test [56]. It might be a similar phenomenon with the potentiostatic polarization behavior shown in Figure 7 in this study, in that the current density of the S specimens underwent a fast decrease followed by a slight increase.

The donor density is known to increase and decrease during the passivation process [60]. Jang et al. [60] reported that the donor density in the passive film of Fe-20Cr-15Ni alloy rapidly increased for the initial 2 h and then decreased slowly. In the same work, the flat band potential was gradually lowered with passivation time. The passive film was thought to grow at a high rate during the initial stage with the generation of many point defects and to reach a steady state when the generation and annihilation of point defects were balanced. The passive film of the S specimens in this work grew faster (Figure 7) and had a lower donor density and flat band potential after the same passivation time as that of the film for the M specimens (Figure 10). It might be due possibly to the high diffusion rates of ions through many grain boundaries in the shot-peened surface as Lv et al. [19] have suggested. After that stage, the transport of vacancies would slow down and reach a steady state, but the high density of dislocations and grain boundaries involved with the residual stress might cause a higher transport rate in the passive film than expected in the passive film formed on normal grains without stress.

4. Conclusions

The effect of shot peening on the corrosion behavior of spring steels was investigated by electrochemical polarization tests and Mott–Schottky analysis.

The compressive stress of 116–155 MPa was induced and the hardness was increased by shot peening. The specimens with compressive residual stress were passivated faster and their density of point defect was lower than the specimens without residual stress. However, the passive current density after 24 h was higher for the specimens with stress than that for the samples without stress, possibly due to the higher diffusion rate involved with the fine and defected grain structure caused by compressive stress.

Author Contributions: Conceptualization: H.J.; Methodology: H.J.; Validation: H.J.; Formal analysis: K.-H.L.; Investigation: K.-H.L. and J.-W.S.; Resources: S.-H.A. and J.-W.S.; Data curation: K.-H.L.; Writing–Original draft preparation: K.-H.L.; Writing–Review and editing: H.J.; Visualization: K.-H.L.; Supervision: H.J. and S.-H.A.; Project administration: H.J. and S.-H.A.; Funding acquisition: H.J. and S.-H.A.

Funding: This work was supported by the "Human Resources Program in Energy Technology" of the Korea Institute of Energy Technology Evaluation and Planning (KETEP), and was granted financial resources from the Ministry of Trade, Industry & Energy, Republic of Korea (No. 20174030201620).

Conflicts of Interest: The authors declare no conflict of interest.

References

1. Park, K.D.; Ki, W.T.; Sin, Y.J. An Evaluation on Corrosion Fatigue Life of Spring Steel by Compressive Residual Stress. *Trans. Korean Soc. Automot. Eng.* **2007**, *15*, 1–7.
2. Jung, J.W.; Park, W.J.; Huh, S.C.; Lee, K.Y.; Ha, K.J. The Effect of Compressive Residual Stress on Fracture Toughness of SUP-9 Spring Steel for Automobile. *Proc. KSAE Conf.* **2003**, 1624–1629.
3. Lim, M.S. Surface treatment method of coil spring. KR10-2011-0116842, 10 November 2011.
4. Bae, D.H.; Lee, G.Y.; Jung, W.S. Stress Analysis of the Automobile's Coil Spring Including Residual Stresses by Shot Peening. *Key Eng. Mater.* **2006**, 459–464. [CrossRef]
5. Akira, T. Coil spring for a suspension of an automobile and manufacturing method thereof. WO2010146898, 23 December 2010.

6. Kim, W.J.; Kim, J.G.; Kim, Y.S.; Ismail, O.; Tsunekawa, Y. Wear-corrosion of cast iron thermal spray coating on Al alloy for automotive components. *Met. Mater. Int.* **2017**, *13*, 317. [CrossRef]
7. Komazaki, I.; Kobayashi, K.; Misawa, T.; Fukuzumi, T. Environmental embrittlement of automobile spring steels caused by wet–dry cyclic corrosion in sodium chloride solution. *Corros. Sci.* **2005**, *47*, 2450–2460. [CrossRef]
8. Ramamurthy, A.C.; Lorenzen, W.I.; Bless, S.J. Stone impact damage to automotive paint finishes: An introduction to impact physics and impact induced corrosion. *Prog. Org. Coat.* **1994**, *25*, 43–71. [CrossRef]
9. Choi, Y.S.; Kim, J.G.; Kim, Y.S.; Huh, J.Y. Corrosion characteristics of coated automotive parts subjected to field and proving ground tests. *Int. J. Automot. Technol.* **2008**, *9*, 625–631. [CrossRef]
10. Matejicek, J.; Brand, P.C.; Drews, A.R.; Krause, A.; Lowe-Ma, C. Residual stresses in cold-coiled helical compression springs for automotive suspensions measured by neutron diffraction. *Mater. Sci. Eng. A* **2004**, *367*, 306–311. [CrossRef]
11. Das, S.K.; Mukhioadhyay, N.K.; Kumar, B.R.; Bhattacharya, D.K. Failure analysis of a passenger car coil spring. *Eng. Fail. Anal.* **2007**, *14*, 158–163. [CrossRef]
12. Del Llano-Vizcaya, L.; Rubio-Gonzalez, C.; Mesmacque, G.; Banderas-Hernandez, A. Stress relief effect on fatigue and relaxation of compression springs. *Mater. Des.* **2007**, *28*, 1130–1134. [CrossRef]
13. Mano, H.; Kondo, S.; Matsumuro, A. Microstructured surface layer induced by shot peening and its effect on fatigue strength. In Proceedings of the International Symposium on Micro-Nano Mechatronics and Human Science, Nagoya, Japan, 5–8 November 2006; IEEE: Nagoya, Japan, 2006.
14. Lim, M.S. Cold coil spring heat treatment method for improvement of inner side residual stress and fatigue life. KR10-2015-0048273, 7 May 2015.
15. Kazuya, I. Compression coil spring and method for producing same. KR10-2015-7009184, 4 October 2015.
16. Go, H.G. Spring manufacturing method. KR10-2014-0133138, 2 October 2014.
17. Macdonald, D.D. Passivity–the key to our metals-based civilization. *Pure Appl. Chem.* **1999**, *71*, 951–978. [CrossRef]
18. Yang, M.Z.; Luo, J.L.; Patchet, B.M. Correlation of hydrogen-facilitated pitting of AISI 304 stainless steel to semiconductivity of passive films. *Thin Soild Films* **1999**, *354*, 142–147. [CrossRef]
19. Lv, J.L.; Luo, H.Y. Comparison of corrosion properties of passive films formed on phase reversion induced nano/ultrafine-grained 321 stainless steel. *Appl. Surf. Sci.* **2013**, *280*, 124–131.
20. Park, K.J.; Ahn, S.J.; Kwon, H.S. Effects of solution temperature on the kinetic nature of passive film on Ni. *Electrochim. Acta* **2011**, *56*, 1662–1669. [CrossRef]
21. Jang, H.J.; Kwon, H.S. Effects of Film Formation Conditions on the Chemical Composition and the Semiconducting Properties of the Passive Film on Alloy 690. *Corros. Sci. Technol.* **2006**, *5*, 141–148.
22. Fattah-Alhosseini, A.; Vafaeian, S. Influence of grain refinement on the electrochemical behavior of AISI 430 ferritic stainless steel in an alkaline solution. *Appl. Surf. Sci.* **2016**, *360*, 921–928. [CrossRef]
23. Jang, H.J.; Kwon, H.S. In situ study on the effects of Ni and Mo on the passive film formed on Fe–20Cr alloys by photoelectrochemical and Mott–Schottky techniques. *J. Electroanal. Chem.* **2006**, *590*, 120–125. [CrossRef]
24. Ge, H.H.; Zhou, G.D.; Wu, W.Q. Passivation model of 316 stainless steel in simulated cooling water and the effect of sulfide on the passive film. *Appl. Surf. Sci.* **2003**, *211*, 321–334. [CrossRef]
25. Li, D.; Wang, J.; Chen, H.; Chen, D. Investigation on electronic property of passive film on nickel in bicarbonate/carbonate buffer solution. *Chin. J. Chem.* **2011**, *29*, 243–253. [CrossRef]
26. Freire, L.; Carmezim, M.J.; Ferreira, M.G.S.; Monremor, M.F. The passive behaviour of AISI 316 in alkaline media and the effect of pH: A combined electrochemical and analytical study. *Electrochim. Acta* **2010**, *55*, 6174–6181. [CrossRef]
27. Sarlak, H.; Atapour, M.; Esmailzadeh, M. Corrosion behavior of friction stir welded lean duplex stainless steel. *Mater. Des.* **2015**, *66*, 209–216. [CrossRef]
28. Lv, J.; Luo, H. Effects of strain and strain-induced α′-martensite on passive films in AISI 304 austenitic stainless steel. *Mater. Sci. Eng. C* **2014**, *34*, 484–490. [CrossRef] [PubMed]
29. Zhang, G.A.; Cheng, Y.F. Micro-electrochemical characterization and Mott–Schottky analysis of corrosion of welded X70 pipeline steel in carbonate/bicarbonate solution. *Electrochim. Acta* **2009**, *55*, 316–324. [CrossRef]
30. Guo, H.X.; Lu, B.T.; Luo, J.L. Study on passivation and erosion-enhanced corrosion resistance by Mott-Schottky analysis. *Electrochim. Acta* **2006**, *52*, 1108–1116. [CrossRef]

31. Yanagisawa, K.; Nakanishi, T.; Hasegawa, Y.; Fushimi, K. Passivity of Dual-Phase Carbon Steel with Ferrite and Martensite Phases in pH 8.4 Boric Acid-Borate Buffer Solution. *J. Electrochem. Soc.* **2015**, *162*, C322–C326. [CrossRef]
32. Liu, C.Y.; Jing, R.; Wang, Q.; Zhang, B.; Jia, Y.Z.; Ma, M.Z.; Liu, R.P. Fabrication of Al/Al_3Mg_2 composite by vacuum annealing and accumulative roll-bonding process. *Mater. Sci. Eng. A* **2012**, *558*, 510–516. [CrossRef]
33. Lee, J.B.; Yoon, S.I. Effect of nitrogen alloying on the semiconducting properties of passive films and metastable pitting susceptibility of 316L and 316LN stainless steels. *Mater. Chem. Phys.* **2010**, *122*, 194–199. [CrossRef]
34. Ningshen, S.; Mudali, U.K.; Mittal, V.K.; Khatak, H.S. Semiconducting and passive film properties of nitrogen-containing type 316LN stainless steels. *Corros. Sci.* **2007**, *49*, 481–496. [CrossRef]
35. Amri, J.; Souier, T.; Malki, B.; Baroux, B. Effect of the final annealing of cold rolled stainless steels sheets on the electronic properties and pit nucleation resistance of passive films. *Corros. Sci.* **2008**, *50*, 431–435. [CrossRef]
36. Yan, M.L.; Zhao, W.Z. Influence of temperature on corrosion behavior of PbCaSnCe alloy in 4.5 M H_2SO_4 solution. *J. Power Sources* **2010**, *195*, 631–637.
37. Cheng, Y.F.; Luo, J.L. A comparison of the pitting susceptibility and semiconducting properties of the passive films on carbon steel in chromate and bicarbonate solutions. *Appl. Surf. Sci.* **2000**, *167*, 113–121. [CrossRef]
38. Oguzie, E.E.; Li, J.; Liu, Y.; Chen, D.; Li, Y.; Yang, K.; Wang, F. The effect of Cu addition on the electrochemical corrosion and passivation behavior of stainless steels. *Electrochim. Acta* **2010**, *55*, 5028–5535. [CrossRef]
39. Jang, H.J.; Oh, K.N.; Ahn, S.J.; Kwon, H.S. Determination of the diffusivity of cation vacancy in a passive film of Ni using Mott-Schottky analysis and in-situ ellipsometry. *Met. Mater. Int.* **2014**, *20*, 277–283. [CrossRef]
40. Jang, H.J.; Kwon, H.S. Effects of Cr on the structure of the passive films on Ni-(15, 30) Cr. *ECS Trans.* **2007**, *3*, 1–11.
41. Liu, B.; Zhang, T.; Shao, Y.; Meng, G.; Wang, F. Effect of hydrostatic pressure on the nature of passive film of pure nickel. *Mater. Corros.* **2011**, *62*, 269–274. [CrossRef]
42. Jang, H.J.; Kwon, H.S. Effects of Mo on the Passive Films Formed on Ni-(15, 30) Cr-5Mo Alloys in pH 8.5 Buffer Solution. *J. Korean Electrochem. Soc.* **2009**, *12*, 258–262. [CrossRef]
43. Park, K.; Kwon, H. Effects of manganese on the passivity of Fe-18Cr-xMn (x = 0, 6, 12). *Korea Adv. Inst. Sci. Technol. Daejeon* **2005**, 32–40.
44. Ahn, S.J.; Kwon, H.S.; Macdonald, D.D. Role of chloride ion in passivity breakdown on iron and nickel. *J. Electrochem. Soc.* **2005**, *152*, B482–B490. [CrossRef]
45. Li, D.G.; Wang, J.D.; Chen, D.R.; Liang, P. The role of passive potential in ultrasonic cavitation erosion of titanium in 1 M HCl solution. *Ultrason. Sonochem.* **2016**, *29*, 279–287. [CrossRef] [PubMed]
46. Hakiki, N.B.; Boudin, S.; Rondot, B.; Belo, M.D.C. The electronic structure of passive films formed on stainless steels. *Corros. Sci.* **1995**, *37*, 1809–1822. [CrossRef]
47. Paola, A.D. Semiconducting properties of passive films on stainless steels. *Electrochim. Acta* **1989**, *34*, 203–210. [CrossRef]
48. Cheng, Y.F.; Luo, J.L. Electronic structure and pitting susceptibility of passive film on carbon steel. *Electrohim. Acta* **1999**, *44*, 2947–2957. [CrossRef]
49. Pourbaix, M. *Atlas of Electrochemical Equilibria in Aqueous Solutions*; National Association of Corrosion Engineers: Houston, TX, USA, 1974.
50. Ahn, S.J.; Kwon, H.S. Effects of solution temperature on electronic properties of passive film formed on Fe in pH 8.5 borate buffer solution. *Electrochim. Acta* **2004**, *49*, 3347–3353. [CrossRef]
51. Smith, W.F. *Structure and Properties of Engineering Alloys*, 2nd ed.; McGraw-Hill: New York, NY, USA, 1990.
52. Han, M.S.; Hyun, G.Y.; Kim, S.J. Effects of Shot Peening Time on Microstructure and Electrochemical Characteristics for Cu Alloy. *J. Korean Soc. Mar. Environ. Saf.* **2013**, *19*, 545–551. [CrossRef]
53. Di Franco, F.; Santamaria, M.; Massro, G.; Di Quarto, F. Photoelectrochemical monitoring of rouging and de-rouging on AISI 316L. *Corros. Sci.* **2017**, *116*, 74–87. [CrossRef]
54. Tranchida, G.; Clesi, M.; Di Franco, F.; Di Quarto, F.; Santamaria, M. Electronic properties and corrosion resistance of passive films on austenitic and duplex stainless steels. *Electrochim. Acta* **2018**, *273*, 412–423. [CrossRef]

55. Di Quarto, F.; Di Franco, F.; Miraghaei, S.; Santamaria, M.; La Mantia, F. The amorphous semiconductor Schottky barrier approach to study the electronic properties of anodic films on Ti. *J. Electrochem. Soc.* **2017**, *164*, C516–C525. [CrossRef]
56. Park, K.D.; Sin, Y.J.; Kim, D.U. The effect of shot peening on the corrosion and fatigue crack to SAE5155 steel. *J. Korean Soc. Mar. Eng.* **2006**, *30*, 731–739.
57. Park, K.D.; Ha, K.J. Influence of shot peening on the corrosion of spring steel. *J. Ocean Eng. Technol.* **2003**, *17*, 39–45.
58. Hosaka, T.; Yoshihara, S.; Amanina, I.; MacDonald, B.J. Influence of grain refinement and residual stress on corrosion behavior of AZ31 magnesium alloy processed by ECAP in RPMI-1640 medium. *Procedia Eng.* **2017**, *184*, 432–441. [CrossRef]
59. Peyre, P.; Scherpereel, X.; Berthe, L.; Carboni, C.; Fabbro, R.; Beranger, G.; Lemaitre, C. Surface modifications induced in 316L steel by laser peening and shot-peening. Influence on pitting corrosion resistance. *Mater. Sci. Eng. A* **2000**, *280*, 294–302. [CrossRef]
60. Jang, H.J.; Park, C.J.; Kwon, H.S. Photoelectrochemical study of the growth of the passive film formed on Fe-20Cr-15Ni in a pH 8.5 buffer solution. *Met. Mater. Int.* **2010**, *16*, 247–252. [CrossRef]

© 2018 by the authors. Licensee MDPI, Basel, Switzerland. This article is an open access article distributed under the terms and conditions of the Creative Commons Attribution (CC BY) license (http://creativecommons.org/licenses/by/4.0/).

Article

Effect of Surface Nanocrystallization on Corrosion Resistance of the Conformed Cu-0.4%Mg Alloy in NaCl Solution

Dan Song [1,2], Jinghua Jiang [1,*], Xiaonan Guan [1], Yanxin Qiao [3], Xuebin Li [4], Jianqing Chen [1], Jiapeng Sun [1] and Aibin Ma [1,2,*]

1. College of Mechanics and Materials, Hohai University, Nanjing 210098, China; songdancharls@hhu.edu.cn (D.S.); guanxiaonanayln123@163.com (X.G.); chenjq@hhu.edu.cn (J.C.); sun.jiap@gmail.com (J.S.)
2. Suqian Research Institute, Hohai University, Suqian 223800, China
3. School of Materials Science and Engineering, Jiansgu University of Science and Technology, Zhenjiang 212003, China; yxqiao@just.edu.cn
4. China Railway Construction Electrification Bureau Group Co., Ltd., Beijing 100043, China; xblee2013@sohu.com
* Correspondence: jinghua-jiang@hhu.edu.cn (J.J.); aibin-ma@hhu.edu.cn (A.M.); Tel.: +86-25-8378-7239 (J.J. & A.M.); Fax: +86-25-8378-6046 (J.J. & A.M.)

Received: 28 August 2018; Accepted: 25 September 2018; Published: 26 September 2018

Abstract: Surface nano-crystallization (SNC) of a conform-extruded Cu-0.4 wt.% Mg alloy was successfully conducted by high-speed rotating wire-brushing to obtain the deformed zone with dislocation cells and nanocrystallines. SNC promotes the anodic dissolution and corrosion rate of the Cu-Mg alloy in the initial stage of immersion corrosion in 0.1 M NaCl solution. The weakened corrosion resistance is mainly attributed to the higher corrosion activity of SNC-treated alloy. With extending the immersion time, the SNC-treated alloy slows the corrosion rate dramatically and exhibits uniform dissolution of the surface. The formation of the dense corrosion products leads to the improvement of overall corrosion performance. It indicates that the SNC-treated Cu-Mg alloy can function reliably for a longer duration in a corrosive environment.

Keywords: Cu-Mg alloy; conform; surface nanocrystallization; corrosion resistance

---

1. Introduction

With the rapid growth of the high-speed railway over the last few decades, more attention has been focused on the development of copper alloys for their high strength, good electrical conductivity, satisfactory resistance to wear, and corrosion for contact wires [1]. Until now, plenty of research has been conducted to enhance the required properties of copper alloys by adding small amounts of alloying elements to them, such as Cr, Zr, Ag, Ni, and Mg [2–6]. Contact wire Cu-Mg alloys with lower production costs demonstrate ideal comprehensive properties and are considered the current preferred material for making contact wires for high-speed trains, of which the operating speed is more than 300 km/h [7,8]. According to the phase diagram of Cu-Mg, a single solid-solution copper alloy containing a small amount of Mg can be obtained at room temperature. Solution strengthening of the Mg element should not cause severe lattice distortion of the copper matrix for their similar atom radius, and should therefore keep the excellent conductivity performance of copper. In China, Cu-Mg contact wires are now widely used in trains running at the speed of ≥300 km [7]. Compared with other copper alloys, single solid-solution Cu-Mg alloys have good comprehensive performance, a simple manufacturing process, and vast application prospects. In their long-term operation, they can hardly be immune from water, humidity, and salts, which affect the lifespan of contact wires. By all appearances, the corrosion control of Cu-Mg alloys is a valuable research issue.

According to the Hall Petch equation, fine grains can improve the strength of materials, and as it is generally believed, grain refinement can simultaneously benefit strength and toughness. Grain refinement of single solid-solution Cu-Mg alloy have been applied to achieve a good combination of strength and conductivity. At present, the China Railway Construction Electrification Bureau Group (Kang Yuan New Materials Co., Ltd.) has developed fine-grained Cu-0.4 wt.% Mg contact wire by using the conform-process, as well as cold drawing. The severe plastic deformation (SPD) procedure is currently one of the most effective ways to produce ultrafine-grained (UFG) alloys, which is gaining an increasing amount of attention [9–11].

Surface nano-crystallization (SNC) can be used to induce severe plastic deformation in the surface layer and obtain a nano-cystallized/ultrafine-grained (NC/UFG) gradient layer with high strength and hardness [12,13]. The enhanced mechanical properties, especially the enhanced surface hardness, will improve the wear resistance of the contact wire and decrease its wear loss induced by the sliding friction with the pantograph. However, due to the complex electrochemical corrosion process and various influencing factors, it still cannot get a unified conclusion to the effect of SNC on the corrosion behavior of treated metals. Some studies reported that the SNC process decreased the corrosion resistance of the metals. Li [14] declared the decreased corrosion resistance of SNC low-carbon steel due to the increased number of active corrosion sites. Others reported the positive effect of SNC on the anti-corrosion performance of the treated metals, such as improved corrosion resistance of the SNC 316L SS steel [15] and AISI 409 SS steel [13] by surface mechanical attrition treatment (SMAT). Our former investigation also found improved passivation ability and corrosion resistance of the SNC-treated low-carbon steel rebar in the Cl^--containing concrete pore solution [16].

As important as it is to develop high-strength, good-conductivity copper contact wires, the present work investigates the corrosion behavior and corrosion resistance of the on-line conformed Cu-0.4wt.% Mg alloy subjected to experimental SNC processes by high-speed wire-brushing [16]. The influential mechanism of the SNC process on the corrosion behavior of this alloy was systematically studied. The SNC-induced special surface microstructure and surface roughness have a close relationship to the evolution in unique corrosion behavior of this alloy.

2. Experiment

The material used was Cu-0.4 wt.% Mg (oxygen ≤ 10 ppm) alloy, which were melted with electrolytic copper and pure magnesium through upward-casting, and then extruded by the conform process of the China Railway Construction Electrification Bureau Group (Kang Yuan New Materials Co., Ltd., Jiangsu, China). Mg atoms of the binary alloy mainly existed at the FCC-structured copper crystal. An illustration of the conform process is presented in Figure 1, which clearly shows it is able to refine the grain size of the alloy, and thus simultaneously improve its strength, plasticity, and conductivity [17]. The conformed round bars were continuously treated by SNC-processing via a high-speed rotating wire brush, which inflicts severe plastic deformation to the sample surface by forceful and repeated scratching (as shown in Figure 2). The detailed SNC processing parameter, such as rotation speed of the wire brush and the feeding speed of the sample, can be found in our former work [16]. Each sample was SNC-processed for four passes to obtain a uniformly modified surface layer and extreme grain refinement on the brushed surface. An optical microscope (Olympus BX51M, Tokyo, Japan) was used to observe the microstructure at the surface of the brushed samples. The composition of the etchant was glacial acetic acid 5 mL, phosphoric acid 11 mL, and nitric acid 4 mL. The etching time was 5 s. Transmission electron microscopy (TEM, JEM-2000EX, Tokyo, Japan) was applied to observe the microstructure and grain size of the Cu-Mg alloy after SNC processing. X-ray diffraction (XRD) analysis of the samples was performed using a Bruker D8 Advance diffractometer (Bruker AXS, Karlsruhe, Germany) with Cu K 1 radiation. The θ–2θ diffraction patterns were scanned from 15° to 85° with a scanning rate of 2° min^{-1}. The laser scanning confocal microscope (Olympus

LEXT OLS4000 3D, Tokyo, Japan) was used to quantitatively analysis the surface roughness after SNC treatment of the alloy.

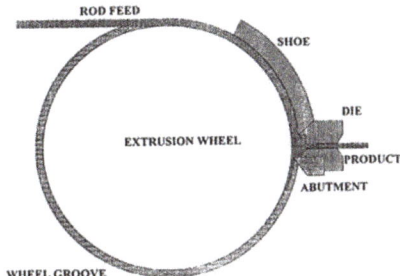

Figure 1. Schematic diagram of the conform.

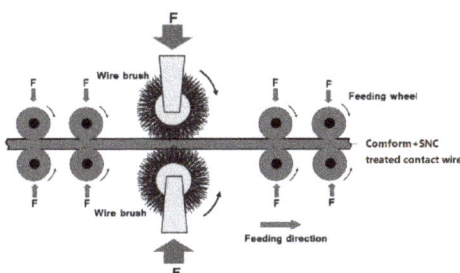

Figure 2. Schematic diagram of surface nanocrystallization.

All SNC samples subjected to corrosion tests were covered by epoxy resins, leaving a columnar exposed area of 9 cm^2. 0.1 M NaCl aqueous solution was used for corrosion tests. Immersion tests were carried out at room temperature for 30 days in an open system. The corrosion morphologies of the samples were observed via a digital microscope (Hirox, KH-7700, Hackensack, NJ, USA) and the scanning electronic microscope (SEM, S-3400N, Hitachi, Tokyo, Japan). The chemical composition of the corrosion product was characterized by the energy-dispersive X-ray spectrometer (EDS, OXFORD instrument, Oxford, Oxfordshire, UK). After the set intervals of immersion, the mass loss of the samples was examined by an electronic balance (accuracy: 0.1 mg) to calculate the corrosion rate (unit: mg·cm^{-2}·h^{-1}) of the SNC-treated alloy. 5 parallel samples were used in this test to get the average corrosion rate of the alloys with and without SNC treatment.

Electrochemical corrosion behavior of SNC samples were evaluated by a CHI660D advanced potentiostat (Huacheng, Shanghai, China) equipped with a saturated calomel electrode (SCE) and a Pt counter electrode. For better repeatability, more than three parallel samples were conducted in each electrochemical test. The samples were freely immersed in the solution for 1000 s to obtain the stable open circuit potential (OCP) values. The frequency of electrochemical impendence spectroscopy (EIS) tests ranged from 10 KHz to 0.01 Hz, and the amplitude of the sinusoidal potential signal was 5 mV with respect to the OCP value of the samples. The potentiodynamic polarization (PDP) tests were performed at a scan rate of 1 mV/s, which started at a potential value 250 mV below the obtained OCP value.

3. Results

3.1. Microstructure Characterization of Cu-0.4%Mg Alloy

Figure 3 presents optical microstructures of the conformed Cu-0.4 wt.%Mg before and after SNC processing. As shown in Figure 3a, the α-Cu grains of the conformed alloy equiaxed and reached an average size of 8 μm. The alloying Mg element in small quantities was solid solutions, thus no second-phase particles exist in the copper matrix. During the continuous conform procedure, the temperature in the die chamber reached as 800 °C and was significantly higher than the recrystallization temperature (about 350 °C) of the Cu-Mg alloy. Therefore, the grain size of the alloy subjected to the conform procedure could not reach nanometers for dynamic recrystallization. As shown in Figure 3b,c, the surface grains after SNC treatment are obviously fine, and there is a gradient microstructure in the longitudinal section of the alloy. The surface area underwent severe plastic deformation during SNC treatment, resulting in the formation of a fibrous, deformed microstructure. More information on the SNC gradient microstructure, especially the nano-grains, can be obtained from TEM observation.

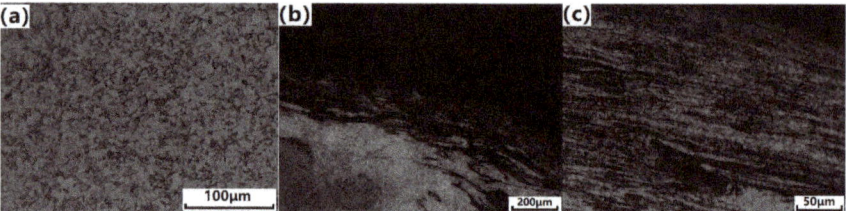

Figure 3. Optical microstructures of the Cu-Mg alloy at (**a**) the as-conformed state; (**b**) and (**c**) are the conform + surface nano-crystallization (SNC) alloy at low and high magnification, respectively.

Figure 4 presents the TEM micrograph of severely deformed α-Cu grains after SNC treatment. It is clear that the α-Cu grains have been further refined into equiaxed nano-grains with an average size of 400 nm. The white areas at the arrowheads in Figure 4a are the low-density zones (LDDZ) of dislocations, which is a typical microstructural characteristic in nanostructured copper samples [18,19]. As is well-known, dislocation tangling is frequently observed in the grain interior of heavily strained alloys. During the SNC procedure, dislocation proliferation which occurred on the alloys' surface generated high-density dislocations, a mass of dislocation cells, and evident dynamic-recrystallization phenomena. Thus, the formation of the LDDZ areas should be attributed to the dynamic equilibrium between the production and annihilation of the dislocations, as well as the dislocation absorption at the grain boundaries with the grain refinement [20,21]. Plenty of approximately equiaxed dislocation cells increasingly formed subgrains during SNC, and eventually, the subgrain boundaries became low-angle grain boundaries (GBs) and even high-angle GBs. In essence, the severely deformed surface reached the nano-scale level after SNC modification. The arrowhead in Figure 4b indicates the deformation twin in particular grains, which is created by the shear stress and severe strain during the SNC process. The stress causes the appearance of partial dislocations at the grain boundary, which react to form the parallel twins.

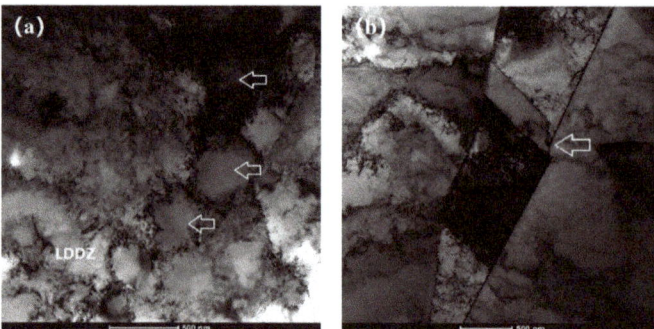

Figure 4. Transmission electron microscopy (TEM) images of the Cu-Mg alloy after SNC: (**a**) at lower magnification of the dislocation tangle; (**b**) at higher magnification of the twin zone.

3.2. Electrochemical Results of Cu-0.4%Mg Alloy

Figure 5a presents the continuous OCP monitoring of the alloy with and without SNC treatment, immersed in 0.1 M NaCl solution for 1000 s. It is clear that the OCP values of both untreated and SNC samples decrease rapidly in the initial 200 s, and then decrease slowly during the rest of the immersion time. The surface of the samples were wet soon after immersing the solution, and an electric double layer was formed at the solid–liquid interface. The lower OCP value of the SNC sample presents the higher corrosion tendency in NaCl solution, compared to the conformed sample without SNC treatment.

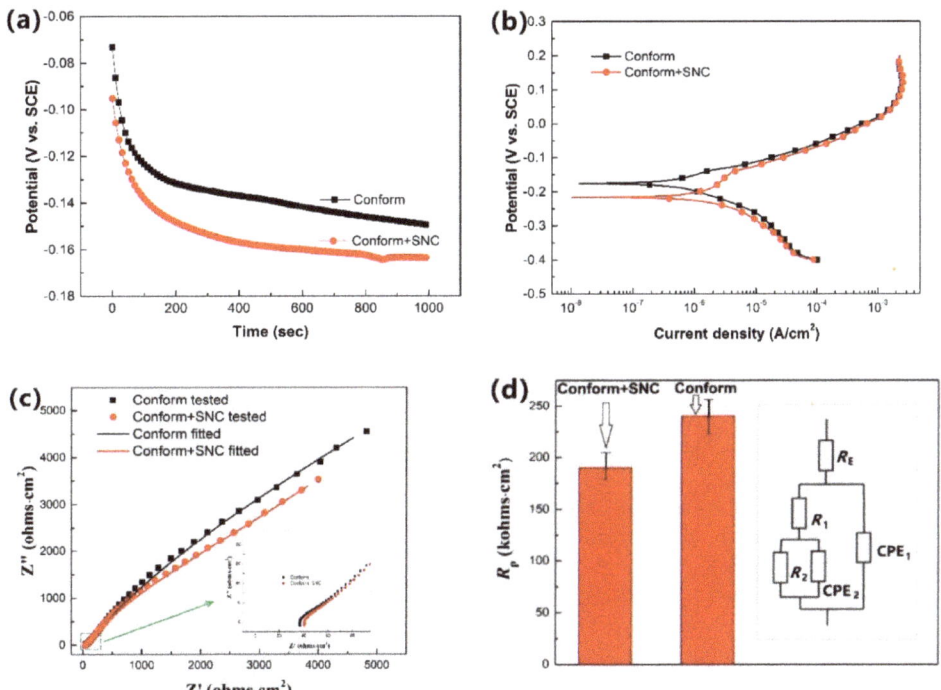

Figure 5. Electrochemical corrosion tests of the conform and conform + SNC Cu-Mg alloy after immersion in 0.1 M NaCl solution for 1000 s. (**a**) Open circuit potential (OCP) curves; (**b**) polarization curves; (**c**) and (**d**) are the EIS Nyquist plots and the relative equivalent circuit and fitted R_p values.

Figure 5b presents the PDP curves of the untreated and SNC-treated samples immersed in 0.1 M NaCl solution. Before the PDP test, the samples were immersed in the solution for 1 h. Table 1 lists the values of corrosion potential (E_{corr}) and corrosion current density (I_{corr}) determined by the Tafel extrapolation procedure from Figure 5b. The similar characteristics of the PDP curves indicates the same corrosion mechanism of the untreated and SNC-treated Cu-Mg samples, which shows up as anodic activation dissolution in the NaCl solution of low concentration. It is well-known that a nobler corrosion potential leads to lower corrosion tendency in thermodynamics, and a higher corrosion current density elucidates a faster corrosion rate in corrosive medium. As seen in Table 1, the SNC sample has a lower E_{corr} value and higher I_{corr} value, in comparison with the untreated sample. It also indicates that the SNC sample is more readily attacked by NaCl solution in the initial.

Table 1. Electrochemical parameters of conform and conform + SNC samples immersed in 0.1 M NaCl solution.

Samples	E_{corr} (V_{SCE})	I_{corr} ($\mu A/cm^2$)
Conform	-0.176 ± 0.01	0.65 ± 0.05
Conform + SNC	-0.216 ± 0.01	0.15 ± 0.03

Figure 5c presents the typical Nyquist impedance plots of the untreated and SNC-treated samples at open-circuit potential after 1000 s immersion in 0.1 M NaCl solution. The similar Nyquist plots represent that the corrosion mechanism of the alloy has not changed after SNC treatment. The line going upwards with slope one in the low-frequency region indicates a diffusion-controlled process. The high-frequency region of the EIS Nyquist plots is shown in the bottom right corner of Figure 5c. The diameter of the capacitive loop is widely accepted to typify the polarization resistance value of the double layer. The SNC samples presents a lower conductive loop diameter, which indicates lower polarization resistance (R_P) after SNC treatment. The $R(Q(R(QR)))$ equivalent circuit was used to fit the EIS plots and to show the used equivalent circuit and fitted R_p values, presented in Figure 5d. Clearly, the R_p value (about 185 kΩ·cm^2) of the SNC sample is smaller than that of the conform sample (about 235 kΩ·cm^2), indicating thqt there was less corrosion resistance in the initial corrosion period.

3.3. Immersion Corrosion Results

A constant immersion test of the samples, followed by a mass-loss measurement and optical microscopy, provided concrete evidence for electrochemical corrosion behavior. Figure 6 presents the mass-loss rate of the untreated and SNC-treated samples after long-term immersion in 0.1 M NaCl. It is clear that the change rule of corrosion rate of the two samples are the same in the solution after a certain time. The mass-loss rate of the samples gradually decreased with the rise in immersion time, and finally returned to a relatively stable value. The higher corrosion rate at the initial stage of immersion is due to anodic activation dissolution without a passivation phenomenon. After immersion for some time, the corrosion product film piled up at the surface conferred a protective effect to reduce the corrosion rate. Figure 7 presents the macro-appearance of corroded regions of the untreated and SNC-treated samples after 2 and 15 days of immersion in 0.1 M NaCl. It indicates that there is no typical pitting corrosion phenomenon, but the uniform corrosion characteristic is presented in the two samples with 2 days of immersion. After 15 days of immersion in 0.1 M NaCl, typical corrosion pits could be located on several sites for the conformed sample, but there was slight corrosion dispersed over an area for the SNC sample.

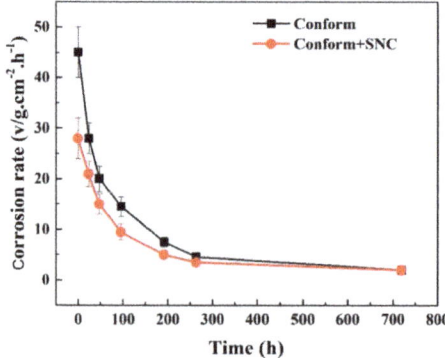

Figure 6. Corrosion rates (with duration) of conform and conform + SNC Cu-Mg alloy immersed in 0.1 M NaCl solutions at different times.

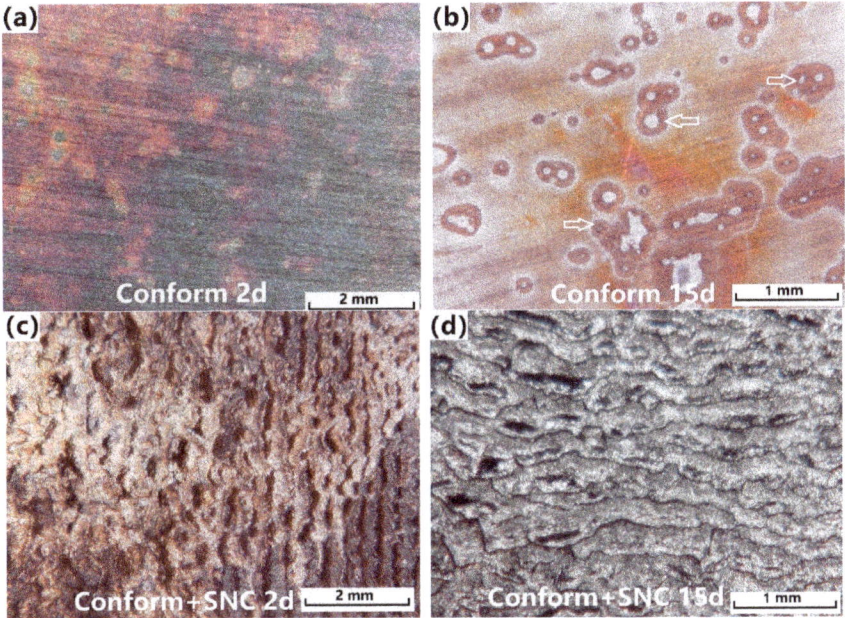

Figure 7. Macro-appearance of the conform and conform + SNC Cu-Mg alloy immersed in 0.1 M NaCl solutions for 2 days and 15 days. (**a**) Conform sample after corrosion for 2 days; (**b**) Conform sample after corrosion for 15 days; (**c**) Conform + SNC sample after corrosion for 2 days; (**d**) Conform + SNC sample after corrosion for 15 days.

3.4. Influential Mechanism of SNC on the Corrosion Behavior of the Cu-0.4%Mg Alloy

Corrosion behavior and corrosion resistance of the materials were strongly influenced by their surface condition, especially the surface microstructure characteristics and the surface roughness [22,23]. Meanwhile, the environmental factors also had an important impact on the corrosion process. In our former investigation on corrosion bahavior of the SNC-treated rebar in a simulated concrete-pore solution, we found that the SNC rebar showed enhanced passivation ability and improved corrosion resistance against Cl$^-$ aggression [16]. Herein, the corrosion behavior of the SNC-treated Cu-Mg alloy seems to be more complicated. It suffered rapid anodic dissolution and showed less corrosion resistance in NaCl solution during the initial corrosion period. However, in long-term corrosion, the SNC-treated alloy showed a decreased corrosion rate and better corrosion resistance compared to the conform alloy. It is possible that the anti-corrosion performance of the SNC-treated alloy was greatly influenced by its SNC microstructure characteristics and surface roughness.

To reveal the effects of grain refinement on the corrosion resistance of the alloy, the XRD analysis was further used to judge the grain size of the alloy before and after SNC modification and corrosion. Before the test, the SNC-modified samples were immersed in 0.1 M NaCl solution for 10 days. From the XRD plots in Figure 8, one can find that all the samples presented with typical copper peaks. Considering the small amount of Mg content, the typical copper peaks can be regarded as the Cu-Mg solid solution. A more detailed difference can be found in the full width at half maxima (FWHM) values of the XRD patterns of the alloys. Many studies have reported that the FWHM values can be used to judge the grain size of the tested materials semi-quantitatively according to the Scherrer equation. Larger FWHM values infers broadening in the diffraction peaks, which denotes a decrease in the surface grain size [24,25]. Herein, the FWHM values of the typical three strongest peaks of the SNC-treated alloy are larger than that of the conform alloy (shown in the Table 2). This phenomenon should be induced by the severe refinement of the surface grains during SNC modification, which will bring a mass of grain boundaries with high energy. As observed in the TEM images above, dislocation proliferation on the alloys' surface generated high-density dislocations, as well as strain-induced twins, during the SNC process. Since the nature of electrochemical corrosion of copper in NaCl solution leads to active dissolution of the anode, the mass of high-energetic crystal defects (such as grain boundaries, dislocations, and twins) may lead to more residual stress and corrosion activity compared to the conform alloy. These energetic crystal defects provide a more active site for corrosion, leading to more rapid anodic dissolution in the initial corrosion period.

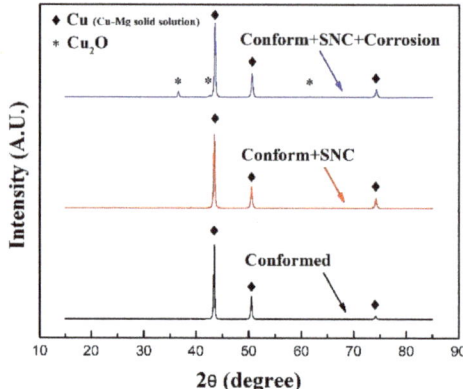

Figure 8. X-ray diffraction (XRD) plots of the conform and conform + SNC Cu-Mg alloys before and after corrosion.

Table 2. Crystallographic parameters of the Cu-0.4%Mg alloys via X-ray diffraction (XRD) analysis.

Peak Angle	FWHM of Conform Alloy	FWHM of Conform + SNC Alloy	FWHM of Conform + SNC + Corrosion alloy
First peak 43.4°	0.248	0.272	0.269
Second peak 50.5°	0.309	0.339	0.338
Third peak 74.2°	0.347	0.389	

As opposed to the conform alloy, there seems to be no change in FWHM values between the SNC samples before and after corrosion. From this phenomenon, one can believe that the corrosion damage of the SNC-treated alloy is quite limited, and the nano-scaled microstructure of this alloy was still kept after corrosion. Besides the copper peaks, the corroded sample also presented with weak Cu_2O peaks. It is generally believed that part of the cathodic process of the electrochemical corrosion of copper is oxygen absorption corrosion, and that the weak Cu_2O peaks should be detected from the corroded products.

It is generally believed that high surface roughness is also harmful to the corrosion resistance of the materials. As shown in Figure 9, the surface roughness of both the conform and conform + SNC samples were quantitatively evaluated by the laser scanning confocal microscope. The typical surface roughness (R_a) of the conform + SNC samples were about 15 μm, nearly 6 times larger than that of the conform sample (R_a = 2.4 μm). The more significant surface roughness of the SNC-treated alloy brought a more exposed area to the aggressive medium, leading to more rapid anodic dissolution during the initial corrosion period.

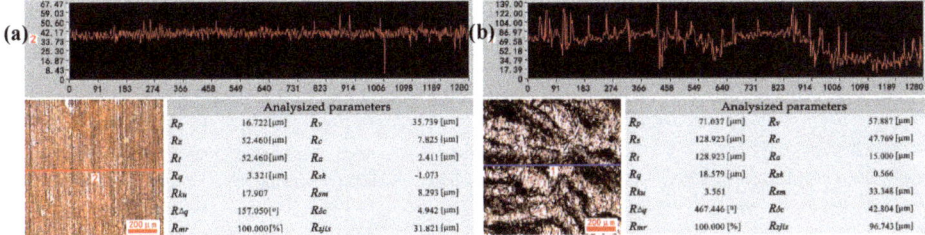

Figure 9. Surface roughness analyzed by the laser scanning confocal microscope. (**a**) Conform + SNC sample; (**b**) conform sample.

SEM analysis of the corroded surfaces of alloys after immersion corrosion in 0.1 M NaCl for 30 days was carried out to disclose the anti-corrosion mechanism during long-term corrosion. As shown in Figure 10, the corroded surfaces of the two samples were covered with corrosion products. It is obvious that the surface of the untreated sample was fully covered with more uniform and loose corrosion products, while the surface of the SNC sample was partially covered with more compact corrosion products. The finer grains with higher-density dislocation after SNC treatment provided more active sites for corrosion reaction to form a more compact corrosion production layer. Table 3 shows EDX results obtained from the corroded surface examination of the same samples in Figure 10c,f. For the two samples, the presence of chloride and oxygen was detected, from which one can deduce that CuCl and Cu_2O should be formed. A higher percentage of the O element was obtained on the surface of the Conform+SNC sample, which indicates that, in this case, CuCl formation was somehow inhibited.

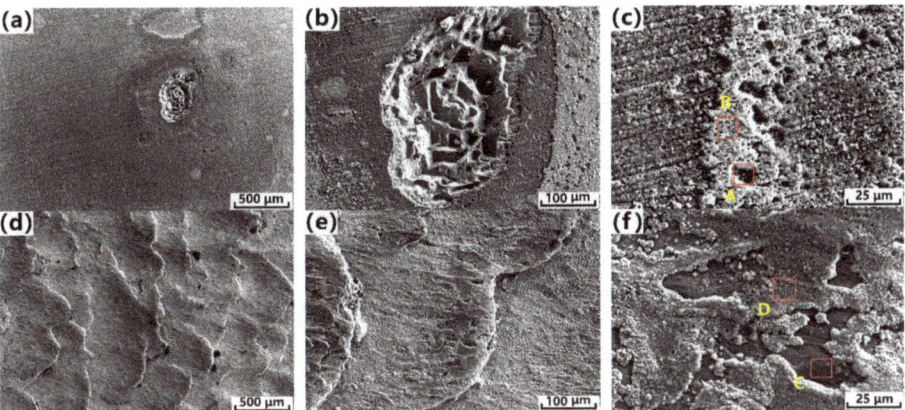

Figure 10. Scanning electronic microscopy (SEM) corrosion morphologies of the Conform (**a**–**c**) and Conform + SNC (**d**–**f**) Cu-Mg alloys' immersion in 0.1 M NaCl solution for 30 days. (**a**) and (**d**) are low-magnification images; (**b**) and (**e**) are medium-magnification images; (**c**) and (**f**) are high-magnification images.

In addition, the conform alloy presented with typical corrosion pits, as shown in Figure 10a,b, which can be related to the corrosion pits observed in the optical corrosion morphologies shown in Figure 7b. It is clear that the corrosion damage in pits has already extended into the deep substrate. Meanwhile, there were no corrosion products covering the pit, which indicates the continuous corrosion damage in the pits during long-term immersion corrosion. Due to the large tensile stress of the contact wire during the service, those corrosion pits on the conform alloy will provide priority nucleation for micro-cracks. Corrosion-induced cracks greatly decrease the fracture toughness of the alloy, which is vital to the safety and lifespan of the contact wire. However, the SNC-treated contact wire has a harder surface, as well as overall corrosion without pitting corrosion risk; these favorable factors will ensure its superior fracture toughness. Fracture toughness can be carefully extracted by the pillar-splitting method, which enables testing of the fracture toughness for sub-micrometer scale materials or in specific zones for bulk materials [26]. This investigation will be the focus of our future work.

Table 3. Energy dispersive X-ray spectrometer (EDS) analysis of corrosion products on the surface of Conform and Conform + SNC specimens immersed in 0.1 M NaCl solutions for 30 days.

Elements	Atomic Percent %			
	Conform Postion A	Conform Postion B	Conform + SNC Postion C	Conform + SNC Postion D
O	50.14	56.88	54.02	56.40
Cl	24.74	15.42	11.82	14.00
Cu	48.73	27.69	33.98	29.60

4. Conclusions

This work has shown the detrimental effect of surface nanocrystalline modification of Cu-0.4%Mg alloy on their electrochemical corrosion behavior, due to the finer grains and more microstructural defects.

The SNC process, via high-speed rotating wire-brushing, forms the severe deformed plastic flow zones and increased surface roughness. The grain size of the surface deformed zone was refined into a nanometer regime, resulting from the formation of plenty of dislocation cells and deformation twins.

Strains-induced grain refinement weakens corrosion resistance of the SNC alloy during the initial corrosion period in 0.1 M NaCl solution, resulting in the lower OCP value and higher I_{corr} values in polarization tests, a smaller capacitive loop and R_p value in EIS tests, higher mass-loss rate, and a partially corroded surface. The SNC sample with a smaller grain size has lower corrosion resistance, indicating that the increased crystal defects and higher surface roughness results in increased corrosion activity.

However, the SNC alloy presented with a gradually decreasing corrosion rate (mass-loss rate) in the long-term immersion corrosion tests. The improved corrosion-resistance performance of the SNC alloy contributed to the formation of more compact corrosion products on the SNC's surface.

Author Contributions: D.S., J.J. and A.M. conceived and designed the experiments; X.G., Y.Q. and X.L. contributed to the sample preparation and corrosion behavior testing; J.C. and J.S. contributed to the data analysis; D.S. and J.J. wrote the paper.

Funding: This research was funded by the Fundamental Research Funds for the Central Universities (2018B57714 and 2018B48414), Natural Science Foundation of China (51878246), Key Research and Development Project of Jiangsu Province of China (BE2017148), "Six talent Peak" project of Jiangsu Province (2016-XCL-196) and the Public Science & Technology Service Platform Program of Suqian City of China (Grant No. M201614).

Acknowledgments: The study was supported by the Fundamental Research Funds for the Central Universities (2018B57714 and 2018B48414), Natural Science Foundation of China (51878246), Key Research and Development Project of Jiangsu Province of China (BE2017148), "Six talent Peak" project of Jiangsu Province (2016-XCL-196) and the Public Science & Technology Service Platform Program of Suqian City of China (Grant No. M201614).

Conflicts of Interest: The authors declare no conflict of interest.

References

1. Peng, L.M.; Mao, X.M.; Xu, K.D.; Ding, W.J. Property and thermal stability of in situ composite Cu-Cr alloy contact cable. *J Mater. Process. Technol.* **2005**, *166*, 193–198. [CrossRef]
2. Zhang, L.; Meng, L.; Liu, J.B. Effects of Cr addition on the microstructural, mechanical and electrical characteristics of Cu-6wt.%Ag microcomposite. *Scr. Mater.* **2005**, *52*, 587–592. [CrossRef]
3. Fernee, H.; Nairn, J.; Atrens, A. Precipitation hardening of Cu-Fe-Cr alloys. *J. Mater. Sci.* **2001**, *36*, 2721–2741. [CrossRef]
4. Liu, J.B.; Zhang, L.; Dong, A.P.; Wang, L.T.; Zeng, Y.W.; Meng, L. Effects of Cr and Zr additions on the microstructure and properties of Cu-6wt.% Ag alloys. *Mater. Sci. Eng. A* **2012**, *532*, 331–338. [CrossRef]

5. Suzuki, S.; Shibutani, N.; Mimura, K.; Isshiki, M.; Waseda, Y. Improvement in strength and electrical conductivity of Cu-Ni-Si alloys by aging and cold rolling. *J. Alloys Compd.* **2006**, *417*, 116–120. [CrossRef]
6. Gao, H.; Wang, J.; Shu, D.; Sun, B. Microstructure and properties of Cu-11Fe-6Ag in situ composite after thermo-mechanical treatments. *J. Alloys Compd.* **2007**, *438*, 268–273. [CrossRef]
7. Zhu, C.C.; Ma, A.B.; Jiang, J.J.; Li, X.B.; Song, D.; Yang, D.H.; Yuan, Y.; Chen, JQ. Effect of ECAP combined cold working on mechanical properties and electrical conductivity of Conform-produced Cu-Mg alloys. *J. Alloys Compd.* **2014**, *582*, 135–140. [CrossRef]
8. Duan, Y.L.; Xu, G.F.; Tang, L.; Li, Z.; Yang, G. Microstructure and properties of the novel Cu-0.30Mg-0.05Ce alloy. processed by equal channel angular pressing. *Mater. Sci. Eng. A* **2015**, *648*, 252–259. [CrossRef]
9. Valiev, R.Z.; Islamgaliev, R.K.; Alexandrov, I.V. Bulk nanostructured materials from severe plastic deformation. *Prog. Mater. Sci.* **2000**, *45*, 103–189. [CrossRef]
10. Sehiotz, J.; Francesco, D.D.T.; Jacobsen, K.W. Softening of nanocrystalline metals at very small grain sizes. *Nature* **1998**, *391*, 561. [CrossRef]
11. McFadden, S.X.; Mishra, R.S.; Valiev, R.Z.; Zhilyaev, A.P.; Mukherjee, A.K. Low-temperature superplasticity in nanostructured nickel and metal alloys. *Nature* **1999**, *398*, 684–686. [CrossRef]
12. Lu, K.; Lu, J. Nanostructured surface layer on metallic materials induced by surface mechanical attrition treatment. *Mater. Sci. Eng. A* **2004**, *38*, 375–377. [CrossRef]
13. Balusamy, T.; Kumar, S.; Sankara Narayanan, T.S.N. Effect of surface nanocrystallization on the corrosion behaviour of AISI 409 stainless steel. *Corros. Sci.* **2010**, *52*, 3826–3834. [CrossRef]
14. Li, Y.; Wang, F.; Liu, G. Grain Size Effect on the Electrochemical Corrosion Behavior of Surface Nanocrystallized Low-Carbon Steel. *Corrosion* **2004**, *60*, 891–896. [CrossRef]
15. Li, N.N.; Shi, S.Q.; Luo, J.L.; Lu, J.; Wang, N. Effects of surface nanocrystallization on the corrosion behaviors of 316L and alloy 690. *Surf. Coat. Technol.* **2017**, *309*, 227–231. [CrossRef]
16. Song, D.; Ma, A.B.; Sun, W.; Jiang, J.H.; Jiang, J.Y.; Yang, D.H.; Guo, G.H. Improved corrosion resistance in simulated concrete pore solution of surface nanocrystallized rebar fabricated by wire-brushing. *Corros. Sci.* **2014**, *82*, 437–441. [CrossRef]
17. Kim, Y.H.; Cho, J.R.; Jeong, H.S. A study on optimal design for CONFORM process. *J. Mater. Process. Technol.* **1998**, *80*, 671–675. [CrossRef]
18. Gong, N.; Wu, H.B.; Yu, Z.C.; Niu, G.; Zhang, D. Studying Mechanical Properties and Micro Deformation of Ultrafine-Grained Structures in Austenitic Stainless Steel. *Metals* **2017**, *7*, 188. [CrossRef]
19. Huang, J.Y.; Zhu, Y.T.; Jiang, H.; Lowe, T.C. Microstructures and dislocation configurations in nanostructured Cu processed by repetitive corrugation and straightening. *Acta Mater.* **2001**, *49*, 1497–1505. [CrossRef]
20. Carlton, C.E.; Ferreira, P.J. What is behind the inverse Hall-Petch effect in nanocrystalline materials. *Acta Mater.* **2007**, *55*, 3749–3756. [CrossRef]
21. Dalla, T.F.; Lapovok, R.; Sandlin, J.; Thomason, P.F.; Davies, C.H.J.; Pereloma, E.V. Microstructures and properties of copper processed by equal channel angular extrusion for 1–16 passes. *Acta Mater.* **2004**, *52*, 4819–4832. [CrossRef]
22. Samaniego, A.; Llorente, I.; Feliu, S., Jr. Combined effect of composition and surface condition on corrosion behaviour of magnesium alloys AZ31 and AZ61. *Corros. Sci.* **2013**, *68*, 66–71. [CrossRef]
23. Lee, S.M.; Lee, W.G.; Kim, Y.H.; Jang, H. Surface roughness and the corrosion resistance of 21Cr ferritic stainless steel. *Corros. Sci.* **2012**, *63*, 404–409. [CrossRef]
24. Lin, Y.; Pan, J.; Zhou, H.F.; Gao, H.J.; Li, Y. Mechanical properties and optimal grain size distribution profile of gradient grained nickel. *Acta Mater.* **2018**, *153*, 279–289. [CrossRef]
25. Rai, P.K.; Shekhar, S.; Mondal, K. Development of gradient microstructure in mild steel and grain size dependence of its electrochemical response. *Corro. Sci.* **2018**, *138*, 85–95. [CrossRef]
26. Ghidelli, M.; Sebastiani, M.; Johanns, K.E.; Pharr, G.M. Effects of indenter angle on micro-scale fracture toughness measurement by pillar splitting. *J. Am. Ceram. Soc.* **2017**, *100*, 5731–5738. [CrossRef]

© 2018 by the authors. Licensee MDPI, Basel, Switzerland. This article is an open access article distributed under the terms and conditions of the Creative Commons Attribution (CC BY) license (http://creativecommons.org/licenses/by/4.0/).

Article

Molybdenum Effects on Pitting Corrosion Resistance of FeCrMnMoNC Austenitic Stainless Steels

Heon-Young Ha [1,*], Tae-Ho Lee [1], Jee-Hwan Bae [2] and Dong Won Chun [2]

[1] Steel Department, Korea Institute of Materials Science, 797 Changwondae-ro, Seongsan-gu, Changwon, Gyeongnam 51508, Korea; lth@kims.re.kr

[2] Advanced Analysis Center, Korea Institute of Science and Technology, 5, Hwarang-ro 14-gil, Seongbuk-gu, Seoul 02792, Korea; jeehwani@kist.re.kr (J.-H.B.); chundream98@kist.re.kr (D.W.C.)

* Correspondence: hyha2007@kims.re.kr; Tel.: +82-55-280-3422; Fax: +82-55-280-3599

Received: 18 July 2018; Accepted: 16 August 2018; Published: 20 August 2018

Abstract: For Fe-based 18Cr10Mn0.4N0.5C(0–2.17)Mo (in wt %) austenitic stainless steels, effects of Mo on pitting corrosion resistance and the improvement mechanism were investigated. Alloying Mo increased pitting and repassivation potentials and enhanced the passive film resistance by decreasing number of point defects in the film. In addition, Mo reduced critical dissolution rate of the alloys in acidified chloride solutions, and the alloy with higher Mo content could remain in the passive state in stronger acid. Thus, it was concluded that the alloying Mo enhanced pitting corrosion resistance of the alloys through increasing protectiveness of passive film and lowering pit growth rate.

Keywords: high interstitial alloy; molybdenum; pitting corrosion; passive film

1. Introduction

FeCrMnNC austenitic stainless steels known as high interstitial alloys (HIAs) are attractive and economical materials to replace conventional FeCrNi austenitic stainless steels [1–10]. The main purpose of using C, N, and Mn for HIA is to stabilize the austenite phase instead of Ni being an expensive austenite stabilizer [1,2,4,11]. In addition, alloying C and N in stainless steels imparts improved mechanical properties including strength and wear resistance. Regarding corrosion properties, the fact that the alloying N improves the resistance to localized corrosion of stainless steels is well known [12,13], and C in solid solution state is also reported to be advantageous to enhance the pitting corrosion resistance [1,5,6,10,14,15]. Thus, new FeCrMnNC alloys have been explored, with comparable and/or superior performances including strength, elongation, and corrosion resistance to the conventional FeCrNi austenite stainless steels; hence, various types of HIAs have been designed and investigated [1–3,5,8,9]. The author group has made an effort to develop new HIAs with high C and N contents (C > 0.3 wt % and N > 0.3 wt %), and we have found that the Fe-based 18Cr10Mn0.4N(0.3–0.5)C (in wt %) alloys exhibit desirable performances [5–7,14]. The alloys have subsequently been modified with various alloying elements such as Mo, Ni, Cu, Nb, and W in order to further improve their mechanical and corrosion properties. Consequently, it has been revealed that the Fe-based 18Cr10Mn0.4N0.5C (in wt %) HIAs with small amount of Mo, Ni, and W (less than 2 wt %) have mechanical properties and resistance to localized corrosion superior to the UNS S30400 and UNS S31603 stainless steels [16,17].

One of the recommended methods to improve the resistance to localized corrosion of stainless steels, including HIAs, is alloying Mo [18–22]. Although Mo is an expensive ferrite former and is able to form a brittle σ phase, which leads to degradation of the physico-chemical properties of stainless steels [23–25], a small amount of Mo (2–4 wt %, sometimes up to 6 wt %) is frequently used in conventional austenitic stainless steels because of its definite advantages to the localized corrosion resistance. Lots of investigations on the mechanism of improved resistance to localized corrosion by Mo addition have been performed [20–22,26–35]. The desirable localized corrosion resistance of Mo-bearing stainless steels is attributed to various factors. Mo is known to promote the protectiveness of the passive film by formation of the Mo- [26,27,36] and/or Cr-enriched [30,37] film and by thickening of the passive film [26]. Mo is also reported to be beneficial to enhance repassivation characteristics [22]. In addition, it is suggested that molybdate ion (MoO_4^{2-}) is formed during dissolution of Mo-bearing metal, which effectively blocks the adsorption of chloride ion (Cl^-) [26,29,33]. The positive influences of Mo on pitting corrosion behavior have been observed in various types of stainless steels, including FeCr-based ferritic stainless steels [22,27,29,30,33,36], FeCrNi-based austenitic stainless steels [22,26,30,35,36], FeCrMnN-based high-nitrogen stainless steels [35], and FeCrNiMo-based duplex stainless steels [36,38,39]. However, Sugimoto [26] reported that the alloying Mo was ineffective to improve the pitting corrosion resistance of FeMo and NiMo binary alloys, and Kaneko [22] reported that the positive effect of Mo was pronounced in FeCrNi austenitic stainless steel in comparison with FeCr ferritic stainless steel. In addition, it is worth mentioning that the beneficial effect of Mo is manifested in the presence of Cr in Fe-based alloys, and Mo exhibits synergistic effects on the corrosion resistance when alloyed with N in stainless steels [35,39]. These observations suggest that the influence of Mo changes depending on the matrix composition, and thus it is worth investigating the Mo effect on the localized corrosion behavior and passivity of newly developed HIAs. Therefore, the objectives of this paper are to investigate the effects of Mo on the resistance to pitting corrosion of Fe-based 18Cr10Mn0.4N0.5C(0–2.17)Mo (in wt %) austenite stainless steels, and to find the reasons for the change in the pitting corrosion resistance.

2. Experimental Section

2.1. Materials and Mechanical Tests

The investigated alloys were Fe-based 18Cr10Mn0.4N0.5C(0–2.17)Mo (in wt %) HIAs, which have been patented recently [16,17]. The detailed chemical compositions of the three alloys are given in Table 1, and were measured using an optical emission spectroscopy (QSN 750-II, PANalytical, Almemo, The Netherlands) and an inductively coupled plasma atomic emission spectroscopy (Optima 8300DV, PerkinElmer, Waltham, MA, USA).

Table 1. Chemical compositions (in wt %) and mechanical properties of the investigated alloys.

Alloys	Fe	Cr	Mn	Mo	N	C	Yield Strength (MPa)	Tensile Strength (MPa)	Elongation (%)
0Mo		18.19	9.72	-	0.36	0.50	502.1	957.2	61.7
1Mo	Balance	17.89	9.81	1.13	0.40	0.47	499.2	897.5	56.4
2Mo		18.10	9.47	2.17	0.38	0.48	529.0	979.9	62.1

The alloy ingots (10 kg) were produced by vacuum induction melting under N_2 atmosphere. The ingots were homogenized at 1250 °C for 1 h under Ar atmosphere. After the homogenization, the ingots were hot-rolled from 40 mm (initial thickness) to 4 mm (final thickness), followed by water quenching. The hot-rolled plates were then solutionized at 1200 °C for 30 min and quenched in water. The temperatures for the thermomechanical processes were determined from equilibrium phase diagrams (Thermo-Calc software version 4.1, TCFE 7.0 database, Solna, Sweden) shown in Figure 1.

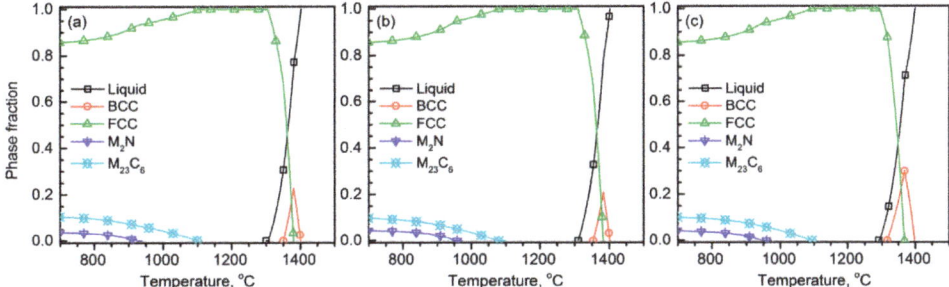

Figure 1. Phase fractions of (**a**) 0Mo, (**b**) 1Mo, and (**c**) 2Mo alloys as a function of temperature calculated using Thermo-Calc software.

Then the microstructures of the alloys were examined. Specimens (15 mm × 10 mm × 4 mm) were cropped from the solutionized plates, mechanically polished using a diamond suspension with a particle size of 1 µm, and chemically etched in acid solution (20 mL HNO$_3$ + 30 mL HCl + 50 mL distilled water) for 1–3 min. A scanning electron microscope (SEM, JSM-5800, JEOL, Tokyo, Japan) was employed to observe the microstructure. The strength and elongation values were measured by tensile tests at 25 °C with a nominal strain rate of 1.67×10^{-3}/s using a servohydraulic machine (INSTRON 5882, Norwood, MA, USA) on tensile specimens (ASTM E8M).

2.2. Electrochemical Tests

The resistance to pitting corrosion of the 0Mo-2Mo alloys was compared with that of S30400 (Fe-based 18.5Cr8.2Ni1.0Mn0.6Si0.04C, in wt %) and S31603 (Fe-based 17.4Cr12.0Ni2.4Mo1.0Mn0.5Si0.02C, in wt %) alloys by linear potentiodynamic polarization tests in a 1 M NaCl solution at 25 °C with a potential sweep rate of 1 mV/s. Then in order to assess the pitting corrosion resistance of 0Mo, 1Mo, and 2Mo alloys more clearly, linear and cyclic potentiodynamic polarization tests were performed in various aqueous solutions containing 4 M NaCl (4 M NaCl, buffered 4 M NaCl, and acidified 4 M NaCl solutions) at 25 °C. The buffer solution was borate-phosphate-citric buffer at pH 8.5 with a composition of 0.2 M boric acid + 0.05 M citric acid + 0.1 M tertiary sodium phosphate. The acidified NaCl solutions were 4 M NaCl with (0.00043–0.1) M HCl solutions, in which both localized and general corrosion behavior could be simultaneously evaluated. The linear polarization was conducted from −0.1 V versus corrosion potential (E_{corr}) to pitting potential (E_{pit}) at a potential sweep rate (dV/dt) of 1 mV/s. For the cyclic polarization tests, the potential was elevated from −0.1 V versus E_{corr} to the potential value at which the current density exceeded 0.1 mA/cm^2, and then lowered to the repassivation potential (E_{rp}) with a dV/dt of 1 mV/s.

The passive behavior and electronic properties of the passive film were investigated in the borate-phosphate-citric buffer solution (pH 8.5) without NaCl. Passive potential range and passive current density ($i_{passive}$) were examined through potentiodynamic polarization tests in the buffer solution at 25 °C with a dV/dt of 1 mV/s, and the resistance of the passive film was investigated by measuring the real part of the impedance (Z'_{real}) during the anodic polarization [35,40–43]. The Z'_{real} values of the passive films were measured by imposing sine-wave voltage perturbation (±10 mV) at a frequency of 0.1 Hz [40,41] during increase in the applied potential from −0.8 to 0.9 V$_{SCE}$. Then the point defect density in the space charge layer of the passive film was measured through Mott-Schottky analysis. For the test, the passive film was formed by applying constant anodic potential of 0.85 V$_{SCE}$ for 3 h in the buffer solution, and then the capacitance of the passivated layer was measured by imposing sine-wave voltage perturbation (±10 mV) at a frequency of 1000 Hz during the negative (cathodic) potential sweep from 0.85 to −0.7 V$_{SCE}$.

The electrochemical tests were controlled by a potentiostat (Reference 600, GAMRY Instruments, Warminster, PA, USA), and performed in a multineck flask (1 L) with three electrodes; a specimen as a working electrode, a Pt plate (50 mm × 120 mm × 0.1 mm) as a counter electrode, and a saturated calomel reference electrode (SCE) as a reference electrode. For the working electrode, specimens (10 mm × 10 mm × 4 mm) were mounted in cold epoxy resin and ground using SiC emery paper up to 2000 grit. The exposed area for the electrochemical tests was 0.2 cm^2, which was controlled using electroplating tape. For each specimen, the polarization tests were conducted 5–6 times, and the resistance and capacitance of the passive layer were measured 4–5 times in order to confirm reproducibility.

3. Results and Discussion

3.1. Microstructure and Tensile Properties

Figure 2a–c exhibits SEM images of the 0Mo, 1Mo, and 2Mo alloys, respectively, after solutionization at 1200 °C for 30 min followed by water quenching. The concentrations of the interstitial alloying elements (C + N) of the alloys were as high as 0.86–0.87 wt %, thus the solution treatment should be conducted at a high temperature range of between 1100 and 1250°C to suppress the formation of $M_{23}C_6$ and/or M_2N (M stands for metal, primarily Cr) as indicated in Figure 1. As shown in Figure 2, the three alloys have an austenite single phase with annealing twins and $M_{23}C_6$ and/or M_2N are not formed even at the grain boundaries. In addition, nonmetallic inclusions such as Mn-oxide and Mn-sulfide are rarely observed in the alloys. Using SEM images taken at 5–6 different locations, the average grain sizes of the samples were measured in accordance with ASTM E112; as a result, the average grain sizes of 0Mo, 1Mo, and 2Mo alloys were 150.2, 148.8, and 142.4 µm, respectively, suggesting that the addition of Mo did not have a significant influence on the grain size.

Figure 2. Microstructures of (**a**) 0Mo, (**b**) 1Mo, and (**c**) 2Mo alloys (SEM).

The tensile properties of the three alloys are summarized in Table 1. The 0Mo–2Mo alloys exhibit yield strength of 499.2–529.0 MPa, tensile strength of 897.5–979.9 MPa, and elongation of 56.4–62.1%. The 2Mo alloy exhibits the best tensile properties among the three alloys. It is worth mentioning that the investigated alloys have better mechanical properties than the commercial austenitic stainless steels. It is reported that the S31603, for example, has yield strength, tensile strength, and elongation values of 170 MPa, 485 MPa and 40%, respectively [44].

3.2. Pitting Corrosion Resistance

Resistance to pitting corrosion of the investigated alloys was compared with the commercial austenitic stainless steels, S30400 and S31603. Figure 3 shows the polarization curves of the alloys measured in a 1 M NaCl solution at 25°C. The five alloys exhibit passive behavior in the potential range from E_{corr} to E_{pit}. The E_{pit} values of S30400 and S31603 alloys were 0.294 and 0.470 V_{SCE}, respectively, and those of 0Mo and 1Mo alloys were 0.390 V_{SCE} and 0.643 V_{SCE}, respectively. The pitting corrosion did not occur in the 2Mo alloy. Figure 3 demonstrates that the resistance to pitting corrosion increased in the order, S30400 < 0Mo < S31603 < 1Mo < 2Mo. It is obvious that the 1Mo and 2Mo exhibit superior corrosion resistance to the S31603, and even 0Mo has better resistance than S30400. Figure 3 confirms the excellent anti-corrosion properties of developed FeCrMnMoNC alloys.

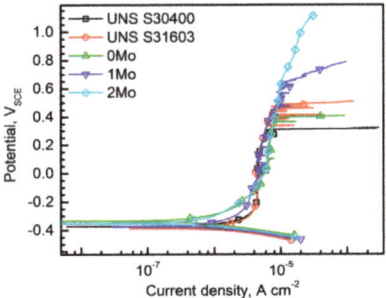

Figure 3. Linear potentiodynamic polarization curves of 0Mo, 1Mo, and 2Mo alloys and UNS S30400 and UNS S31603 alloys measured in 1 M NaCl solution at 25 °C (dV/dt = 1 mV/s).

The pitting corrosion resistance of the 0Mo, 1Mo, and 2Mo alloys was evaluated more clearly. Figure 4a shows cyclic potentiodynamic polarization curves of 0Mo, 1Mo, and 2Mo alloys measured in a 4 M NaCl solution at 25 °C. The E_{corr} value of 0Mo alloy was -0.399 V_{SCE} and it slightly increased to -0.321 V_{SCE} for 2Mo alloy. The alloys were in a passive state under open circuit conditions in this solution, and the passivity appeared in a limited potential range from the E_{corr} to the E_{pit}, at which an abrupt and irreversible increase in the current density began. The average E_{pit} and E_{rp} values obtained from the repetitive polarization tests (5–6 times) in the 4 M NaCl solution were plotted versus the Mo content in Figure 4b. The average E_{pit} increased linearly from 0.213 to 0.940 V_{SCE} with an increase in the Mo content from 0 to 2.17 wt %, indicating the improved pitting corrosion resistance of the HIAs by alloying Mo. The E_{rp} is also shifted to a higher value as the Mo content increases. For 0Mo alloy, the stable pit can repassivate below -0.344 V_{SCE}, which is close to its E_{corr} value, while the average E_{rp} of 2Mo alloy is 0.840 V_{SCE}. It is worth mentioning that the difference between the E_{pit} and E_{rp} values, $\Delta(E_{pit} - E_{rp})$ significantly decreases from 0.557 to 0.099 V as the Mo content increases.

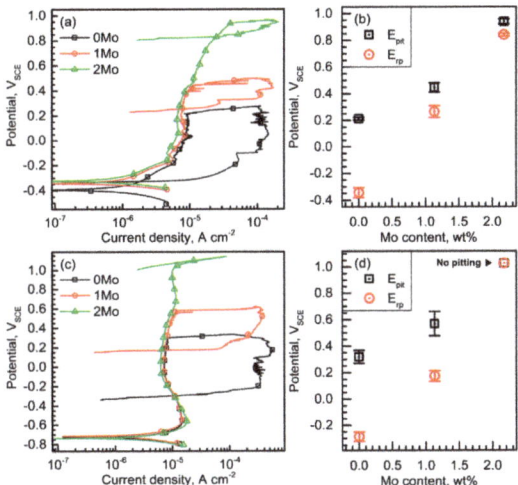

Figure 4. Cyclic potentiodynamic polarization curves of 0Mo, 1Mo, and 2Mo alloys measured in (**a**) 4 M NaCl and (**c**) borate-phosphate-citric buffer containing 4 M NaCl (pH 8.5) solutions at 25 °C. (dV/dt = 1 mV/s). Average pitting and repassivation potentials with standard deviation values (scatter band) of the alloys measured in (**b**) 4 M NaCl and (**d**) buffered 4 M NaCl solutions.

The polarization curves in Figure 4a show changes in other features by addition of Mo, such as $i_{passive}$ and number of metastable pitting corrosion events. The lowest $i_{passive}$ value is observed in the 2Mo alloy in the entire potential range, and the $i_{passive}$ decreases from 8.3 to 5.8 µA/cm^2 (at 0 V$_{SCE}$, for example) as the Mo content increases. Moreover, small current spikes indicating the initiation and repassivation of metastable pits are more frequently observed in the polarization curve of the alloys with lower Mo content.

Figure 4c exhibits the cyclic potentiodynamic polarization curves of the alloys measured in a borate-phosphate-citric buffer solution (pH 8.5) with 4 M NaCl at 25°C. In this solution, the E_{corr} of the alloys was approximately −0.73 V$_{SCE}$. The alloys also exhibited passivity at E_{corr}, and pitting corrosion occurred under sufficient anodic polarization, except for the 2Mo alloy. The average E_{pit} values of 0Mo and 1Mo alloys were 0.319 and 0.568 V$_{SCE}$, respectively, and the E_{rp} of 0Mo alloy was −0.287 V$_{SCE}$ and that of 1Mo alloy was 0.173 V$_{SCE}$. Figure 4d confirms again that the alloying Mo raises both E_{pit} and E_{rp}. Moreover, the $\Delta(E_{pit} - E_{rp})$ also decreases from 0.606 to 0.395 V$_{SCE}$ as the Mo content increased from 0 to 1.13 wt %.

The E_{pit} values obtained in the buffered NaCl solution (Figure 4d) were higher than those obtained in the simple NaCl solution (Figure 4b). Besides, comparing the polarization curves measured in the buffered and simple NaCl solutions, it was found that the occurrence of metastable pitting corrosion during the polarization was restrained in the buffered NaCl solution. The better resistance to pitting corrosion obtained in the buffer solution is considered to be due to two possible reasons. First, anions in the borate-phosphate-citric buffer solution, $H_2BO_3^-$, $H_2PO_4^-$, and $H_2C_6H_5O_7^-$, act as inhibitors to adsorption of Cl^- on the passivated electrode, and second, the buffer solution containing the anions is helpful to form protective passive film as reported [40,45].

It is obvious from Figure 4 that the alloying Mo enhances the resistance to pitting corrosion of the Fe-based 18Cr10Mn0.4N0.5C(0–2.17)Mo (in wt %) HIAs. More specifically, alloying Mo improves the resistance to pit initiation, which is supported by the decrease in the number of metastable pitting corrosion events, and accelerates repassivation kinetics, which is proven by the rise in the E_{rp}; thus, the Mo alloying consequently shifts the E_{pit} to the higher level. Moreover, it is considered that the decrease in the $\Delta(E_{pit} - E_{rp})$ value along with increase in the Mo alloying demonstrates that the alloying Mo is more effective in accelerating repassivation kinetics than the pit initiation probability, which is related to the passive film protectiveness. Thus, in order to explain the role of Mo alloying in the improvement of resistance to stable pitting corrosion, passive and general corrosion behavior of the three HIAs was investigated. The former is correlated with the pit initiation probability and the latter is related to the pit growth rate [41].

3.3. Passive Behavior

The passive behavior and passive film properties were investigated. For this, a borate-phosphate-citric buffer solution (pH 8.5) without NaCl was used because a thick and protective passive film can be formed on Fe-based alloys in this solution, as reported [40,45]. The passive behavior of the alloy in the buffer solution was examined through potentiodynamic polarization tests (Figure 5a). In this mild basic buffer solution, the polarization curves of the three alloys are almost identical in shape, and the passivity was observed in the potential range from E_{corr} (approximately −0.82 V$_{SCE}$) to transpassive potential (approximately 0.95 V$_{SCE}$), where the oxygen evolution began. In the passive potential range, three current peaks are observed at −0.66, −0.47, and 0.64 V$_{SCE}$, respectively. It is known that the first current peak at −0.66 V$_{SCE}$ is due to the oxidation of Fe to Fe^{2+}, and the second current peak at −0.47 V$_{SCE}$ is attributed to re-oxidation of Fe^{2+} to Fe^{3+}. The third current peak at 0.64 V$_{SCE}$ reflects the re-oxidation reaction of Cr^{3+} to Cr^{6+} [40,41,46,47]. The polarization curves, however, do not clearly show a distinct influence of Mo on the passive behavior. Thus, the Z'_{real} value of the passive film was measured during the polarization [40–43].

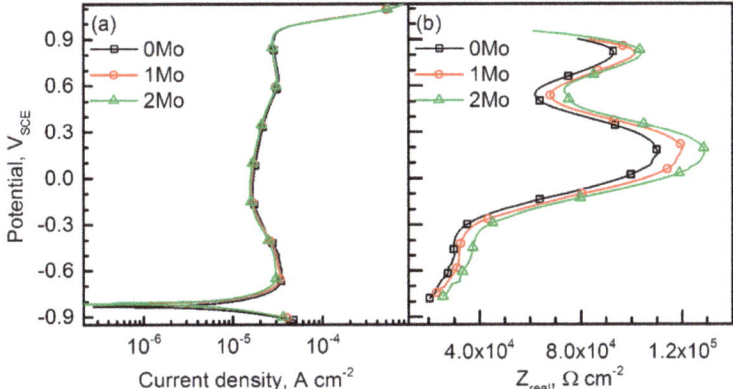

Figure 5. (a) Potentiodynamic polarization curves of 0Mo, 1Mo, and 2Mo alloys measured in a borate-phosphate-citric buffer solution (pH 8.5) at 25 °C (dV/dt = 1 mV/s). (b) Graphs of real part of the impedance (Z'_{real}) versus applied potential of the alloys measured in the same solution by imposing sinusoidal voltage perturbation (±10 mV) at a frequency of 0.1 Hz.

The graph of the measured Z'_{real} value versus the applied potential is presented in Figure 5b. In the three alloys, the Z'_{real} values at E_{corr} are approximately 2.5×10^4 Ω/cm^2, and it begin to steeply increase at approximately -0.33 V_{SCE}. The maximum Z'_{real} values of the three alloys are obtained at approximately 0.15 V_{SCE}, and those of 0Mo, 1Mo, and 2Mo alloys are 1.10×10^5, 1.19×10^5, and 1.28×10^5 Ω/cm^2, respectively. Then the Z'_{real} values decreased to 6.2–7.5×10^5 Ω/cm^2 at 0.55 V_{SCE}, which is due to the re-oxidation reaction of Cr^{3+} to Cr^{6+} as mentioned above. Good correlation between Figure 5a,b confirms that the Z'_{real}-potential curve is useful to understand the polarization behavior more clearly. Figure 5b obviously shows the influence of Mo on the passive behavior; that is, the alloying Mo in the HIAs improved the resistance to passive film in the entire passive potential range.

Investigation on the point defect density of the passive film can explain the change in the film resistance. The passive film generally contains a large number of point defects; thus, the passive film behaves as an extrinsic semiconductor [48,49]. The semiconductive parameters of the passive film such as point defect density and flat band potential can be obtained using the measurement of specific interfacial capacitance (C_{total}) as a function of the applied potential (E_{app}), that is, Mott-Schottky analysis [50–52]. The C_{total} can be obtained using the relation of $C_{total} = 1/\omega Z''_{imag}$, where ω is the angular frequency and Z''_{imag} is the imaginary part of the specific impedance, and the C_{total} is a series combination of the double layer capacitance (Helmholtz layer capacitance, C_H) and space charge layer capacitance (C_{SC}). For n-type semiconductors, the relation between the C_{SC}, C_H, and C_{total} values and applied potential (E_{app}) is given as follows;

$$\frac{1}{C_{SC}^2} = \frac{1}{C_{total}^2} - \frac{1}{C_H^2} = \left(\frac{2}{\varepsilon \varepsilon_0 e N_D}\right)\left(E_{app} - E_{fb} - \frac{\kappa_B T}{e}\right), \quad (1)$$

where ε_0 is the vacuum permittivity (8.85×10^{14} F/cm), ε is the dielectric constant of the passive film (taken as 15.6 [5,41,51]), κ_B is the Boltzmann constant (1.38×10^{-23} J/K), N_D is a density of point defect (donor in this case), and e is the electron charge (1.60×10^{-19} C). In the Mott-Schottky relation, C_H can be neglected, because it is sufficiently higher than the C_{SC}. In accordance with Equation (1), the reciprocal of the C_{SC}^{-2} and the E_{app} exhibits a linear relationship, thus the N_D in the space charge layer can be estimated from the slope of the graph of C_{SC}^{-2} versus E_{app}.

Figure 6a shows Mott-Schottky plots for the passive films of the investigated HIAs, which were formed by applying a constant anodic potential of 0.85 V_{SCE} for 3 h in the buffer solution. The Mott-Schottky plots of the HIAs exhibit two potential sections showing linear increase, region I (from −0.4 to 0 V_{SCE}) and region II (from 0.3 to 0.7 V_{SCE}). The dominant and detective point defects of the n-type passive film are oxygen vacancy (V_O^{2+}, shallow donor) and Cr^{6+} (deep donor), and the shallow and deep donor densities can be calculated using the positive slopes ($\Delta C^{-2}/\Delta V$) of the region I and II in the Mott-Schottky plot, respectively.

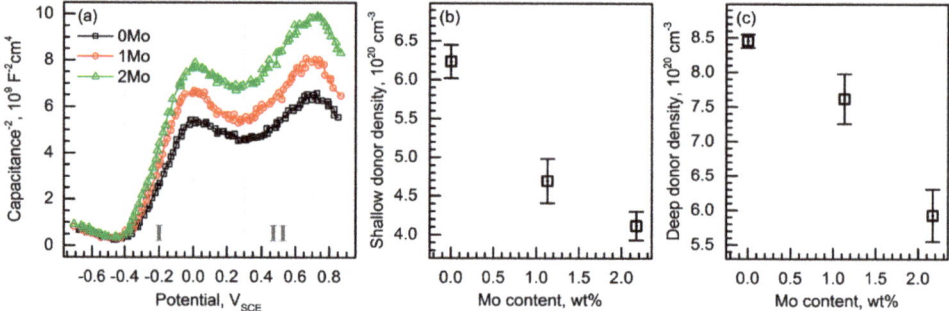

Figure 6. (a) Mott-Schottky plots of 0Mo, 1Mo, and 2Mo alloys measured in a borate-phosphate-citric buffer solution (pH 8.5) at 25 °C with decreasing applied potential by imposing sinusoidal voltage perturbation (±10 mV) at a frequency of 1000 Hz. Average (b) shallow donor (V_O^{2+}) density and (c) deep donor (Cr^{6+}) density with standard deviation (scatter band) in the passive films of the alloys.

The average density values of the shallow and deep donors in the passive films formed on the 0Mo–2Mo alloys are presented in Figure 6b,c, respectively, as a function of the Mo content. Shallow donor density for 0Mo alloy was 6.23×10^{20} /cm^3, and that for the 2Mo alloy was 4.12×10^{20} /cm^3. In addition, the deep donor densities for the 0Mo and 2Mo alloys were 8.45×10^{20} and 5.93×10^{20} /cm^3, respectively. Figure 6b,c shows that the numbers of both shallow and deep donors linearly decrease as the Mo content increases. Point defects in the passive film function as charge carriers; thus, Figure 6 well explains the increased resistance of the passive film by addition of Mo. It can be concluded that the improved resistance against pitting corrosion initiation of the investigated HIAs by alloying Mo (Figure 4a,c) is partly attributed to the formation of more resistant passive film with less point defects.

3.4. General Corrosion Behavior

Figure 7a shows the linear potentiodynamic polarization curves of the HIAs measured in 4 M NaCl + 0.01 M HCl (pH 1.21) solution at 25 °C. In this strongly acidic chloride solution, typical active-passive transition and pitting corrosion occurred during the polarization. Similar to Figure 4a, the E_{pit} values obtained in this acidified NaCl solution also increased by addition of Mo (Figure 7b). More importantly, the polarization curves in Figure 7a demonstrate the change in the general corrosion behavior of the HIAs by addition of Mo. The average E_{corr} values of the 0Mo, 1Mo, and 2Mo alloys are −0.667, −0.633, and −0.597 V_{SCE}, respectively (Figure 7c), which increases with the Mo content. In this solution, metal dissolution actively occurred between the E_{corr} and primary passive potential (E_{pp}). The E_{pp} of the alloys are gradually lowered from −0.499 V_{SCE} for 0Mo alloy to −0.530 V_{SCE} for 2Mo alloy. In addition, the critical anodic current density (i_{crit}) values of 0Mo, 1Mo, and 2Mo alloys are 13.09, 2.71, and 0.66 mA/cm^2, respectively, as shown in Figure 7d. To sum up, the alloying Mo raised the E_{corr}, reduced the i_{crit}, and lowered the E_{pp} of the HIAs, thus it can be concluded that the alloying Mo made the HIA matrix noble and improved the general corrosion resistance.

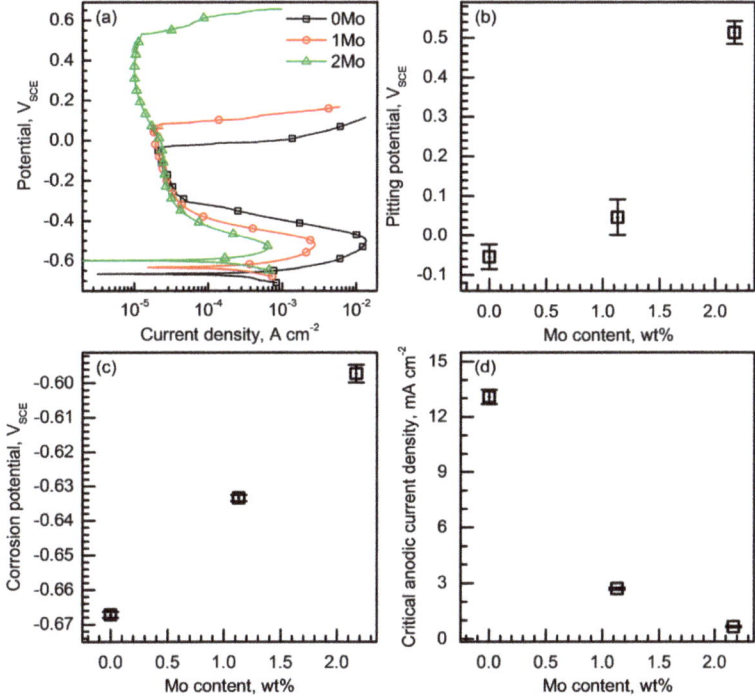

Figure 7. (a) Potentiodynamic polarization curves of 0Mo, 1Mo, and 2Mo alloys measured in a 4 M NaCl + 0.01 M HCl solution at 25 °C (dV/dt = 1 mV/s). Average (b) pitting and (c) corrosion potentials, and (d) critical anodic current density values with standard deviation (scatter band) of the alloys.

The general corrosion resistance of the matrix can affect the pit propagation [19,21,41]. Inside the pit cavity formed in stainless steel, the confined chloride solution becomes acidified due to a hydrolysis of the metal ions [19], and the bare metal surface without a passive film is directly exposed to the acidified chloride solution. In this situation, general corrosion occurs on the bare matrix surface in the pit cavity. Therefore, the general corrosion behavior in the acidified NaCl solution shown in Figure 7 can reflect the matrix dissolution behavior inside the pit cavity. Since the Mo alloying induced the decrease in the i_{crit} and E_{pp}, it is conceivable that the pit propagation rate is lowered and the pit extinction (that is, repassivation) is accelerated in the HIAs as the Mo content increases.

Figure 8a–c shows the potentiodynamic polarization curves of the 0Mo, 1Mo, and 2Mo alloys, respectively, measured in the acidified 4 M NaCl solutions with various solution pHs. The pH values of the solutions were adjusted by addition of HCl ranging between 0.44 (4 M NaCl + 0.1 M HCl) and 2.88 (4 M NaCl + 0.00043 M HCl). For all the alloys, the i_{crit} values increase as the solution is acidified. In the solution with pH 2.88, the polarization curves of the three alloys do not show distinct active-passive transition, and the maximum anodic current density values (at approximately −0.45 V$_{SCE}$) of the 0Mo, 1Mo, and 2Mo alloys are 52.65, 33.91, and 24.45 µA/cm^2, respectively, showing slight decrease along with the alloyed Mo content. On the other hand, in the solution with pH 0.44, the i_{crit} values of the alloys are as high as several tens of mA/cm^2, and the polarization curves exhibit typical active-passive transition. In the strongest acid, the average i_{crit} values of the 0Mo, 1Mo, and 2Mo alloys are 71.21, 32.61, and 11.50 mA/cm^2, respectively, which also decrease as the Mo content increases.

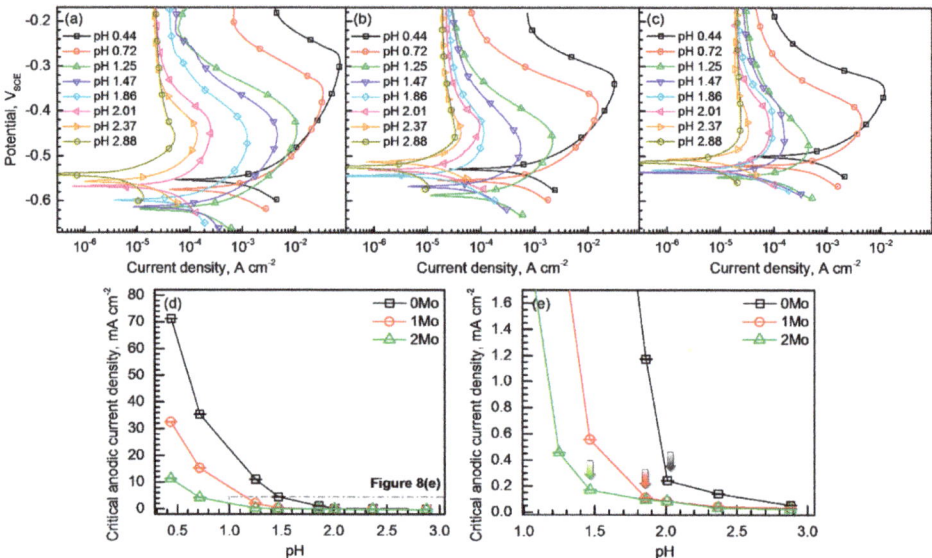

Figure 8. Potentiodynamic polarization curves of (**a**) 0Mo, (**b**) 1Mo, and (**c**) 2Mo alloys measured in 4 M NaCl + HCl solutions with different solution pHs of 0.44–2.88 at 25 °C (dV/dt = 1 mV/s). (**d**) Graphs of average critical anodic current density values with standard deviation (scatter band) of the alloys versus solution pH. (**e**) A magnified part of graphs in Figure 7d.

For the three HIAs, the i_{crit} values measured in each solution are plotted in Figure 8d as a function of the solution pH. The i_{crit} steadily decreases as the pH increases, and the 0Mo alloy exhibits the highest i_{crit} values in the entire pH range. In Figure 8d, the pH value at which the i_{crit} value abruptly changed requires further attention. Figure 8e magnifies the part of Figure 8d ranging from pH 1 to 3. For 0Mo alloy, the i_{crit} value slightly increases in the range of 50–250 µA/cm² as the solution pH decreases from 2.88 to 2.01, but in the acidic solution with pH lower than 2, the i_{crit} abruptly increases to 1.17 mA/cm². For 1Mo alloy, the i_{crit} value also negligibly increases with decrease in the solution pH in the pH range of 1.86–2.88, but it begins to rapidly increases at the pH lower than 1.47. In the 2Mo alloy, the rapid increase in the i_{crit} is observed in the solution with pH lower than 1.25. The pH values at which the abrupt increase in the i_{crit} begins are marked with arrows in Figure 8e. This result demonstrates that the HIA with lower Mo is more likely to undergo active dissolution even in less acidified chloride solution inside the pit, thus even the small pits tend to grow more easily to stable pits. The results obtained Figures 7 and 8 simultaneously indicate that the elevated E_{rp} and the decreased $\Delta(E_{pit} - E_{rp})$ value by addition of Mo shown in Figure 4b,d are due to the lowered pit propagation rate and accelerated repassivation kinetics by addition of Mo. In addition, because the alloying Mo imparts a higher resistance to active dissolution in a stronger acid solution, the pit embryos become more difficult to grow in the alloy with higher Mo content, which consequently leads to increase in the E_{pit}.

Based on the findings of this investigation, it can be concluded that the alloying Mo is effective to improve the corrosion properties of Fe-based 18Cr10Mn0.4N0.5C(0–2.17)Mo (in wt %) alloys. The literature presents examples of pitting corrosion of UNS S30400 and/or S31603 alloys in various chloride solutions [1,5,6,53–56]. Thus, in aqueous environments containing Cl⁻, it can be considered to replace UNS S30400 and/or S31603 alloys by usage of the Ni-free Fe-based 18Cr10Mn0.4N0.5C(0–2.17)Mo (in wt %) alloys especially 2Mo grade, which presents superior pitting corrosion resistance.

4. Conclusions

For Fe-based 18Cr10Mn0.4N0.5C(0–2.17)Mo (in wt %) HIAs, the effects of Mo on pitting corrosion resistance and the improvement mechanism were investigated. The investigated alloys have been patented due to their corrosion resistance and mechanical properties, which are superior to commercial austenitic stainless steels such as UNS S30400 and S31603. The following points summarize the findings of this research.

(1) Potentiodynamic polarization tests in chloride solutions revealed that the alloying Mo suppressed metastable pitting corrosion and raised both E_{pit} and E_{rp} of the alloys. In addition, it was found that the difference between the E_{pit} and E_{rp} decreased as the Mo content increased.
(2) Passive film analysis through a resistance measurement and Mott-Schottky analysis indicated that the alloyed Mo increased the film resistance by decreasing the number of point defects in the passive film.
(3) The alloyed Mo reduced the critical dissolution rate of the alloys in acidified chloride solutions, and the alloy with higher Mo content was able to resist active dissolution in stronger acid.
(4) It is concluded that the alloying Mo enhanced pitting corrosion resistance of the alloy through increasing protectiveness of passive film and lowering pit propagation rate.

Author Contributions: Conceptualization, H.-Y.H. and T.-H.L.; Methodology, H.-Y.H. and T.-H.L.; Investigation, H.-Y.H., T.-H.L., J.-H.B. and D.W.C.; Validation, J.-H.B. and D.W.C.; Writing-Original Draft Preparation, H.-Y.H. and J.-H.B.; Writing-Review & Editing, D.W.C. and T.-H.L.

Funding: This study was financially supported by Fundamental Research Program (grant number: PNK5850) of the Korea Institute of Materials Science (KIMS). This study also supported by the Ministry of Trade, Industry & Energy (MI, Korea) under Strategic Core Materials Technology Development Program (No. 10067375).

Conflicts of Interest: The authors declare no conflicts of interest.

References

1. Thomann, U.I.; Uggowitzer, P.J. Wear-corrosion behavior of biocompatible austenitic stainless steels. *Wear* **2000**, *239*, 48–58. [CrossRef]
2. Gavrilyuk, V.G.; Berns, H. High-strength Austenitic Stainless Steels. *Met. Sci. Heat Treat.* **2007**, *49*, 566–568. [CrossRef]
3. Rawers, J.C. Alloying effects on the microstructure and phase stability of Fe–Cr–Mn steels. *J. Mater. Sci.* **2008**, *43*, 3618–3624. [CrossRef]
4. Gavriljuk, V.G.; Shanina, B.D.; Berns, H. A physical concept for alloying steels with carbon + nitrogen. *Mater. Sci. Eng.* **2008**, *481–482*, 707–712. [CrossRef]
5. Ha, H.-Y.; Lee, T.-H.; Oh, C.-S.; Kim, S.-J. Effects of combined addition of carbon and nitrogen on pitting corrosion behavior of Fe-18Cr-10Mn alloys. *Scr. Mater.* **2009**, *61*, 121–124. [CrossRef]
6. Ha, H.-Y.; Lee, T.-H.; Oh, C.-S.; Kim, S.-J. Effects of carbon on the corrosion behaviour in Fe-18Cr-10Mn-N-C stainless steels. *Steel Res. Int.* **2009**, *80*, 488–492.
7. Lee, T.-H.; Shin, E.; Oh, C.-S.; Ha, H.-Y.; Kim, S.-J. Correlation between stacking fault energy and deformation microstructure in high-interstitial-alloyed austenitic steels. *Acta Mater.* **2010**, *58*, 3173–3186. [CrossRef]
8. Schymura, M.; Stegemann, R.; Fischer, A. Crack propagation behavior of solution annealed austenitic high interstitial steels. *Int. J. Fatigue* **2015**, *79*, 25–35. [CrossRef]
9. Seifert, M.; Siebert, S.; Huth, S.; Theisen, W.; Berns, H. New Developments in Martensitic Stainless Steels Containing C + N. *Steel Res. Int.* **2015**, *86*, 1508–1516. [CrossRef]
10. Niederhofer, P.; Richrath, L.; Huth, S.; Theisen, W. Influence of conventional and powder–metallurgical manufacturing on the cavitation erosion and corrosion of high interstitial CrMnCN austenitic stainless steels. *Wear* **2016**, *360–361*, 67–76. [CrossRef]
11. Gavriljuk, V.G.; Berns, H. *High Nitrogen Steels*, 1st ed.; Springer: Berlin, Germany, 1999; Chapter 1–2.
12. Levey, P.R.; van Bennekom, A. A mechanistic study of the effects of nitrogen on the corrosion properties of stainless steels. *Corrosion* **1995**, *51*, 911–921. [CrossRef]

13. Jargelius-Pettersson, R.F.A. Electrochemical investigation of the influence of nitrogen alloying on pitting corrosion of austenitic stainless steels. *Corros. Sci.* **1999**, *41*, 1639–1664. [CrossRef]
14. Ha, H.-Y.; Lee, T.-H.; Kim, S.-J. Effect of C fraction on corrosion properties of high interstitial alloyed stainless steels. *Metall. Mater. Trans. A* **2012**, *43*, 2999–3005. [CrossRef]
15. Speidel, M.O. Nitrogen containing austenitic stainless steels. *Mater. Sci. Eng. Technol.* **2006**, *37*, 875–880. [CrossRef]
16. Lee, T.-H.; Kim, S.-J.; Oh, C.-S.; Ha, H.-Y. High Strength and High Corrosion Coal Nitrogen Combined Addition Austenitic Stainless Steel and a Manufacturing Method Thereof (In Japanese). Patent JP 5272078 B2, 17 May 2013.
17. Lee, T.-H.; Kim, S.-J.; Oh, C.-S.; Ha, H.-Y. High Strength/Corrosion Resistant Austenitic Stainless Steel with Carbon-Nitrogen Complex Additive, and Method for Manufacturing Same. Patent EP 2455508 B1, 23 November 2016.
18. Lo, K.H.; Shek, C.H.; Lai, J.K.L. Recent developments in stainless steels. *Mater. Sci. Eng.* **2009**, *65*, 39–104. [CrossRef]
19. Frankel, G.S. Pitting Corrosion of Metals—A Review of the Critical Factors. *J. Electrochem. Soc.* **1998**, *145*, 2186–2198. [CrossRef]
20. Newman, R.C. The dissolution and passivation kinetics of stainless alloys containing molybdenum—1. Coulometric studies of Fe-Cr and Fe-Cr-Mo alloys. *Corros. Sci.* **1985**, *25*, 331–339. [CrossRef]
21. Newman, R.C. The dissolution and passivation kinetics of stainless alloys containing molybdenum—II. Dissolution kinetics in artificial pits. *Corros. Sci.* **1985**, *25*, 341–350. [CrossRef]
22. Kaneko, M.; Isaacs, H.S. Effects of molybdenum on the pitting of ferritic- and austenitic-stainless steels in bromide and chloride solutions. *Corros. Sci.* **2002**, *44*, 1825–1834. [CrossRef]
23. Villanueva, D.M.E.; Junior, F.C.P.; Plaut, R.L.; Padilha, A.F. Comparative study on sigma phase precipitation of three types of stainless steels: austenitic, superferritic and duplex. *Mater. Sci. Technol.* **2006**, *22*, 1098–1104. [CrossRef]
24. Sourmail, T. Precipitation in creep resistant austenitic stainless steels. *Mater. Sci. Technol.* **2001**, *17*, 1–14. [CrossRef]
25. Weiss, B.; Stickler, R. Phase Instabilities During High Temperature Exposure of 316 Austenitic Stainless Steel. *Metall. Mater. Trans. B* **1972**, *3*, 851–866. [CrossRef]
26. Sugimoto, K.; Sawada, Y. The role of molybdenum additions to austenitic stainless steels in the inhibition of pitting in acid chloride solutions. *Corros. Sci.* **1977**, *17*, 425–445. [CrossRef]
27. Hashimoto, K.; Asami, K.; Teramoto, K. An X-ray photo-electron spectroscopic study on the role of molybdenum in increasing the corrosion resistance of ferritic stainless steels in HCl. *Corros. Sci.* **1979**, *19*, 3–14. [CrossRef]
28. Clayton, C.R.; Lu, Y.C. A Bipolar Model of the Passivity of Stainless Steel: The Role of Mo Addition. *J. Electrochem. Soc.* **1986**, *133*, 2465–2473. [CrossRef]
29. Landolt, D.; Mischler, S.; Vogel, A.; Mathieu, H.J. Chloride Ion Effects on Passive Films on FeCr and FeCrMo Studied by AES, XPS and SIMS. *Corros. Sci.* **1990**, *31*, 431–440. [CrossRef]
30. Montemor, M.F.; Simoes, A.M.P.; Ferreira, M.G.S.; Da Cunha Belo, M. The role of Mo in the chemical composition and semiconductive behaviour of oxide films formed on stainless steels. *Corros. Sci.* **1999**, *41*, 17–34. [CrossRef]
31. Ilevbare, G.O.; Burstein, G.T. The role of alloyed molybdenum in the inhibition of pitting corrosion in stainless steels. *Corros. Sci.* **2001**, *43*, 485–513. [CrossRef]
32. Bastidas, J.M.; Torres, C.L.; Cano, E.; Polo, J.L. Influence of molybdenum on passivation of polarised stainless steels in a chloride environment. *Corros. Sci.* **2002**, *44*, 625–633. [CrossRef]
33. Tobler, W.J.; Vertanen, S. Effect of Mo species on metastable pitting of Fe18Cr alloys—A current transient analysis. *Corros. Sci.* **2006**, *48*, 1585–1607. [CrossRef]
34. Li, D.G.; Wang, J.D.; Chen, D.R.; Liang, P. Molybdenum addition enhancing the corrosion behaviors of 316 L stainless steel in the simulated cathodic environment of proton exchange membrane fuel cell. *Int. J. Hydrogen Energy* **2015**, *40*, 5947–5957. [CrossRef]
35. Loable, C.; Vicosa, I.N.; Mesquita, T.J.; Mantel, M.; Nogueira, R.P.; Berthome, G.; Chauveau, E.; Roche, V. Synergy between molybdenum and nitrogen on the pitting corrosion and passive film resistance of austenitic stainless steels as a pH-dependent effect. *Mater. Chem. Phys.* **2017**, *186*, 237–245. [CrossRef]

36. Mesquita, T.J.; Chauveau, E.; Mantel, M.; Nogueira, R.P. A XPS study of the Mo effect on passivation behaviors for highly controlled stainless steels in neutral and alkaline conditions. *Appl. Surf. Sci.* **2013**, *270*, 90–97. [CrossRef]
37. Vignal, V.; Olive, J.M.; Desjardins, D. Effect of molybdenum on passivity of stainless steels in chloride media using ex situ near field microscopy observations. *Corros. Sci.* **1999**, *41*, 869–884. [CrossRef]
38. Mesquita, T.J.; Chauveau, E.; Mantel, M.; Kinsman, N.; Roche, V.; Nogueira, R.P. Lean duplex stainless steels—The role of molybdenum in pitting corrosion of concrete reinforcement studied with industrial and laboratory castings. *Mater. Chem. Phys.* **2012**, *132*, 967–972. [CrossRef]
39. Olsson, C.-O.A. The influence of nitrogen and molybdenum on passive films formed on the austenoferritic stainless steel 2205 studied by AES and XPS. *Corros. Sci.* **1995**, *37*, 467–479. [CrossRef]
40. Ha, H.-Y.; Lee, T.-H.; Kim, S.-J. Role of nitrogen in the active–passive transition behavior of binary Fe-Cr alloy system. *Electrochim. Acta* **2012**, *80*, 432–439. [CrossRef]
41. Ha, H.-Y.; Jang, M.-H.; Lee, T.-H. Influences of Mn in solid solution on the pitting corrosion behaviour of Fe-23 wt. %Cr-based alloys. *Electrochim. Acta* **2016**, *191*, 864–875. [CrossRef]
42. Krakowiak, S.; Darowicki, K.; Slepski, P. Impedance investigation of passive 304 stainless steel in the pit pre-initiation state. *Electrochim. Acta* **2005**, *50*, 2699–2704. [CrossRef]
43. Nagarajan, S.; Rajendran, N. Crevice corrosion behaviour of superaustenitic stainless steels: Dynamic electrochemical impedance spectroscopy and atomic force microscopy studies. *Corros. Sci.* **2009**, *51*, 217–224. [CrossRef]
44. American Society for Testing and Materials (ASTM). *ASTM A276-06, Standard Specification for Stainless Steel Bars and Shapes*; ASTM: West Conshohocken, PA, USA, 2016.
45. Ha, H.-Y.; Kwon, H.-S. Effects of pH levels on the surface charge and pitting corrosion resistance of Fe. *J. Electrochem. Soc.* **2012**, *159*, C416–C421. [CrossRef]
46. Piao, T.; Park, S.-M. Spectroelectrochemical studies of passivation and transpassive breakdown reactions of stainless steel. *J. Electrochem. Soc.* **1997**, *144*, 3371–3377. [CrossRef]
47. Cho, E.A.; Kwon, H.S.; Macdonald, D.D. Photoelectrochemical analysis on the passive film formed on Fe–20Cr in pH 8.5 buffer solution. *Electrochim. Acta* **2002**, *47*, 1661–1668. [CrossRef]
48. Macdonald, D.D. The Point Defect Model for the Passive State. *J. Electrochem. Soc.* **1992**, *139*, 3434–3449. [CrossRef]
49. Macdonald, D.D. The history of the Point Defect Model for the passive state: A brief review of film growth aspects. *Electrochim. Acta* **2011**, *56*, 1761–1772. [CrossRef]
50. Dean, M.H.; Stimming, U. The electronic properties of disordered passive films. *Corros. Sci.* **1989**, *29*, 199–211. [CrossRef]
51. Fattah-alhosseini, A.; Vafaeian, S. Comparison of electrochemical behavior between coarse-grained and fine-grained AISI 430 ferritic stainless steel by Mott–Schottky analysis and EIS measurements. *J. Alloy. Compd.* **2015**, *639*, 301–307. [CrossRef]
52. Ahn, S.J.; Kwon, H.S. Effects of solution temperature on electronic properties of passive film formed on Fe in pH 8.5 borate buffer solution. *Electrochim. Acta* **2004**, *49*, 3347–3353. [CrossRef]
53. Sedek, P.; Brozda, J.; Gazdowicz, J. Pitting corrosion of the stainless steel ventilation duct in a roofed swimming pool. *Eng. Fail. Anal.* **2008**, *15*, 281–286. [CrossRef]
54. Szala, M.; Beer-Lech, K.; Walczak, M. A study on the corrosion of stainless steel floor drains in an indoor swimming pool. *Eng. Fail. Anal.* **2017**, *77*, 31–38. [CrossRef]
55. Alfonsson, E.; Mameng, S.H. The possibilities & limitations of austenitic and duplex stainless steels in chlorinated water systems. *Nucl. Exch.* **2012**, 30–34.
56. Olsson, J.; Snis, M. Duplex—A new generation of stainless steels for desalination plants. *Desalination* **2007**, *205*, 104–113. [CrossRef]

© 2018 by the authors. Licensee MDPI, Basel, Switzerland. This article is an open access article distributed under the terms and conditions of the Creative Commons Attribution (CC BY) license (http://creativecommons.org/licenses/by/4.0/).

Review

Effects of Different Parameters on Initiation and Propagation of Stress Corrosion Cracks in Pipeline Steels: A Review

M.A. Mohtadi-Bonab

Department of Mechanical Engineering, University of Bonab, Velayat Highway, Bonab 5551761167, Iran; m.mohtadi@bonabu.ac.ir; Tel.: +98-9144203460

Received: 27 March 2019; Accepted: 20 May 2019; Published: 22 May 2019

Abstract: The demand for pipeline steels has increased in the last several decades since they were able to provide an immune and economical way to carry oil and natural gas over long distances. There are two important damage modes in pipeline steels including stress corrosion cracking (SCC) and hydrogen induced cracking (HIC). The SCC cracks are those cracks which are induced due to the combined effects of a corrosive environment and sustained tensile stress. The present review article is an attempt to highlight important factors affecting the SCC in pipeline steels. Based on a literature survey, it is concluded that many factors, such as microstructure of steel, residual stresses, chemical composition of steel, applied load, alternating current (AC) current and texture, and grain boundary character affect the SCC crack initiation and propagation in pipeline steels. It is also found that crystallographic texture plays a key role in crack propagation. Grain boundaries associated with {111}||rolling plane, {110}||rolling plane, coincidence site lattice boundaries and low angle grain boundaries are recognized as crack resistant paths while grains with high angle grain boundaries provide easy path for the SCC intergranular crack propagation. Finally, the SCC resistance in pipeline steels is improved by modifying the microstructure of steel or controlling the texture and grain boundary character.

Keywords: stress corrosion cracking; residual stress; AC current density; crystallographic texture; intergranular and transgranular cracks

1. Introduction

The demand for energy has increased in recent decades which forced the industry to develop high resistance pipeline steels [1–3]. Such steels show better mechanical properties and a higher corrosion resistance compared with normal carbon steels. However, these steels still suffer from two important failure modes including hydrogen induced cracking (HIC) and stress corrosion cracking (SCC) [4–6]. There are numerous studies in the literature focused on these failure modes. The SCC has been recognized as one of the main important failure modes in humid environments and causes a huge amount of economical loss and environmental disasters all around the world. The SCC susceptibility in pipeline steels depends on various factors such as the microstructure of steel, distribution of inclusions and precipitates inside the steel, texture and micro-texture of steel, chemical composition of steel, pH of the oil and gas which is transported, the pH of soil and environment where the pipeline steel is buried, and many other factors. Importance of the SCC in pipeline failure motivated us to write this review paper. This paper concentrates on different factors affecting the SCC crack nucleation and propagation in pipeline steel and looks for new ways to increase the resistance of pipeline steels to the SCC. Tables 1 and 2 show the chemical composition and mechanical properties of common used pipeline steels (API X60, API X65, API X70, API X80 and L360NS).

Table 1. Chemical composition of API X60, X65, X70, X80 and L360NS pipeline steel (wt. %) [7–9].

Pipeline Steel	C	Mn	Si	Nb	Mo	Ti	Cr	Cu	Ni	V
X60	0.052	1.50	0.15	0.067	0.096	0.022	0.07	0.18	0.19	0.001
X65	0.081	1.54	0.33	0.04	-	0.002	-	0.18	-	0.001
X70	0.025	1.65	0.26	0.068	0.175	0.015	0.07	0.21	0.08	0.001
X80	0.056	1.90	0.31	0.046	0.213	0.018	-	0.044	0.221	-
L360NS	0.12	1.50	0.25	-	0.02	-	0.07	0.08	0.04	-

Table 2. Mechanical properties of API X60, X65, X70, X80 and L360NS pipeline steels [8–10].

Pipeline Steel	Yield Strength (MPa)	Tensile Strength (MPa)	Elongation (%)
X60	520	610	33
X65	568	650	32
X70	615	720	29
X80	640	780	25
L360NS	380	510	41

The microstructure of API X60 and X70 pipeline steels has been mainly composed of polygonal and acicular ferrite. Moreover, there are some particles of martensite in the microstructure of both steels [7]. The microstructure of X65 steel includes mostly ferrite and some pearlite [10]. When the strength of pipeline steel increases, the microstructure becomes different from other types of steels. For example, the microstructure of X80 and X100 pipeline steels is mainly formed from ferrite and bainite [11,12]. There are also some martensite particles in the microstructure of both X80 and X100 steels. The microstructure of L360NS pipeline steel has been composed of white blocky polygonal ferrite, gray irregular blocky quasi-polygonal ferrite and black blocky pearlite colony [10].

2. Explanation of SCC and HIC

The HIC and the SCC are categorized as two types of corrosion that occur in pipeline steels. Since they have a close correlation, it is necessary to define both. In order to have an accurate definition of the HIC and the SCC, it would be better to explain the corrosion concept. Corrosion is the material degradation due to environmental effects. During the corrosion process, electrons are released due to the metal dissolution at anodic site [13]. Such electrons transfer to the cathode, where oxygenated water is reduced to hydrogel ions. The following overall reactions occur during the metal corrosion.

$$\text{Anodic dissolution: Fe} \rightarrow \text{Fe}^{+2} + 2e^- \qquad (1)$$

Figure 1 shows how a rust begins with the oxidation of iron to ferrous ions. The rust formation is a very complicated process, which begins with the oxidation of iron.

$$\text{Oxidation at anode: } 2H_2O(l) \rightarrow O_2(g) + 4H^+(aq) + 4e^- \qquad (2)$$

$$\text{Oxygen reduction in neutral or alkalis media: } O_2 + 2H_2O + 4e^- \rightarrow 4OH^- \qquad (3)$$

$$\text{Oxygen reduction in acidic media: } O_2 + 4H + 4e^- \rightarrow 2H_2O \qquad (4)$$

$$\text{Overall corrosion reaction: Fe} + 2H^+ \rightarrow Fe^{+2} + H^2 \qquad (5)$$

The SCC cracks are cracks that are induced due to the combined effect of the corrosive environment and sustained tensile stress. The tensile stress can be directly applied inside the pipeline or can be in the form of residual tensile stress. Therefore, three parameters including, a susceptible material (pipeline steel), a specific chemical species (environment), and tensile stress are required for crack nucleation and propagation. Therefore, the SCC is a type of environmentally assisted cracking (EAC), which is of great interest to the oil and gas pipeline manufactures. Recently, thousands of colonies of the SCC cracks have been observed in pipeline steels. Such cracks usually become dormant at depth of 1 mm. However, sometimes these cracks result in failure of pipeline by crack propagation [14]. Figure 2 shows effective factors influencing SCC crack initiation in pipeline steels.

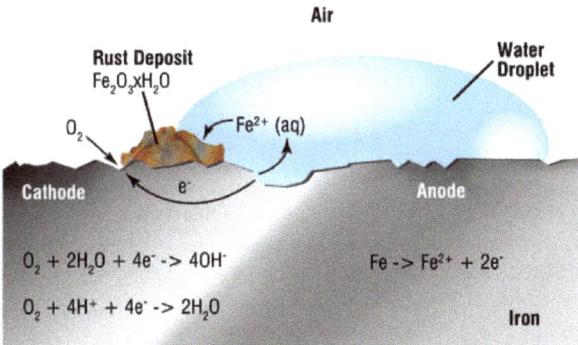

Figure 1. Formation of rust by the oxidation of iron to ferrous ions [13]. Reproduced with permission from [13], Noria Corporation and Machinery Lubrication, 2018.

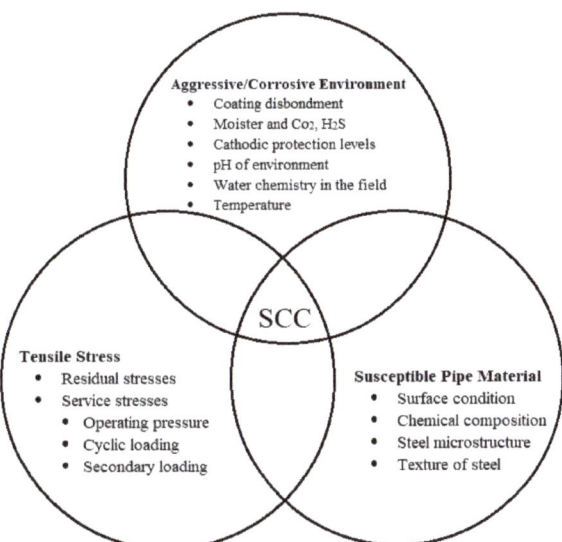

Figure 2. Effective factors for the stress corrosion cracking (SCC) crack initiation in pipeline steels.

Several factors such as microstructure, chemical composition, residual stress, texture of steel, water chemistry in the field, applied stress, pH of environment and AC current density may affect the SCC crack nucleation and propagation in pipeline steels, see Figure 2 [15–20]. Two types of corrosion happen in pipeline steels [21]. The first one is the sweet corrosion, which happens due to the presence of CO_2. The sweet corrosion [21] in carbon steels is formed in an acidic solution by mixing the CO_2 and water. The corrodant material is H^+ which is derived from H_2CO_3. The CO_2 gas is entered during some processes such as injection of CO_2 gas into the steel during the recovery operation. The sweet corrosion starts with the reaction of Fe and CO_2. This reaction can be written as follows:

$$Fe + CO_2 \rightarrow FeCO_2 \tag{6}$$

The reaction between the adsorbed surface complex with water produces Fe^{2+} (aq) and H_2CO^3. This reaction provides the cathodic reactant H^+ during dissociation. The cathodic reaction can be written as follows:

$$2H^+(aq) + 2e^- \rightarrow H_2 \tag{7}$$

$$2H_2CO_3 + 2e^- \rightarrow H_2 + 2HCO_3^- \tag{8}$$

The dissociation of H_2CO_3 in solution creates hydrogen ion for cathodic reaction.

$$H_2CO_3 + e^- \rightarrow H^+ + HCO_3^- \tag{9}$$

Several studies have been focused on sweet corrosion [21], however, the mechanism of a cathodic reaction has not been fully understood. It was shown that when a pH value is lower than 4, the hydrogen reduction is the dominant mode for the corrosion. However, when the pH value varies between 4 and 7, the adsorbed H_2CO_3 reduction is considerable. This type of corrosion is called sweet corrosion since it occurs with the absence of hydrogen sulphide or high levels of hydrogen sulphide. Carbon dioxide or carbonic acid are the main causes of sweet corrosion.

The second type and more common type of corrosion occurs owing to the presence of hydrogen sulphide (H_2S). Hydrogen sulphide which is present in oil and natural gas is decomposed to H^+ and HS^-. HS ion acts as a hydrogen recombination poison and avoid hydrogen molecule formation [7]. The following reactions occur:

$$H_2S \rightarrow HS^- + H^+ \tag{10}$$

$$HS^- \rightarrow S^{-2} + H^+ \tag{11}$$

Hydrogen atoms in the forms of protons get electrons from the iron and converted to the hydrogen atoms based on the following equations:

$$H^+ + e^- \rightarrow H_{ads} \tag{12}$$

$$H_{ads} + H_{ads} \rightarrow H_2 \tag{13}$$

It is worth mentioning that the hydrogen atoms are accumulated at microstructural defects such as empty spaces between inclusions and precipitates and metal matrix. The hydrogen atoms are combined at these regions and create a high amount of pressure. When this pressure reaches a critical value, the cracks initiate. Such cracks are known as hydrogen-induced cracks. The cavities or empty spaces are formed between inclusions and the metal matrix due to the difference between their thermal expansion coefficients. These cavities are formed during solidification of slabs or hot rolling process and can capture hydrogen atoms due to their small sizes. When the hydrogen atoms are accumulated in these areas, they combine to make hydrogen molecules, which make a high amount of pressure. The following equation shows the Gibbs free energy for hydrogen atoms combination.

$$\Delta_r G_H = \Delta_r G_H^\ominus + RT \ln \frac{p_{H_2}/p^\ominus}{C_H^2} \tag{14}$$

In the above equation, $\Delta_r G_H^\ominus$ is the reaction standard Gibbs free energy, T and R are the reaction temperature and gas constant and C_H and $p^\ominus$ re the concentration of hydrogen atoms near the inclusion and standard atmospheric pressure. When the concentration of H atoms around the inclusions reach a certain value, the reaction will occur and hydrogen molecules are formed. Based on the above equation, the increase of C_H at the reaction interface will decrease the reaction Gibbs free energy and further leads to the production of H_2 molecules. Hydrogen molecule formation creates a high amount of pressure and this results in hydrogen-induced cracks.

3. Role of Microstructure on the SCC Cracks

Microstructure of pipeline steel plays an important role in the SCC crack propagation. In general, hard and brittle phases ease the SCC crack propagation. There are a few papers in the literature focused on the role of microstructure on the SCC crack propagation. Most of the papers have discussed the role of microstructure on the nature of crack. For instance, Zhu et al. [22] investigated the mechanism of failure by the SCC in X80 pipeline steel in high pH carbonate and bicarbonate solution. These authors concluded that the nature of the SCC cracks mainly depends on the microstructure of steel. They observed that when the local microstructure of steel has been formed from bulky polygonal ferrite and granular bainite in a high pH solution, the nature of the crack is intergranular. However, when the local microstructure changes to the fine acicular ferrite and granular bainite, both intergranular and trangranular types of the SCC cracks are observed. Moreover, they implied that the microstructure of steel has a decisive role in transgranular SCC cracking and the probability of transgranular crack propagation increases with a decrease of pH solution. As seen in Figure 3a,b, both types of cracks are seen in the cross section of X80 pipeline steel after slow strain rate tensile (SSRT) test in high pH solution. At the crack initiation stage, the crack type is intergranular; however, both types of cracks are observed at the later stage.

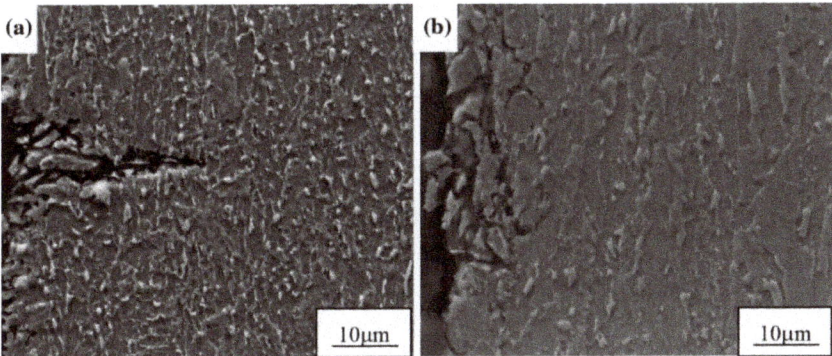

Figure 3. (a,b) SCC cracks in the cross section of X80 pipeline steel after SSRT test in high pH solution [22]. Reproduced with permission from [22], Springer Nature, 2014.

In another study, Gonzalez et al. [23] studied the effects of microstructure on the SCC behavior in HSLA steel. Their results documented that the microstructure of steel play a key role in the SCC behavior. Based on their findings, the type of microstructure is very important in transgranular crack propagation, while other metallurgical parameters including grain size and grain boundary character become determinative factors during intergranular cracking. It has been reported that the SSC susceptibility directly depends on the hardness, microstructure and chemical composition of steel [24]. Typically, the SSC cracks are internally propagated in intergranular manner and show little crack branching; however, some cracks have been also reported in transgranular mode [25]. Roffey et al. [26] investigated the SCC in an austenitic stainless steel hydrocarbon gas pipeline and concluded that transgranular SCC cracks initiate from corrosion pits from the internal and external surfaces and are divided to some branches, see Figure 4a,b.

Elboujdaini et al. [27] carried out research on the role of metallurgical factors on the SCC susceptibility. They found that the SCC cracks usually initiate from corrosion pits and pits nucleate from sulfides. Therefore, sulfide precipitates, inclusions and stringers can affect the SCC susceptibility in pipeline steel. Figure 5a–c shows a SCC crack initiating from a corrosion pit in X65 pipeline steel.

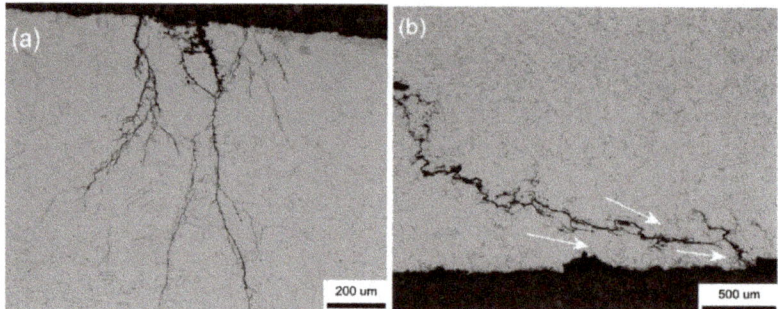

Figure 4. Transgranular type of cracks from corrosion pits on (**a**) internal surface, and (**b**) external surface [26]. Reproduced with permission from [26], Elsevier, 2014.

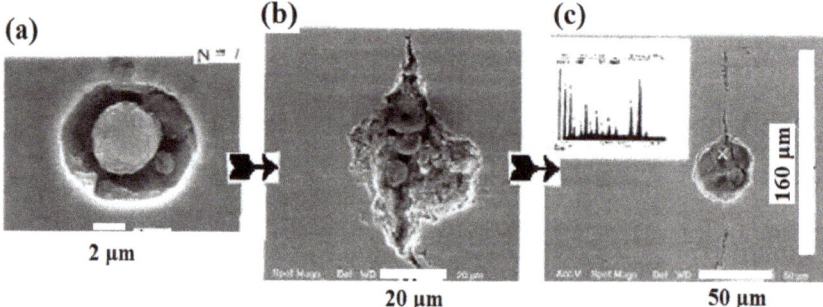

Figure 5. (**a**) Pit formation from an inclusion in X65 pipeline steel, (**b**) SCC crack initiation from a pit and (**c**) EDS analysis on the inclusion showing the existence of Na, Mg, Al, Si, P, S, Cl, K, and Ca elements [27]. Reproduced with permission from [27], Springer Nature, 2009.

4. Role of Stress Intensity Factor (SIF) in SCC and HIC Crack Propagation

SCC crack initiation and propagation in pipeline steel occur in three stages, see Figure 6 [28]. In the first step, small SCC cracks continuously initiate and coalesce. Initiation of these small cracks occurs in a large proportion of lifetime of pipeline steel. In the second step, the initiated small cracks propagate rapidly. This step is characterized by a function of stress intensity factor (SIF). SIF which is used in materials with small scale yielding at a crack tip represents the state of stress near the crack tip. In linear elastic fracture mechanics, it is used to explain the intensification of applied stress at the crack tip and is an important factor in characterizing the driving force and the crack propagation rate [29]. It basically depends on the geometry of the crack, location of the crack and applied load. In the third step, the material fails by rapid crack growth. Crack branching is one of the main characteristics of SCC cracks in pipeline steels. Two types of branching including micro-branching and macro-branching are observed in pipeline steels. In macro-branching, the main crack is divided to two running cracks and crack growth rate is independent from the crack length. The critical SIF for crack branching should exceed $\sqrt{2}$ two times of SIF for the crack propagation. In micro-branching, a main SCC crack splits into several small cracks at intervals of the order of one grain diameter. This phenomenon occurs when the critical SIF reaches at least $\sqrt{2}$ times of the subcritical value of the stress intensity [30]. The lifetime of pipeline steel is clearly determined by the second step. In fatigue corrosion, according to figure below, one process of fast crack propagation in step two is started when the maximum SIF at the crack tip exceeds the threshold SIF for SCC (K_{ISCC}). Another process of fast crack propagation occurs when the maximum stress intensity range (ΔK) exceeds the fatigue threshold (ΔK_{th}).

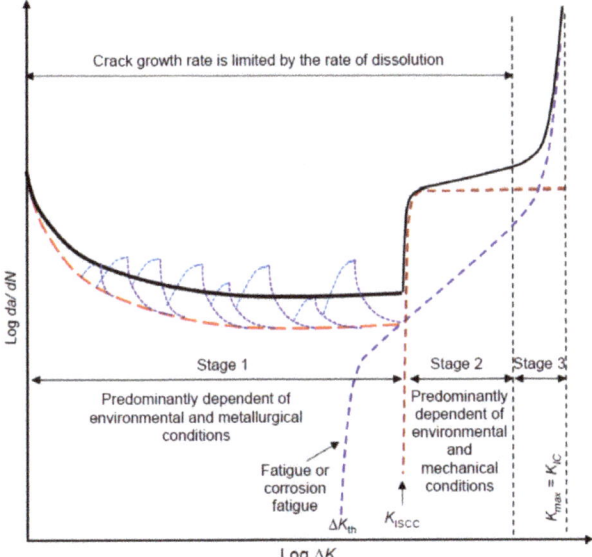

Figure 6. Effect of ΔK upon SCC velocity of pipeline steel exposed to carbonate, bicarbonate solution [28]. Reproduced with permission from [28], Elsevier, 2017.

Besides the SCC phenomenon, the SIF plays a key role in HIC crack propagation in pipeline steels. Costin et al. [31] reported that the threshold SIF (K_{th}) range for crack propagation is between 1.56 MPa $\sqrt{m}$ and 4.36 MPa $\sqrt{m}$. This range is lower than the K_{th} value which is calculated for ferrous alloys. This shows that subcritical HIC cracks propagate at micro-scale at lower SIFs.

5. Role of Surface Films in SCC

One of the important factors affecting SCC in pipeline steels is surface films. During the corrosion reaction, metallic iron is oxidized to the Fe^{2+} due to the electrochemical potential. There are some cracks at this oxide layer and at the interface between the oxide and the steel [32]. Figure 7 shows that SCC cracks initiated from the oxide film. The nature of these cracks is from transgarular. When the oxide films are cracked, the SCC cracks penetrated into the steel in transgraular manner. Therefore, two types of SIFs can be considered for SCC cracks in pipeline steels [28]. The first one is related to the external load while the oxide film induces the SIF in the second type. When the critical condition for film-induced SIF, the crack initiation and propagation will occur. The film-induced SIF can be considered as a driving force for SCC crack propagation.

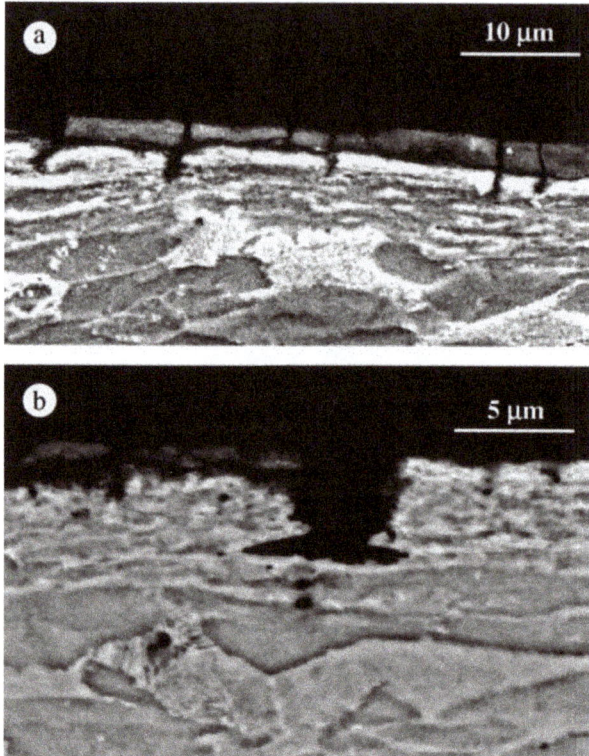

Figure 7. Initiation of transgranular SCC cracks from the surface oxide: (**a**) region 1, and (**b**) region 2 [32]. Reproduced with permission from [32], Elsevier, 2003.

6. Effect of Residual Stress on the SCC Crack Propagation

Beside the role of microstructure of pipeline steel on the SCC crack propagation in pipeline steel, the effect of tensile and compressive residual stresses on the SCC crack propagation is undeniable. Residual stresses are mainly developed during the manufacturing of forming and welding process in pipeline steels. Such stresses are high and sometimes reach yield stress of steel. Tensile residual stresses provide a driving for the SCC crack propagation while the compressive residual stresses have an opposite effect. In this field of study, Chen et al. [33] investigated the role of residual stresses on the SCC susceptibility in pipeline steel in neutral pH solution. These authors found that tensile residual stress gives a high amount of mechanical driving force for crack initiation and short crack propagation. These authors also showed that the SCC cracks are blunted by plastic deformation when the crack is propagated in a 45° where the shear stress for the plastic deformation is highest, see Figure 8a,b.

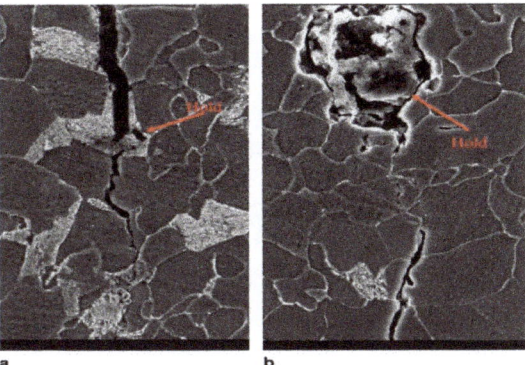

Figure 8. (a,b) Crack tip morphology after the pipeline specimens subjected to constant load for 7 days in different soil solutions [33]. Reproduced with permission from [33], Elsevier, 2007.

7. Role of AC Current on the SCC Susceptibility

Nowadays, pipeline steels are usually buried parallel with the electric power lines or electrified railways due to the limitation space [34]. Therefore, there is a possibility for the flowing of the AC current to the soil encompassed pipeline. Such AC flow results in AC corrosion of a pipeline where the coating is disbonded [35]. There are some investigations showing that AC interference increases the corrosion rate in metal alloys [36–38]. It has also been reported that AC current may break down the insulation layer of pipeline steel and destroy the cathodic protection system [39–41]. Zhu et al. [34] studied the role of AC current on the SCC susceptibility in carbonate/bicarbonate solution in X80 pipeline steel using SSRT experiment. They concluded that AC current plays a significant role on the SCC behavior. They observed that when there is no AC current, the SCC cracks propagate in intergranular manner and its mechanism is from the anodic dissolution. However, with the presence of AC current, the SCC susceptibility is increased and the SCC cracks propagate in transgranular manner at high AC current densities. In another study, Wan et al. [42] investigated the effect of alternating current on the SCC mechanism in X80 pipeline steel in near-neutral solution. The results of their research showed that crack propagation in X80 steel does not depend on the AC current and its mechanism shows a transgranular fracture feature. However, they observed that the depth of crack propagation is enhanced with the increase of AC current. Figure 7 depicts the fracture surfaces of X80 steel after SSRT experiment. As shown in the Figure 9a, there is no crack in the specimen tested in the air. However, there are small cracks in specimens tested in NS4 solution without AC current indicating SCC susceptibility, as shown in the Figure 9b. NS_4 solution or near-neutral soil solution (pH = 7) is used as a test solution for SCC experiment. It is prepared by mixing distilled water and pure chemical reagent. The chemical composition of NS_4 solution is shown in Table 3.

It is important to note that the number and the length of the secondary cracks increase with the increase of AC current density see the Figure 9c,d. When the AC current density reaches 30 and 50 A/m^2, as shown in the Figure 9e,f, the secondary cracks become wide signifying an increased SCC susceptibility. As seen in these figures, some cracks nucleated from pitting illustrating that anodic dissolution affected the SCC in X80 steel [42].

Table 3. NS_4 solution composition (g/L) [43].

NS_4 solution	KCl	$NaHCO_3$	$CaCl_2 \cdot 2H_2O$	$MgSO_4 \cdot 7H_2O$
-	0.122	0.483	0.181	0.131

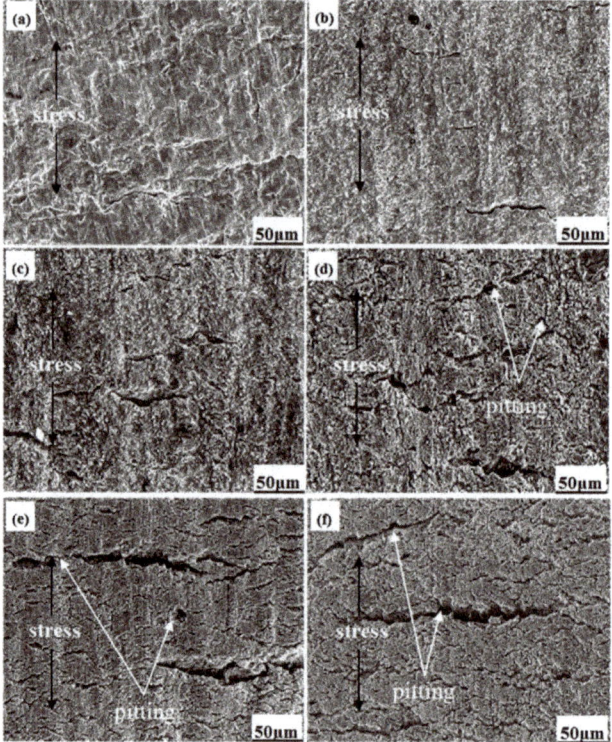

Figure 9. SEM image of fracture surfaces of X80 pipeline steel in air and NS4 solution in (**a**) in air, (**b**) 0 A/m², (**c**) 5 A/m², (**d**) 10 A/m², (**e**) 30 A/m² and (**f**) 50 A/m² [42]. Reproduced with permission from [42], Elsevier, 2017.

It is worth mentioning that the mechanism of AC current on SCC has not been fully understood. However, AC current transfers between pipeline steels and soil when there is a coating defect on pipeline surface and this leads to the AC corrosion [44]. The corrosion potential is negatively shifted by AC current and this phenomenon degrades the steel passivity in carbonate/bicarbonate solution. When the AC current increases, the corrosion type will change from uniform to pitting. Zhu et al. [45] studied the effect of short term AC current on SCC susceptibility in pipeline steel and observed several intergranular SCC cracks and the anodic dissolution of steel was the main mechanism for SCC. They concluded that when the AC current increases, the rate of intergranular crack nucleation from corrosion pits will increase as well.

8. Type of Environment on the SCC Behavior

Pipeline steels are usually used in various environments. For example, API 5L X60, X65 and X70 pipeline steels carry oil and natural gas in sour environments having a high amount of H_2S and CO_2 with low pH value. An acidic environment, containing H_2S, CO_2, and water, promotes corrosion and the H_2S enhances the absorption of hydrogen into the pipeline steel by the corrosion reaction as well. Some types of pipeline steels pass from sea water which has salt. Chloride ions are considered as one of the common atmospheric corrosive agent [46,47]. One of the environments is the deep sea water which applies a hydrostatic pressure to the pipeline steel. Even though there are few papers focused in this field, the deep sea water has a crucial effect on SCC susceptibility in pipeline steel. It has been reported that the SCC in land soil are dived into two types. The first type happens due to the effect of anodic dissolution in high pH solution among the grain boundaries [11,48,49]. The second one, however, occurs due to the combination effect of anodic dissolution and hydrogen embrittlement in near neutral pH solution [50,51]. In subsea environments, when the coating on the pipeline surface is destroyed, the permeation of seawater in the gaps will promote the peeling of the coating resulting in a complicated and more severe corrosion in pipeline steel [52].

Sun et al. [53] carried out an interesting study on SCC susceptibility in deep and shallow sea water on X70 pipeline steel. They showed that the SCC susceptibility first decreased, reached minimum amount at 15 MPa and then increased with the increase of hydrostatic pressure. In other word, the SCC susceptibility is the lowest at 1500 m deep sea environment; however, it reaches the its highest amount at 3000 m.

Moreover, the electrolyte has a considerable effect on SCC susceptibility in pipeline steels. It has been reported that the electrolyte pH has a key role in determining the nature of SCC crack in pipeline steels [54]. The type of SCC crack is intergranular when concentrated carbonate electrolytes with high pH is present as a composition of environment. It is worth-mentioning that transgranular type of SCC crack occurs in pipeline steels with dilute electrolytes [55]. One should consider that the main components of ground electrolytes contain carbonate/bicarbonate, sulfate, chloride and nitrate onions. Moreover, some soils are acidic with different types of compositions and concentrations. The effect of each on SCC should be considered separately.

One of the factors affecting the SCC susceptibility is the produced hydrogen inside the pipeline steel. The correlation between the hydrogen effect and anodic dissolution has not been completely recognized. However, it appears that hydrogen may have a destructive effect on SCC susceptibility due to the hydrogen embrittlement phenomenon. Lu et al. [56,57] studied the effect of hydrogen on SCC susceptibility and reported that hydrogen restricts pipeline steel dissolution in near-neutral pH water. However, unexpected results were proposed by Liu et al. [58]. These authors investigated the SCC susceptibility in X70 steel under cathodic protection in an acidic and near-neutral solution by using SSRT experiment. The results of their findings proved that hydrogen induced plasticity has a positive impact on SCC susceptibility by releasing the stress concentration at the crack nucleation sites and reducing the stress intensity. However, they reported that in order to postpone the crack initiation and propagation, the applied potential should be within the high SCC susceptibility potential range.

The sulphur element has a key role on SCC susceptibility in pipeline steels. One can consider its effect in two different ways. First, it plays an important role inside the microstructure of steel. It has been reported that the sulfur element has a detrimental effect on HIC susceptibility by creating MnS inclusion [8,59,60]. This type of inclusion usually has an elongated shape and provides regions with high stress concentration. Therefore, this inclusion is considered as the HIC crack initiation site. From the second viewpoint, sulphur element provides a corrosive environment for pipeline steel. Fan et al. [9] investigated the SCC susceptibility in L360NS pipeline steel in sulfur environment and concluded that the existence of sulfur element considerably increases the SCC susceptibility. As shown in Figure 10, these authors modeled the SCC behavior in L360NS pipeline steel in sulfur melting cladding condition and observed that H^+ ion permeates into the substrate in region with high stress concentration. This permeation breaks the corrosion product film and the surface of pipe steel is corroded continuously. The combined effects of corrosion and hydrogen diffusion degrade the mechanical properties of steel and the pipeline steel fractures under the stress.

Moreover, an applied load plays an important role during crack initiation and propagation on the SCC phenomenon. Three types of stress may affect the SCC susceptibility in pipeline steel. The first one is the constant load such as the pressure inside the pipe. The second one is the preload and the last one is the fluctuating loads such as wind or ground movements. Pipe bending or welding process can apply preload as a residual stress in pipeline steel. There is stress fluctuation in pipeline steel and the effect of such stress on crack initiation and propagation should not be neglected. It is reported that the SCC crack propagation near-neutral pH solution does not occur under constant loading and fluctuation loading is required to initiate and propagate the cracks [61–64]. Jia. et al. [65] studied the effect of constant loading and preloading on the SCC in X80 pipeline steel in near-neutral pH environment. They observed that preload do not change the SCC behavior in X80 steel near-neutral solution. However, the time of crack initiation near-neutral environment under constant load is decreased when preload is applied.

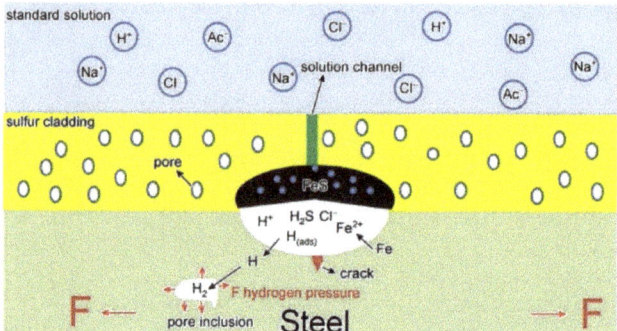

Figure 10. The SCC model for L360NS pipeline steel in sulfur melting cladding condition [9]. Reproduced with permission from [9], Elsevier, 2017.

9. Effect of Different Elements on SCC Susceptibility

There are several studies focused on the effects of different elements on the HIC and the SCC susceptibility in pipeline steel. No doubt, different elements play different roles on the SCC and the HIC susceptibility. The main elements playing a key role on SCC susceptibility are carbon, manganese, and phosphorus elements. These elements are recognized as segregation elements during solidification of steels such as pipeline steels. The mentioned elements are segregated at the center of thickness of pipeline slabs and create some hard phases and structures such as bainite and martensite. It has been reported that the manganese segregation ratio depends on carbon content [66]. It is important to note that low strength carbon steels can tolerate higher phosphorus segregation than high strength alloy steels [67]. Calcium element is usually added to pipeline steel due to its role on controlling the sulphur element [68,69]. Calcium element is combined with the sulphur and makes sulphide based inclusion which has a spherical shape. Therefore, the sulphur element is consumed by combining with the calcium and this phenomenon avoids the formation of more manganese sulphide inclusions. It is notable that manganese sulphide inclusion has an elongated shape and is considered as crack nucleation site [10,70].

It has been reported that copper has a beneficial effect on both increasing the strength of steel and HIC resistance by producing Cu-enriched fine precipitates [71]. Copper also makes a protective layer on the surface of steel and decrease the hydrogen diffusion inside the steel. Baba et al. [72] investigated the effect of copper addition on the prevention of hydrogen permeation in sour environment in pipeline steel. These authors concluded that copper creates an inner layer of corrosion product with 100 nm on the surface of steel preventing hydrogen entry inside the microstructure of steel.

10. Effect of Crystallographic Texture and Meso-Texture on SCC Susceptibility

There are several traditional methods, such as using micro-alloying elements, adding some elements, using some special heat treatments and reducing number of inclusions, to increase the SCC susceptibility in pipeline steels. However, most of such methods are not effective enough to increase the SCC susceptibility. Crystallographic texture has been recently considered as a novel technique which can reduce the SCC susceptibility in pipeline steels. There are a few studies focused on the role of texture on the SCC susceptibility in pipeline steel [60–62]. Arafin et al. [12] investigated the role of grain boundary character and crystallographic texture on the SCC susceptibility in X65 pipeline steel by electron backscatter diffraction (EBSD) and X-ray texture measurements. Figure 11 shows one SCC crack in X65 pipeline steel. As shown in this figure, the SCC crack gets two branches when it approaches to grains number 7 and 8. It is important to note that the misorientation angle between grains (7, 13) and (7, 10) are higher than 15° falling in the classification of high angle grain boundaries. Therefore, the boundaries between these grains with high energies provide an easy path for crack propagation. As shown with red circle, the SCC crack is deflected about 45° when it reaches grain number 30. Calculation of misorientation between grains number (30, 31) and (31, 32) proves that the misoriatation angles between them are 12.3° and 36.1°, respectively. This proves why the SCC crack deflects when it reaches grain number 30. Coincidence site lattice boundaries are categorized as low angle boundaries having a low energy. Therefore, such boundaries are not favorable for the SCC crack propagation. One should expect that the crack follows the grain boundary between grains number 37 and 38; however, the EBSD analysis shows that such boundary is from $\Sigma 11$ type boundary with a low energy. An interesting result was observed when the SCC crack reaches the grain number 39. The grain boundaries between grains number (39, 40), and (40, 41) are from low angle grain boundary (LAGB) and high angle grain boundary (HAGB), respectively. Surprisingly, the SCC crack has propagated though the grain boundary between grains number 39 and 40 which has a low energy. The EBSD analysis indicates that these grains have orientations close to the {110}||rolling plane with <110> boundary rotation axis. This might be the main reason for resisting of mentioned path for crack propagation.

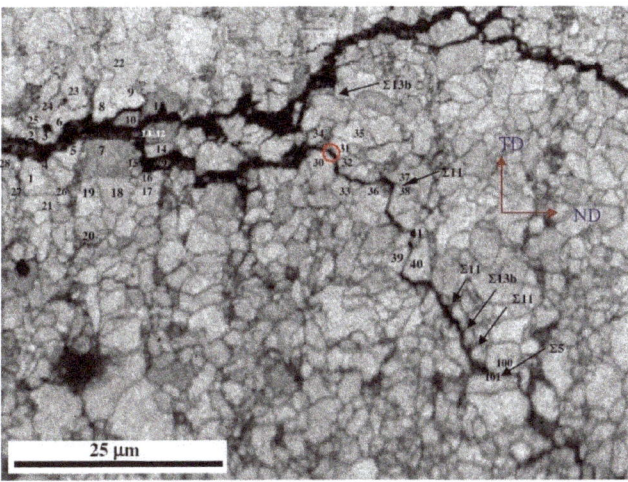

Figure 11. Electron backscatter diffraction (EBSD) map of the SCC crack propagation in X65 pipeline steel [12]. Reproduced with permission from [12], Elsevier, 2009.

The results of Arafin et al. [12] illustrated that the grain boundary character plays an important role in intergranular SCC. They also observed that CSL boundaries beyond the $\Sigma 13b$ do not provide the resistance path for the crack propagation. Finally, they showed that the macro and miro-texture significantly affect the intergraular SCC. It is worth-mentioning that the study of texture on the SCC crack propagation is a novel method and there are a few researches in the literature. Control of texture and grain boundary character may be used to produce new pipeline steels with a higher resistance to the HIC and the SCC in the near future.

11. Improvement of SCC Resistance

There are some methods that can be considered as new ways to improve the pipeline steel performance against SCC cracking. Crystallographic texture and grain boundary engineering have a key role in increasing the SCC resistance in pipeline steel. Arafin et al. [12] investigated the role of texture on X65 pipeline steel and concluded that a new understanding of texture and grain boundary character will help to produce new pipeline steels with superior intergranular SCC resistance. They concluded that a large amount of CSL boundaries and LAGBs can avoid intergranular SCC crack propagation in pipeline steels. Such boundaries are provided by modifying the surface texture. It has been reported that some phases and microstructures have a better SCC resistance in pipeline steels. Bulger et al. [73] investigated the effect of microstructure in pipeline steels on SCC resistance in near-neutral pH and concluded that there is a possibility to improve the SCC resistance by providing the fine-grained bainite and ferrite microstructure. These authors also documented that the pipeline steels with ferrite and pearlite microstructure have a poor SCC resistance. Moreover, the role of inclusions on SCC crack initiation is considerable. Wang et al. [74] investigated role of inclusions on pitting corrosion and SCC in X70 pipeline steels in near-neutral pH environment. They concluded that oxide and silicon enriched inclusions are considered as SCC micro-crack initiation sites. Based on their observations, most of SCC micro-cracks propagated through ferrite grains in transgranular manner. Therefore, manufacturing a pipeline steel with low number of inclusions or with uniform distribution of inclusions would be desirable for a high SCC resistance in pipeline steels.

12. Conclusions

Based on the above-discussion, several factors playing a significant role on the SCC susceptibility in pipeline steels are as follow:

(1) Microstructure of steel plays a key role on the SCC crack initiation and propagation. Nature of the SCC cracks, specially transgranular crack propagation, highly depends on the microstructure of steel. Other metallurgical factors including grain size and grain boundaries become determinative factors during intergranular cracking.
(2) The effect of residual stresses in crack initiation and propagation should not be neglected. Tensile residual stress gives a high amount of mechanical driving force for crack initiation and short crack propagation.
(3) AC current density may affect the crack propagation by two ways. First, AC current may break down the insulation layer of pipeline steel and destroy the cathodic protection system. Secondly, AC current density affects the nature of the SCC crack. When there is no AC current, the SCC cracks propagate through intergranular manner and its mechanism is the anodic dissolution. However, with the presence of AC current, the SCC susceptibility is increased and the SCC cracks propagate through transgranular manner at high current densities.
(4) In a subsea environment, the permeation of sea water in the gaps will promote the peeling of the coating resulting in a complicated and more severe corrosion in pipeline steel by destroying the coating of pipeline steel.
(5) Addition of some elements to the pipeline steel have a substantial effect on the SCC susceptibility. For instance, Carbon, manganese and phosphorus elements are segregated at the center of thickness of pipeline slabs and make some hard phases and structures such as bainite and martensite. Copper also makes a protective layer on the surface of steel and decrease the hydrogen diffusion inside the steel.
(6) The SCC crack propagation near-neutral pH solution does not occur under constant loading and fluctuation loading is required to initiate and propagate the cracks. The time of crack initiation near-neutral environment under constant loading is decreased when preload is applied.
(7) Crystallographic texture plays a key role in SCC crack propagation. Grains with {111}||rolling plane and {110}||rolling plane, coincidence site lattice boundaries and low angle grain boundaries are recognized as crack resistant paths while grains with high angle grain boundaries provides easy path for intergranular SCC crack propagation.

Acknowledgments: We would like to thank the Research Center of University of Bonab for the financial support of this study.

Conflicts of Interest: The author declares no conflict of interest.

References

1. Liu, Z.Y.; Li, X.G.; Du, C.W.; Lu, L.; Zhang, Y.R.; Cheng, Y.F. Effect of inclusions on initiation of stress corrosion cracks in X70 pipeline steel in an acidic soil environment. *Corros. Sci.* **2009**, *51*, 895–900. [CrossRef]
2. Shi, X.B.; Yan, W.; Wang, W.; Zhao, L.Y.; Shan, Y.Y.; Yang, K. HIC and SSC behavior of high-strength pipeline steels. *Acta Metall. Sinica* **2015**, *28*, 799–808. [CrossRef]
3. Hara, T.; Asahi, H.; Ogawa, H. Conditions of hydrogen-induced corrosion occurrence of X65 grade linepipe steels in sour environments. *Corros. Sci.* **2004**, *60*, 1113–1121. [CrossRef]
4. Shi, X.B.; Yan, W.; Wang, W.; Zhao, L.Y.; Shan, Y.Y.; Yang, K. Effect of microstructure on hydrogen induced cracking behavior of a high deformability pipeline steel. *J. Iron Steel Res.* **2015**, *22*, 937–942. [CrossRef]
5. Kim, W.K.; Koh, S.U.; Yang, B.Y.; Kim, K.Y. Effect of environment and metallurgical factors on hydrogen induced cracking. *Corros. Sci.* **2008**, *50*, 3336–3342. [CrossRef]
6. Maciejewski, J. The effects of sulfide inclusions on mechanical properties and failures of steel components. *J. Fail. Anal. Prev.* **2015**, *15*, 169–178. [CrossRef]
7. Mohtadi-Bonab, M.A.; Szpunar, J.A.; Razavi-Tousi, S.S. A comparative study of hydrogen induced cracking behavior in API 5L X60 and X70 pipeline steels. *Eng. Fail. Anal.* **2013**, *33*, 163–175. [CrossRef]
8. Mohtadi-Bonab, M.A.; Eskandari, M.; Karimdadashi, R.; Szpunar, J.A. Effect of different microstructural parameters on hydrogen induced cracking in an API X70 pipeline steel. *Met. Mater. Int.* **2017**, *23*, 726–735. [CrossRef]
9. Fan, Z.; Hu, X.; Liu, J.; Li, H.; Fu, J. Stress corrosion cracking of L360NS pipeline steel in sulfur environment. *Petroleum* **2017**, *3*, 377–383. [CrossRef]
10. Mohtadi-Bonab, M.A.; Eskandari, M.; Ghaednia, H.; Das, S. Effect of microstructural parameters on fatigue crack propagation in an API X65 pipeline steel. *J. Mater. Eng. Perform.* **2016**, *25*, 4933–4940. [CrossRef]
11. Arafin, M.A.; Szpunar, J.A. Effect of bainitic microstructure on the susceptibility of pipeline steels to hydrogen induced cracking. *Mater. Sci. Eng. A* **2011**, *528*, 4927–4940. [CrossRef]
12. Arafin, M.A.; Szpunar, J.A. A new understanding of intergranular stress corrosion cracking resistance of pipeline steel through grain boundary character and crystallographic texture studies. *Corros. Sci.* **2009**, *51*, 119–128. [CrossRef]
13. Wright, J. Inhibiting rust and corrosion to prevent machine failures. In Proceedings of the Machinary Lubrication Conference and Exhibition, Houston, TX, USA, 6–8 November 2018.
14. National Energy Board. *Stress Corrosion Cracking on Canadian Oil and Gas Pipelines*; Report No. MH-2-95; National Energy Board: Calgary, AB, Canada, 1996.
15. Aly, O.F.; Neto, M.M. *Stress Corrosion Cracking, Developments in Corrosion Protection*; Aliofkhazraei, M., Ed.; IntechOpen Limited: London, UK, 2014.
16. Sutcliffe, J.M.; Fessler, R.R.; Boyd, W.K.; Parkins, R.N. Stress corrosion cracking of carbon steel in carbonate solutions. *Corrosion* **1972**, *28*, 313–320. [CrossRef]
17. Beavers, J.A.; Harle, B.A. Mechanisms of high-pH and nearneutral-pH SCC of underground pipelines. *Offshore Mech. Arct. Eng.* **2003**, *123*, 147–151. [CrossRef]
18. Charles, E.A.; Parkins, R.N. Generation of stress corrosion cracking environments at pipeline surfaces. *Corrosion* **1995**, *51*, 518–527. [CrossRef]
19. Fang, B.Y.; Atrens, A.; Wang, J.Q.; Han, E.H.; Zhu, Z.Y.; Ke, W. Review of stress corrosion cracking of pipeline steels in "low" and "high" pH solutions. *J. Mater. Sci.* **2003**, *38*, 127–132. [CrossRef]
20. Mohtadi-Bonab, M.A.; Ghesmati-Kucheki, H. Important Factors on the Failure of Pipeline Steels with Focus on Hydrogen Induced Cracks and Improvement of Their Resistance: Review Paper. *Met. Mater. Int.* **2019**. [CrossRef]
21. Kadhim, M.G.; Albdiry, M. A critical review on corrosion and its prevention in the oilfield equipment. *J. Petrol. Res. Stud.* **2017**, *14*, 162–189.

22. Zhu, M.; Du, C.; Li, X.; Liu, Z.; Wang, S.; Zhao, T.; Jia, J. Effect of strength and microstructure on stress corrosion cracking behavior and mechanism of X80 pipeline steel in high pH carbonate/bicarbonate solution. *J. Mech. Eng. Perform.* **2014**, *23*, 1358–1365. [CrossRef]
23. Gonzalez, J.; Gutierrez-Solana, F.; Varona, J.M. The effects of microstructure, strength level, and crack propagation mode on stress corrosion cracking behavior of 4135 steel. *Met. Mater. Trans. A* **1994**, *27*, 281–290. [CrossRef]
24. Masouri, D.; Zafari, M.; Araghi, A. Sulfide stress cracking of pipeline-case history. In Proceedings of the NACE International, Corrosion 2008 Proceedings, New Orleans, LA, USA, 16–20 March 2008.
25. Lancsater, J. *Handbook of Structural Welding: Processes, Materials and Methods in the Welding of Major Structures, Pipelines and Process Plant*; Woodhead Publishing: Sawston, UK, 2003.
26. Roffey, P.; Davies, E.H. The generation of corrosion under insulation and stress corrosion cracking due to sulphide stress cracking in an austenitic stainless steel hydrocarbon gas pipeline. *Eng. Fail. Anal.* **2014**, *44*, 148–157. [CrossRef]
27. Elboujdaini, M.; Revie, R.W. Metallurgical factors in stress corrosion cracking (SCC) and hydrogen-induced cracking (HIC). *J. Solid State Electrochem.* **2009**, *13*, 1091–1099. [CrossRef]
28. Chen, W. *Modeling and prediction of stress corrosion cracking of pipeline steels, Trends in Oil and Gas Corrosion Research and Technologies*; Woodhead Publishing Series in Energy; Woodhead Publishing: Sawston, UK, 2017; pp. 707–748.
29. Hongliang, Y.; He, X.; Fuqiang, Y.; Lingyan, Z. Effect of film-induced stress on mechanical properties at stress corrosion cracking tip. *Rare Met. Mater. Eng.* **2017**, *46*, 3595–3600. [CrossRef]
30. Austen, I.M. Effective stress intensities in stress corrosion cracking. *Int. J. Fract.* **1976**, *12*, 253–263. [CrossRef]
31. Costin, W.L.; Lavigne, O.; Kotousov, A.; Ghomashchi, R.; Linton, V. Investigation of hydrogen assisted cracking in acicular ferrite using site-specific micro-fracture tests. *Mater. Sci. Eng. A* **2016**, *651*, 859–868. [CrossRef]
32. Wang, J.Q.; Atrens, A. SCC initiation for X65 pipeline steel in the high pH carbonate/bicarbonate solution. *Corros. Sci.* **2003**, *45*, 2199–2217. [CrossRef]
33. Chen, W.; van Boven, G.; Rogge, R. The role of residual stress in neutral pH stress corrosion cracking of pipeline steels—Part II: Crack dormancy. *Acta Mater.* **2007**, *55*, 43–53. [CrossRef]
34. Zhu, M.; Du, C.; Li, X.; Liu, Z.; Li, H.; Zhang, D. Effect of AC on stress corrosion cracking behavior and mechanism of X80 pipeline steel in carbonate/bicarbonate solution. *Corros. Sci.* **2014**, *87*, 224–232. [CrossRef]
35. Hosokawa, Y.; Kajiyama, F.; Fukuoka, T. Alternating current corrosion risk arising from alternating current-powered rail transit systems on cathodically protected buried steel pipelines and its measures. *Corrosion* **2004**, *60*, 408–413. [CrossRef]
36. Tan, T.C.; Chin, D.T. Ac corrosion of nickel in sulphate solutions. *J. Appl. Electrochem.* **1988**, *18*, 831–838. [CrossRef]
37. Goidanich, S.; Lazzari, L.; Ormellese, M.; Pedeferri, M. *Influence of AC on Corrosion Kinetics for Carbon Steel, Zinc and Copper*; CORROSION/2005, Paper No. 05189; NACE International: Houston, TX, USA, 2005.
38. Wendt, J.L.; Chin, D.T. The A.C. corrosion of stainless steel – II. The breakdown of passivity of ss304 in neutral aqueous solutions. *Corros. Sci.* **1985**, *25*, 889–900.
39. Muralidharan, S.; Kim, D.K.; Ha, T.H.; Bae, J.H.; Ha, Y.C.; Lee, H.G.; Scantlebury, J. Influence of alternating, direct and superimposed alternating and direct current on the corrosion of mild steel in marine environments. *Desalination* **2007**, *216*, 103–115. [CrossRef]
40. Vasudevan, S.; Lakshmi, J. Effects of alternating and direct current in electrocoagulation process on the removal of cadmium from waterea novel approach. *Sep. Purif. Technol.* **2011**, *80*, 643–651. [CrossRef]
41. Jiang, Z.; Du, Y.; Lu, M.; Zhang, Y.; Tang, D.; Dong, L. New findings on the factors accelerating AC corrosion of buried pipeline. *Corros. Sci.* **2014**, *81*, 1–10. [CrossRef]
42. Wan, H.; Song, D.; Liu, Z.; Du, C.; Zeng, Z.; Yang, X.; Li, X. Effect of alternating current on stress corrosion cracking behavior and mechanism of X80 pipeline steel in near-neutral solution. *J. Nat. Gas Sci. Eng.* **2017**, *38*, 458–465. [CrossRef]
43. Luo, J.; Zhang, L.; Li, L.; Yang, F.; Ma, W.; Wang, K.; Zhao, X. Electrochemical corrosion behaviors of the X90 linepipe steel in NS4 solution. *Nat. Gas. Ind.* **2016**, *3*, 346–351. [CrossRef]
44. Zhu, M.; Du, C.W. A new understanding on AC corrosion of pipeline steel in alkaline environment. *J. Mater. Eng. Perform.* **2017**, *26*, 221–228. [CrossRef]

45. Zhu, M.; Du, C.; Li, X.; Liu, Z.; Wang, S.; Li, J.; Zhang, D. Effect of AC current density on stress corrosion cracking behavior of X80 pipeline steel in high pH carbonate/bicarbonate solution. *Electrochem. Acta* **2014**, *117*, 351–359. [CrossRef]
46. Rodríguez, J.J.S.; Hernández, F.J.S.; González, J.E.G. The effect of environmental and meteorological variables on atmospheric corrosion of carbon steel, copper, zinc and aluminium in a limited geographic zone with different types of environment. *Corros. Sci.* **2003**, *45*, 799–815. [CrossRef]
47. Iakovleva, E.; Mäkilä, E.; Salonen, J.; Sitarz, M.; Sillanpää, M. Industrial products and wastes as adsorbents for sulphate and chloride removal from synthetic alkaline solution and mine process water. *Chem. Eng. J.* **2015**, *259*, 364–371. [CrossRef]
48. Mustapha, A.; Charles, E.A.; Hardie, D. Evaluation of environment-assisted cracking susceptibility of a grade X100 pipeline steel. *Corros. Sci.* **2012**, *54*, 5–9. [CrossRef]
49. Oskuie, A.A.; Shahrabi, T.; Shahriari, A.; Saebnoori, E. Electrochemical impedance spectroscopy analysis of X70 pipeline steel stress corrosion cracking in high pH carbonate solution. *Corros. Sci.* **2012**, *61*, 111–122. [CrossRef]
50. Yan, M.C.; Xu, J.; Yu, L.B.; Wu, T.Q.; Sun, C.; Ke, W. EIS analysis on stress corrosion initiation of pipeline steel under disbonded coating in near-neutral pH simulated soil electrolyte. *Corros. Sci.* **2016**, *110*, 23–34.
51. Kang, Y.W.; Chen, W.X.; Kania, R.; Boven, G.V.; Worthingham, R. Simulation of crack growth during hydrostatic testing of pipeline steel in near-neutral pH environment. *Corros. Sci.* **2011**, *53*, 968–975. [CrossRef]
52. Liu, Y.; Wang, J.W.; Liu, L.; Li, Y.; Wang, F.H. Study of the failure mechanism of an epoxy coating system under high hydrostatic pressure. *Corros. Sci.* **2013**, *74*, 59–70. [CrossRef]
53. Suna, F.; Ren, S.; Li, Z.; Liu, Z.; Li, X.; Du, C. Comparative study on the stress corrosion cracking of X70 pipeline steel in simulated shallow and deep sea environments. *Mater. Sci. Eng. A* **2017**, *685*, 145–153. [CrossRef]
54. Marshakov, A.I.; Ignatenko, V.E.; Bogdanov, R.I.; Arabey, A.B. Effect of electrolyte composition on crack growth rate in pipeline steel. *Corros. Sci.* **2014**, *83*, 209–216. [CrossRef]
55. Parkins, R.N.; Blanchard, W.K., Jr.; Delanty, B.S. Transgranular stress corrosion cracking of high-pressure pipelines in contact with solutions of near-neutral-pH. *Corrosion* **1994**, *50*, 394–408. [CrossRef]
56. Lu, B.T.; Luo, J.L.; Norton, P.R. Environmentally assisted cracking mechanism of pipeline steel in near-neutral pH groundwater. *Corros. Sci.* **2010**, *52*, 1787–1795. [CrossRef]
57. Lu, B.T.; Luo, J.L.; Norton, P.R.; Ma, H.Y. Effects of dissolved hydrogen and elastic and plastic deformation on active dissolution of pipeline steel in anaerobic groundwater of near-neutral pH. *Acta Mater.* **2009**, *57*, 41–49. [CrossRef]
58. Liu, Z.Y.; Wang, X.Z.; Du, C.W.; Li, J.K.; Li, X.G. Effect of hydrogen-induced plasticity on the stress corrosion cracking of X70 pipeline steel in simulated soil environments. *Mater. Sci. Eng. A* **2016**, *658*, 348–354. [CrossRef]
59. Mohtadi-Bonab, M.A.; Eskandari, M.; Szpunar, J.A. Role of cold rolled followed by annealing on improvement of hydrogen induced cracking resistance in pipeline steel. *Eng. Fail. Anal.* **2018**, *91*, 172–181. [CrossRef]
60. Mohtadi-Bonab, M.A.; Eskandari, M.; Szpunar, J.A. Effect of arisen dislocation density and texture components during cold rolling and annealing treatments on hydrogen induced cracking susceptibility in pipeline steel. *J. Mater. Res.* **2016**, *31*, 3390–3400. [CrossRef]
61. Fang, B.Y.; Han, E.H.; Zhu, Z.Y.; Wang, J.Q.; Ke, W. Stress corrosion cracking of pipeline steels. *J. Mater. Sci. Technol.* **2002**, *18*, 3–6.
62. Lu, B.T.; Luo, J.L. Crack initiation and early propagation of X70 Steel in simulated near-neutral pH groundwater. *Corrosion* **2006**, *62*, 723–731. [CrossRef]
63. Chen, W.X.; Kania, R.; Worthingham, R.; van Boven, G. Transgranular crack growth in the pipeline steels exposed to near-neutral pH soil aqueous solutions-The role of hydrogen. *Acta Mater.* **2009**, *57*, 6200–6214. [CrossRef]
64. Chen, W.; Wang, S.H.; Chu, R.; King, F.; Jack, T.R.; Fessler, R.R. Effect of precyclic loading on stress-corrosion-cracking initiation in an X-65 pipeline steel exposed to near-neutral pH soil environment. *Metall. Mater. Trans. A* **2003**, *34*, 2601–2608. [CrossRef]

65. Jia, Y.Z.; Wang, J.Q.; Han, E.H.; Ke, W. Stress corrosion cracking of x80 pipeline steel in near-neutral pH environment under constant load tests with and without preload. *J. Mater. Sci. Technol.* **2011**, *27*, 1039–1046. [CrossRef]
66. Ghosh, G.; Rostron, P.; Garg, R.; Panday, A. Hydrogen induced cracking of pipeline and pressure vessel steels: A review. *Eng. Fract. Mech.* **2018**, *199*, 609–618. [CrossRef]
67. Dayal, R.K.; Grabke, H.J. Hydrogen induced stress corrosion cracking in low and high strength ferritic steels of different phosphorus content in acid media. *Mater. Corros.* **1987**, *38*, 409–416. [CrossRef]
68. Mohtadi-Bonab, M.A.; Szpunar, J.A.; Basu, R.; Eskandari, M. The mechanism of failure by hydrogen induced cracking in an acidic environment for API 5L X70 pipeline steel. *Int. J. Hydrogen Energy* **2015**, *40*, 1096–1107. [CrossRef]
69. Mohtadi-Bonab, M.A.; Eskandari, M. A focus on different factors affecting hydrogen induced cracking in oil and natural gas pipeline steel. *Eng. Fail. Anal.* **2017**, *79*, 351–360. [CrossRef]
70. Mohtadi-Bonab, M.A.; Eskandari, M.; Sanayei, M.; Das, S. Microstructural aspects of intergranular and transgranular crack propagation in an API X65 steel pipeline related to fatigue failure. *Eng. Fail. Anal.* **2018**, *94*, 214–225. [CrossRef]
71. Shi, X.; Yan, W.; Wang, W.; Shan, Y.; Yang, K. Novel Cu-bearing high-strength pipeline steels with excellent resistance to hydrogen-induced cracking. *Mater. Des.* **2016**, *92*, 300–305. [CrossRef]
72. Baba, K.; Mizuno, M.D.; Yasuda, K.; Nakamichi, H.; Ishikawa, N. Effect of Cu addition in pipeline steels on prevention of hydrogen permeation in mildly sour environments. *Corrosion* **2016**, *72*, 1107–1115. [CrossRef]
73. Bulger, J.T.; Lu, B.T.; Luo, J.L. Microstructural effect on near-neutral pH stress corrosion cracking resistance of pipeline steels. *J. Mater. Sci.* **2006**, *41*, 5001–5005. [CrossRef]
74. Wang, L.; Xin, J.; Cheng, L.; Zhao, K.; Sun, B.; Li, J.; Wang, X.; Cui, Z. Influence of inclusions on initiation of pitting corrosion and stress corrosion cracking of X70 steel in near-neutral pH environment. *Corros. Sci.* **2019**, *147*, 108–127. [CrossRef]

 © 2019 by the author. Licensee MDPI, Basel, Switzerland. This article is an open access article distributed under the terms and conditions of the Creative Commons Attribution (CC BY) license (http://creativecommons.org/licenses/by/4.0/).

MDPI
St. Alban-Anlage 66
4052 Basel
Switzerland
Tel. +41 61 683 77 34
Fax +41 61 302 89 18
www.mdpi.com

Metals Editorial Office
E-mail: metals@mdpi.com
www.mdpi.com/journal/metals

www.ingramcontent.com/pod-product-compliance
Lightning Source LLC
LaVergne TN
LVHW070430100526
838202LV00014B/1569